Ecology and Tropical Biology

Ian Deshmukh

Blackwell Scientific Publications

Palo Alto   Oxford   London
Edinburgh   Boston   Victoria

Editorial Offices

667 Lytton Avenue, Palo Alto,
California 94301, USA

Osney Mead, Oxford, OX2 OEL

8 John Street, London, WC1N 2ES

23 Ainslie Place, Edinburgh, EH3 6AJ

52 Beacon Street, Boston,
Massachusetts 02108, USA

107 Barry Street, Carlton,
Victoria, 3053 Australia

Distributors

USA and Canada
Blackwell Scientific Publications
P.O. Box 50009
Palo Alto, California 94303

United Kingdom and Europe
Blackwell Scientific Publications, Ltd.
Osney Mead, Oxford, OX2 OEL

Australia
Blackwell Scientific Publications
(Australia) PTY. LTD.
107 Barry Street
Carlton, Victoria
3053 Australia

Editor: John Staples
Production Coordinator: Deborah Gale
Artist: Robin Mouat
Interior and Cover Design: John Edeen
Composition: Skillful Means Press

First published 1986.

Library of Congress Cataloging-in-Publication Data

Deshmukh, Ian.
    Ecology and tropical biology.

    Bibliography; p.
    Includes index.
    1. Ecology—Tropics.   2. Ecology—Developing
countries.   3. Ecology.   I. Title.
QH84.5.D48   1986   574.5′2623   85–32551
ISBN 0–86542–316–4

54,588

Preface

In writing *Ecology and Tropical Biology,* I had four major objectives:

1. To provide an up-to-date textbook covering all the concepts of basic ecology at the undergraduate level. The texts currently available are showing signs of age, with limited coverage of some recent topics of interest. Also the better of these books are too large for many undergraduate courses.

2. To give the book a different flavor by illustrating principles of basic ecology mainly with examples of tropical organisms and tropical ecosystems. As Janzen (1977) points out, most texts are strongly biased toward the temperate zone. In the ten books that he reviewed, only 9% of the bibliographies referred to tropical studies, and of these, tropical studies composed a maximum of 16% of the listings. I hope that my approach will prove useful and stimulating to students of tropical ecology worldwide. In addition to the wealth of new examples, the comparative approach (see below) and the fact that many tropical ecosystems are less disturbed than many temperate ones, make the book valuable to all ecology students.

3. To identify and evaluate differences between tropical and extratropical (mainly temperate) ecology. Many such differences have been proposed at various times, and some of these have gained the status of "conventional wisdoms." I have assessed

current evidence and suggest which of these ideas hold up (and to what extent), which need more evidence, and which are without substantial support at the moment.

4. To give a broad view of human ecology in tropical countries. Concepts developed earlier in the book for natural systems are used to examine the roles of people in ecological systems in the tropics.

Although the geographical scope of the book covers the whole of the tropical zone, only terrestrial ecology is considered. Aquatic systems are referred to only where they interact with terrestrial systems, or for specific examples that apply equally well to terrestrial situations. To cover both types of systems in a book of this length would, I believe, lead to inadequate coverage of both. Although all tropical regions are covered, I have tried to pick the best examples rather than to parcel them out equally between regions or countries.

Because of the comparative approach of this book, some sections read more like short reviews than is often the case with introductory text books. I feel that this approach is an important feature of the book. Many of the ideas collated here have not been juxtaposed in this way before, and it is important that readers can trace the origin of the conclusions presented. When I began to write in 1982, the time seemed ripe for a straightforward synthesis of the tropical literature. This conviction has not changed, but my appreciation of the gaps in our knowledge has. Many notions that seemed well-founded a few years ago are actually based upon flimsy evidence (sometimes none at all), and many recent studies contradict rather than confirm what seemed like established ideas.

Applied ecology has a vital role to play in developing countries. The environmental impact of rapid cultural changes in these countries can only be fully understood in an ecological context. I have, therefore, tried to supply this context in the last three chapters. (A few other aspects of human ecology are dealt with in appropriate sections of earlier chapters.) I do not attempt to provide a comprehensive account of problems and their solutions as this would require thorough expositions of political, social, and economic phenomena. Neither have I written a tract in support of the various so-called environmentalist or ecological movements that have arisen in Europe and North America. Rather I stress the importance of rational assessment of the options available in different countries as they pursue development. Particular attention is payed to ecological and cultural contexts. I hope this approach will prove useful to resource planners and managers as well as those interested from the purely ecological point of view.

Tropical ecological research burgeoned in the 1960s and 1970s. Much of the work was done by people from former colonial powers or by ecologists from the United States. There are also many able ecologists within the tropical countries themselves. However, there are indications that tropical studies have declined in the 1980s. A perusal of *Current Advances in Ecological Research* (a "scanning" journal) reveals a drop of 60% in tropical titles. Such papers now represent only 1% of the ecological literature (Cole 1984). This trend is unfortunate given the enormous need for more information stressed throughout this book. I hope that this text will provide sufficient stimulation to stem this decline in some small way. Even with modest expenditure, the potential rewards in both "pure" and applied ecology are enormous.

Acknowledgments

I would like to thank John Staples of Blackwell Scientific Publications and his associates for reviewing, editing, and publishing the manuscript so efficiently. Particular thanks is owed to John for his encouragement and enthusiasm when my faith in the publishing trade was low.

The following group of friends and colleagues reviewed draft chapters (or parts of chapters) indicated by the numbers in parentheses following their names. There is no doubt that the book has been markedly improved by their comments. Any remaining mistakes or heresies are probably where I ignored their advice. Many thanks to Willem Brakel (4,5,6,7), Oliver A. Chadwick (3), Robert K. Colwell (7), Tony Diamond (8,11), Diane Gifford-González (9), Jeremy Greenwood (6), Alan Hildrew (4, 5), Katharine Milton (6, 8), David Mouat (8, 11), John Phillipson (2), Thelma Rowell (9, 10) and Peter M. Vitousek (3).

I am particularly indebted to Christopher Dunford, who reviewed the whole manuscript in draft from both technical and editorial viewpoints. His rapid review was much appreciated. Many thanks, also, to Robin Mouat who somehow transformed my appalling sketches into the fine illustrations found throughout the book.

The Animal Ecology Research Group (Department of Zoology), Oxford University, and the Zoology Department at the University of California in Berkeley provided "homes" during the writing. John Phillipson and Thelma Rowell, respectively, were especially helpful in those two institutions.

Partial financial support during the first four months of writing came from an Inter-University Council Fellowship (British Council), which is gratefully acknowledged.

Ian Deshmukh

Contents

ECOLOGY OF NATURAL SYSTEMS

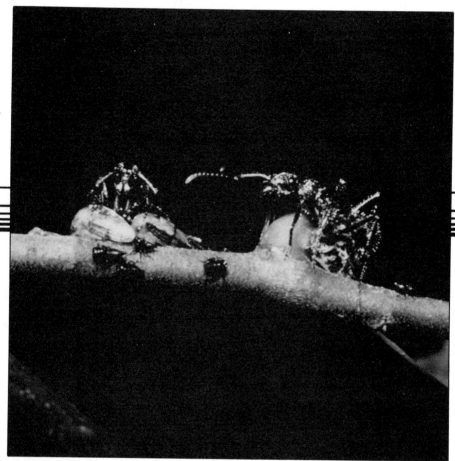

A liquid food chain on Barro Colarado Island, Panama. Plant sap is sucked by the scale insects on top of the twig. In turn, the ants "milk" the scale insects for their honeydew excreta. (Photograph by Eldridge S. Adams.)

CHAPTER 1

Introduction

1.1   What Is Ecology?

The word ecology comes from the Greek *oikos* meaning house or home, and was first used by the German biologist Haeckel in 1896. He regarded the ecology of an organism as ". . . . . the knowledge of the sum of the relations of organisms to the surrounding outer world, to organic and inorganic conditions of existence. . . ." Ecology as a distinct discipline grew out of natural history early in this century as natural historians began to collect their observations into a body of theory. Pivotal in this process was the work of Charles Darwin. Although his book *On the Origin of Species* was published in 1859, before the term ecology was coined, it contains many seeds that have grown to dominate modern ecology.

    Krebs (1985) gives the following simple, modern, and comprehensive definition of ecology: *Ecology is the scientific study of the interactions that determine the distribution and abundance of organisms.*

1.1.1   Ecological Terminology

Any specialized discipline inevitably develops its own terminology. Most of these terms are introduced and defined in the relevant chapters. This section is an introduction to the broader concepts and types of study that constitute ecology.

In Haeckel's definition of ecology, he refers to the "surrounding outer world," which we now call the **environment** of an organism. His "organic and inorganic conditions," we call biotic and abiotic environmental factors, respectively. **Biotic factors** are the other organisms encountered, whether of the same or different species. **Abiotic factors** are the physical and chemical conditions such as temperature, moisture, respiratory gases, and substrate.

Biology attempts to define and explain patterns within and among organisms at each of the following hierarchy of levels of organization:

LEVEL OF ORGANIZATION	SUBDISCIPLINE OF BIOLOGY
Organic molecules	Biochemistry
Subcellular organelles	Cell biology
Cells	
Tissues	Histology
Organs	
Organ systems	Anatomy, physiology
Organisms	
Populations	
Communities	Ecology
Ecosystems	

Each level of organization has properties peculiar to it that are not identifiable at the levels below, but studies at higher levels must take account of the lower levels. For example, a random collection of tissues—say blood, muscle, and connective tissue—do not make a heart. These tissues must interact in particular ways to form that organ. Similarly, a community may be a coherent unit (albeit much less tightly organized than a heart), which can be understood only by a consideration of the constituent populations and their interactions.

The three ecological levels of organization are somewhat distinct from one another because of the types of questions that are posed by ecologists. Each ecological level and examples of the questions posed at each level and the approaches used to answer them are described next.

A **population** is a group of individual organisms of the same species in a given area. Population ecologists ask, "Why is this population of a particular density?" Answers lie in the inherent biology of the species (constraints imposed by the levels of organization below), and the ways in which members of the population interact with their environment.

A **community** is a group of populations of different species in a given area. It may include all the populations in that area—all plants, animals, and microorganisms—or may be defined more narrowly as a particular group such as the fern community or the seed-eating bird community of that area. A major concern of community ecologists is the question, "Why is this community of a particular diversity?" **Diversity** is a combination of the number of species and the number of individuals of each species in a community. Another important question is, "Why does a particular community occur at a given location?" Answers are sought in the influences of the abiotic environment, how communities interact, and how communities change through time. An important point stemming from evolutionary biology is that communities are much looser assemblages than are populations, and that answers to problems at the community level must pay due regard to the evolutionary histories of the constituent populations.

An **ecosystem** is the whole biotic community in a given area plus its abiotic environment. It therefore includes the physical and chemical nature of the sediments, water, and gases as well as all the organisms. Ecosystem ecology emphasizes the movements of energy and nutrients (chemical elements) among the biotic and abiotic components of ecosystems. A major concern is, "How much and at what rates are energy and nutrients being stored and transferred between components of a given ecosystem?"

Because the ecosystem is the highest level of biological organization, all ecological concepts can be set within its framework. The biotic components of any ecosystem are linked as **food chains**. For example, the population in an African grassland may be characterized by their feeding relationships in the following two food chains:

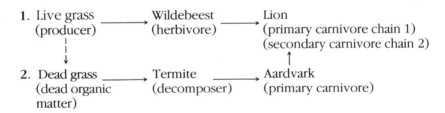

The broken arrows show how the two chains are linked into a **food web**. Real food webs usually have hundreds of species interlinked by their feeding habits. The terms producer, herbivore, primary carnivore, secondary carnivore and decomposer refer to **trophic levels**. Notice that a particular population may exist simultaneously on more than one trophic level as do lions in this example. Similarly, the termites are simultaneously decomposers feeding on dead organic matter and microbivores because they inevitably feed on living microorganisms present in the dead grass.

Population ecologists study **competition**, usually between populations from the same trophic level (such as herbivores competing for the same grass), and **prey-predator** interactions between members of adjacent trophic levels. For example, a lion population influences that of the wildebeest and vice versa. In tropical ecosystems **mutualistic** interactions between populations, in which all participants benefit, are also important. Termite populations have mutualistic associations with fungi or Protozoa to help them digest dead grass.

Food webs are basic units of ecosystem ecology because it is around them that energy and nutrient transfers take place. In addition to the organisms, there are also exchanges with the abiotic environment. Figure 1-1 shows the basic patterns of energy and nutrient transfers in a generalized ecosystem. Herbivores and carnivores are combined as **consumers** (or biophages) feeding on living organisms to contrast them with **decomposers** (or saprophages), which eat dead organic matter. The patterns of energy and nutrient movements differ significantly in their relationships with the abiotic environment and with the ecosystem boundary. *Energy flows through ecosystems,* being acquired from *outside* as light energy from the sun and being ultimately lost from the ecosystem as heat dissipated by the respiration of all community members. *Nutrients are cycled within ecosystems* to a much greater extent. Plants acquire their nutrients from an environmental inorganic pool *inside* their ecosystem in the atmosphere, water, soil, or sediment. These nutrients are passed around the food web in organic molecules, but most are eventually

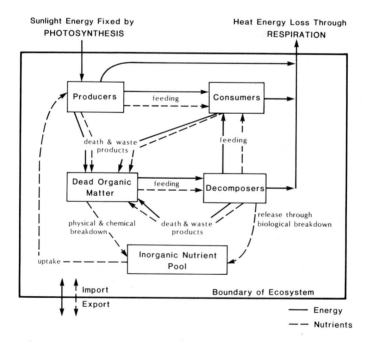

Figure 1-1   Patterns of energy and nutrient exchanges in a hypothetical ecosystem.

returned to the inorganic pool by the breakdown of dead organic matter. While organic and inorganic matter are transferred between ecosystems, such exports and imports are usually compared with transfers within ecosystems.

Most ecosystems change through time, sometimes quite rapidly. A fresh volcanic lava flow is colonized by plants and animals and may develop into a rain forest if the climate is suitable. Such changes are called **successions**. During successions the biota changes in species composition, and the abiotic environment is modified by the interactions between physical and chemical factors and the organisms. For example, a soil may develop where there was previously rock. Inevitably changes in patterns and magnitudes of energy and nutrient transfers occur during these changes.

Two other terms commonly encountered in ecology are **habitat** and **niche**. Elton (1927) likened habitat to the "address" and niche to the "profession" of an organism (or population) in an ecosystem. In the food web on p. 5, the habitat of termites would be dead grass and soil, and their niche would be decomposition of dead grass and food for aardvarks. In this book, niche is given a more formal definition introduced in section 5.1.4. Habitat has acquired two distinct meanings; the first, as just described, is the locale occupied by a population. This meaning is adequately replaced by the environment of a population. The second meaning, and the one adopted here, is the spatial subdivision of the environment within an ecosystem into convenient units. These units do not always coincide with the area occupied by specific populations. For example, the spatial unit (habitat) "upper tree canopy" in a forest may be occupied by both leaf-eating monkeys and sap-sucking insects. However, the former occupy a large area of forest, while the latter may be restricted to a few leaves on a single tree.

Applied ecology (or **human ecology**) is the use of ecological concepts to describe human activities and the determination of ways in which people can best obtain their needs from ecosystems. Ecosystems that are substantially altered by human activities are said to be **managed**, while those free from such disturbances are referred to as **natural**. This subdivision is arbitrary, since most ecosystems have been occupied by people at some time. It is best to think of natural ecosystems as those in which human populations play no greater role than other dominant populations in the community. The last three chapters are devoted to applied ecology. Other aspects, such as harvesting of populations and pollution are dealt with in other chapters.

1.1.2   The Problem of Boundaries

The definitions of population, community, and ecosystem are deliberately vague in referring to the "given area" occupied. This lack of precision reflects the difficulty of setting clear boundaries. Where, for example, does one population end and an adjacent one begin? Sometimes the answer is straightforward; the populations of bamboo plants on Mounts Kenya and Kilimanjaro in East Africa are separated by 300 km and clearly constitute distinct populations with negligible interactions between them. More frequently, no precise answer is possible. A forest community may intergrade with an adjacent savanna at a particular place. Indeed, they may change one into the other through time.

In many types of study the problem is avoided rather than overcome. Sometimes, particularly in population ecology, laboratory microcosms are established in which the container sets the boundary. Field ecologists often take representative samples and express population density or rates of processes on the basis of units of area; for example, 50 wildebeest pre square kilometer (50 km^{-2}) or an energy fixation rate by plants of 1500 kilojoules per square meter per annum (1500 kJ m^{-2}a^{-1}). This problem of defining boundaries only becomes acute in ecological mapping where ecologists use a mixture of subjective approaches (common sense and experience) and objective techniques, including statistical manipulations and instrumental environmental sensing.

1.2   Where Are the Tropics?

Geographically, the tropics are the highest latitudes at which the sun can be observed vertically overhead. This corresponds to 23.5° north and south of the equator. Such a definition is inadequate ecologically, since the terms "tropical," "temperate," etc., have climatic connotations and climate does not change uniformly as latitude increases. Ecologists have long recognized that the vegetation of the earth can be conveniently divided into **biomes**, recognizable by structural attributes and species composition, and that the distribution of these biomes is related to climate and latitude. For example, forest biomes can be subdivided into tropical, temperate, and boreal types. The major types of biome are defined and described in Chapter 8. In simple terms, a forest refers to an area with a closed canopy of trees, a savanna has a broken canopy of trees (or no trees) and a grassy herb layer, and a desert has a low woody or herb stratum interspersed with a large proportion of bare ground. For climatic correlates (such as rain forest, semiarid savanna) and other vegetation types (such as woodland, alpine savanna) see Chapter 8 for precise meanings.

Figure 1-2 shows an ecological boundary between tropical and other environments based on the division between characteristically tropical and extratropical biomes. Climatically, the average temperature at sea level is generally above 18°C during the coolest month in the tropics, but there are anomalies (see Figure 1-2). At higher altitudes the lowest monthly temperature may be much less, but throughout the tropics, the annual range of temperatures is rarely greater than the largest

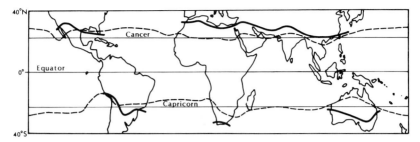

Figure 1-2   The delimitation of tropical environments; the heavy line depicts the limits of tropical vegetation types. Also shown are the geographic tropics and the 18°C mean isotherm at sea level for the coolest month (dashed line). (Modified from Walter [1979] and Bolin et al. [1979].)

diurnal fluctuation. In contrast, the seasonal temperature range is greater than the daily one at higher latitudes.

Over much of the temperate zone low winter temperatures greatly reduce, or preclude, plant growth and consequently the activity of other organisms. In the tropics, seasonal fluctuations in temperature do not limit biological activity. However, seasonal changes in rainfall are of great significance as discussed next.

1.2.1   Rainfall Seasonality in the Tropics

A relatively small proportion of the tropical zone receives sufficient rainfall for unrestricted plant growth throughout the year. Figure 1-3 is a classification of seasonal rainfall regimes, with a typical example from each. Areas with low mean annual rainfall usually experience great fluctuations from year to year. In arid areas the fluctuations may be so great that the amount of rainfall in any one year is unpredictable (Figure 1-4). High annual rainfall is usually correlated with short dry seasons. For example, in India, regions with 3000 mm annual rainfall generally have less than three dry months, while those with 1500 mm may have up to seven (Walter 1979).

Plants obtain moisture from the soil. The amount of water in a soil results from the balance between several factors. Import from rain is counteracted by runoff (ultimately to oceans or lakes) and **evapotranspiration,** which includes evaporation directly from the soil surface and transpiration by the plant canopy. In seasonal climates, soils lose their moisture during dry periods and are recharged during the rainy seasons. Any surplus during the rains is lost as runoff. Because of their high temperatures and solar radiation, tropical regions have higher potential evapotranspiration than most temperate regions. This means that in general the same amount of precipitation is less effective in maintaining a moist soil in the tropics. In fact, over much of the tropics, rainfall during dry seasons is less than the potential evapotranspiration dictated by solar

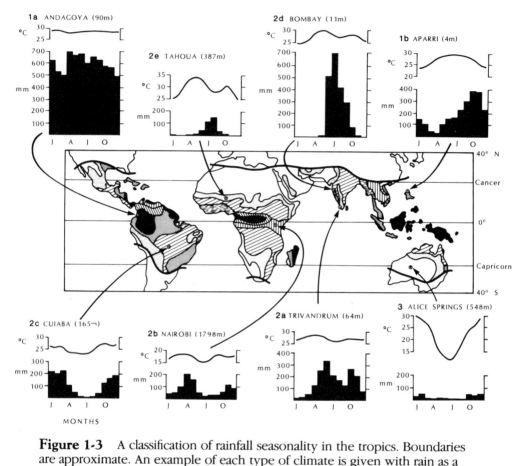

Figure 1-3 A classification of rainfall seasonality in the tropics. Boundaries are approximate. An example of each type of climate is given with rain as a bar chart and mean monthly temperature as a line graph. Note the small temperature variation through the year but lower temperatures at higher altitude (given next to place names). (Modified from Jackson [1977].)

1. Humid:
   ■ 1a. Annual precipitation (AP) > 2000 mm; no month < 100 mm
   ▨ 1b. AP   < 2000 mm; some months < 100 mm
2. Wet and Dry (moist to semiarid):
   ▨ 2a. AP   1000–2000 mm; two wet seasons; some months < 50 mm
   ▥ 2b. AP   650–1500 mm; two short wet seasons; one long dry season; some months < 25 mm
   ▨ 2c. AP   650–1500 mm; one long wet season (three to five months > 75 mm); one long dry season
   ▨ 2d. AP   < 1500 mm; one very wet season; one long dry season
   □ 2e. AP   250–650 mm; one short wet season (three to four months > 50 mm); one long dry season
3. Dry (arid):
   □   AP   < 250 mm; no pronounced wet season; annual rainfall often occurs during short periods, with very little or none at other times.

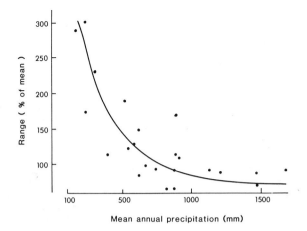

Figure 1-4   Actual annual rainfall in any year is much more variable compared with the annual mean in drier areas. Results from 23 rainfall stations in East Africa each with more than 20 years of continuous records are plotted. Mean rainfall (all years) is compared with the range (wettest minus driest year) as a percentage of the mean. (Data from Pratt and Gwynne [1977].)

radiation. As a result, there is a temporary water deficit as soil reserves are depleted. In humid climates these deficits occur only briefly, if at all, but in drier climates they occur for much of the year. Most of the time, arid climates experience only a brief period of soil moisture recharge, negligible runoff, and short periods of plant growth following rainfall.

Further Reading

For more detailed accounts of tropical climates and water balance, the following may be consulted:

Jackson, I. J. (1977). *Climate, Water and Agriculture in the Tropics.* London: Longman.
Nieuwolt, S. (1977). *Tropical Climatology.* London: Wiley.

Tropical rain forests have higher primary production than other natural terrestrial ecosystems. Shown here is the undisturbed canopy of Corcovado National Park, Costa Rica. (Photograph by Robert K. Colwell.)

CHAPTER 2

Energy Flow in Ecosystems

All organisms require energy to move, grow, reproduce, and perform bodily functions. In most ecosystems this energy derives entirely from sunlight, which is converted to chemical energy by the process of photosynthesis in green plants. The light energy converted to growth of new plant tissues then enters food webs in forms available to organisms requiring organic food (animals, fungi, many bacteria). All the organisms in an ecosystem respire and eventually release the energy supplied by the plants as heat. The increase in energy content of an individual, a population, or a community per unit of time is called **production**. In most studies the rate of production is expressed on an annual basis (a^{-1}). Such quantitative information is important in understanding the relative importance of different trophic groups within an ecosystem and in comparing patterns of energy flow in different ecosystems.

Many of the ideas that have developed into production ecology can be found in Charles Elton's *Animal Ecology* (1927), but intensive studies did not begin until the work of the American aquatic ecologists Juday and Lindemann in the 1940s. Particular emphasis and impetus was given in the 1960s and 1970s by the International Biological Program (IBP), whose aim was to investigate "the biological basis of productivity and human welfare." To give easy access to the literature, IBP terminology is used here, although it is not an entirely consistent system for all organisms.

Figure 2-1 shows energy flow in a Caribbean rain forest. Two of the major concerns of production ecology are shown: the estimation of **standing crop** of energy present in each part of the food web, and the **rate of flow** of energy between these components. Some organisms have a much higher rate of flow relative to their standing crop than others. This can be illustrated by the ratio of energy dissipation (respiration) to standing crop. For microorganisms the ratio is 47, 450:1; for the animal community, 270:1; and for plants, 4:1. These differences are clearly related to body size. Small organisms such as bacteria process energy more rapidly than do large trees. Also shown in Figure 2-1 is the subdivision of terrestrial ecosystems into above-ground and below-ground parts. Relatively little is known about root production and its dependent food web in most ecosystems. Much of this chapter concentrates, therefore, on the production and utilization of above-ground production.

The chapter opened with an illustration of energy flow in one ecosystem. The nature of energy transformations in ecology is described next, followed by a dissection of ecosystems to look at **primary production** by plants (autotrophs) and **secondary production** by all other organisms (consumers and decomposers, or heterotrophs). An understanding of these processes allows for the construction of a model of a hypothetical ecosystem and the subsequent examination of patterns of energy flow in the distinct herbivore and decomposer food webs.

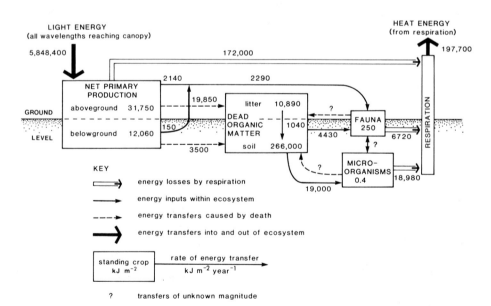

Figure 2-1   Energy flow in a Puerto Rican rain forest. (Data from Odum [1970].)

2.1   Energy Transformation in Ecology

Energy transformations of ecological significance occur at the subcellular level. They comprise photosynthesis (energy fixation) and respiration (degradation of chemical energy to heat). Photosynthesis is the use of light energy to build carbohydrate molecules from inorganic sources of carbon dioxide and water in the chloroplasts of green plants. The overall reaction is:

$$6\,CO_2 + 6\,H_2O + 2964\ \text{kilojoules} \longrightarrow C_6H_{12}O_6 + 6\,O_2$$

carbon   water   light energy   carbohydrate   oxygen
dioxide   (glucose)

The light energy input is determined by the thermodynamics of the reaction that produces glucose. Plants do not capture all the light energy available, and they also degrade some of the energy fixed to produce more complex carbohydrates and other compounds. The energy assimilated in photosynthesis is degraded by plants and by heterotrophs in their respiration, and the energy released is used in their life processes. Although the detailed biochemistry is entirely different, the overall equation for respiration of carbohydrates is the exact reverse of that for photosynthesis. Oxygen and glucose combine to produce water and carbon dioxide, and energy is released as heat. Fats and, more rarely, proteins may also be respired by yet different metabolic pathways, with the former yielding more and the latter less energy on average than carbohydrates on combustion (Table 2-1).

Some bacteria (chemoautotrophs) are able to obtain their energy by **chemosynthesis**. For example, sulfur bacteria synthesize organic compounds using the energy released when they oxidize hydrogen sulfide. In most ecosystems, chemosynthesis is negligible compared with photosynthesis.

Chemical reactions are governed by the laws of thermodynamics, but the application of these laws directly to ecological energetics is not straightforward. Ecological systems (from individual to ecosystem) are open thermodynamic systems because both energy and matter are able to enter or leave the system. The classical laws of thermodynamics describe closed systems in which only energy can cross the system boundary. Ecological energy transformations are governed by the first law of dynamics (Wiegert 1968). This law can be stated as follows: "Energy may be transformed from one form to another, but is neither created nor destroyed." This simply means that the same total amount of energy must be present after the transformation as before, even though it may be in different forms. In other words, equations involving energy must balance.

Table 2-1   Energy content of organic materials and organisms. Note that differences in energy values of organisms (Nos. 2 to 4) result from differences in body composition (No. 1). Units = kJ g^{-1} (dry weight).

1. ORGANIC NUTRIENTS

Carbohydrate	15.5–17.6
Fat	39.7
Protein	16.3–17.3

2. GRAND MEANS FOR TERRESTRIAL ORGANISMS (Cummins and Wuycheck 1971)

Plants	19.9	Invertebrates	23.7
Vertebrates	27.3		

3. COMPARISON OF TROPICAL AND TEMPERATE VEGETATION (Golley and Leith 1972)

	Tropical	*Temperate*
Forests	17.1–17.6	19.2–20.1
Grasslands	16.7	16.7

4. VARIATIONS IN ENERGY CONTENT WITHIN ANIMAL SPECIES

a. Migrant birds moving from United States to tropical America and back (Odum et al. 1965):

Before migration	33.0–34.7
After return	23.8–30.9

b. Castes of the termite *Trinervitermes geminatus* (Baroni-Urbani et al. 1978):

Reproductives	29.2
Eggs	24.7
Major soldier	26.0
Workers	20.8

The energy exchanges between an animal and its environment can be characterized as follows (Wiegert 1968):

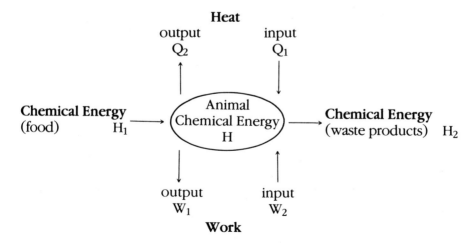

According to the first law of thermodynamics, any growth in this animal (ΔH) must result from a positive overall balance between work, heat, and matter exchanges with its environment:

$$\begin{array}{ccccccc} \Delta H & = & (H_1 - H_2) & + & (Q_1 - Q_2) & + & (W_1 - W_2) \\ \text{change in} & & \text{matter} & & \text{heat} & & \text{work} \\ \text{energy content} & & \text{exchanges} & & \text{exchanges} & & \text{exchanges} \\ \text{of animal} & & & & & & \end{array}$$

Work exchanges (W_1, W_2) between animals and their environment do occur (for example, output when termites build a mound of soil; input when birds soar on convection currents). However, such exchanges are negligible compared with other terms in the equation and can be ignored. Heat exchanges (Q_1, Q_2) are effectively equal to the heat dissipated in respiration. Other heat exchanges such as when animals bask in the sun (input) or dive into cold water (output) balance out over the long term. If this were not so, the animal would get continually hotter or colder.

The equation can therefore be rewritten as:

$$\begin{array}{ccccccc} \text{Production} & = & \text{Consumption} & - & \text{Waste products} & - & \text{Respiration} \\ (\Delta H) & & (H_1) & & (H_2) & & (Q_1 - Q_2) \end{array}$$

Converting to IBP terminology, the equation is (see Figure 2-6):

$$\begin{array}{ccccccc} P & = & C & + & FU & + & R \\ \text{production} & & \text{consumption} & & \text{waste products} & & \text{respiration} \end{array}$$

The same type of reasoning can be applied to populations of all types of organism or to whole ecosystems.

The energy unit used is the **joule** (J) and its multiples such as the kilojoule (kJ). The earlier ecological literature uses the calorie (1 joule = 0.239 calories). Such universal units are essential if direct comparisons are to be made between different forms of energy such as light, heat, and organic materials. Where production of only organic matter is of interest, **biomass** (weight per unit area) is often used instead of energy units. In many cases this unit is the best available. However, biomass is an inaccurate index for energy budgets if different components have different energy values. For example, plant food usually contains less energy per unit weight than does herbivore tissue. Such differences are illustrated in Table 2-1. Note that the energy content of organisms is restricted to a fairly narrow range on average. Golley (1969) suggests that tropical forest plants have lower energy values than their temperate equivalents, but there is no evidence to indicate similar divergences in other organisms or ecosystems (Table 2-1). However, in some populations, energy values vary by as much as 30% within and between individuals. Examples include migratory birds and termites (Table 2-1). Migratory birds store energy from their food as fat deposits prior to migrating thousands of miles to avoid the temperate winter. During the flight, they feed little and deplete the fat deposits. Alate termites, the potentially reproductive individuals, are also rich in fat. This energy is necessary for the swarming nuptial flight and the establishment of a new colony. Such differences in energy content of individuals must be accounted for if accurate and balanced energy budgets for populations are to be drawn up.

Because all organisms dissipate energy through respiration, production of one trophic level is always less than that of the preceding level. As a result, a whole ecosystem can be illustrated as a pyramid with plant production as the broad base and production of top-level consumers as the apex. Figure 2-2 shows such a pyramid for a West African savanna. The idea of ecological pyramids comes from Elton (1927), who envisaged a pyramid of numbers in which, ''a) smaller animals are preyed upon by larger animals and b) the smaller animals can increase faster than the larger ones and so are able to support the latter.'' Sometimes numbers are not adequate to account for differences in abundance between successive trophic levels. For example, one tree may support many insect herbivores; one host, many parasites. Biomass is an alternative measure that accounts for differences in size; a tree weighs more than its insects, and a buffalo, more than its ticks. Even biomass can create anomalies, however. A population of low biomass but high production can support a less productive predator population of greater biomass. Such a situation arises in some planktonic communities and is likely in decomposition systems where a small mass of productive microorganisms supports a larger mass of microbivores. The pyramid of production (or some other measure of energy flow) is the only one that *must* take a pyramidal shape.

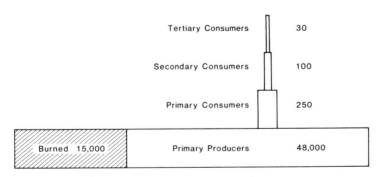

Figure 2-2   Pyramid of production in the Lamto savanna of the Ivory Coast. Producers are mainly herbaceous plants. Consumer production is the aggregate of herbivore and decomposer food webs at each level. Herbivores are primarily grasshoppers and rodents; decomposers are earthworms and termites. Secondary and tertiary consumers include ants, spiders, birds, lizards, and frogs. Units = kJ m^{-2}a^{-1}. (Data from Lamotte [1979].)

2.2   Primary Production

Table 2-2 shows average net primary production (in terms of dry weight) for terrestrial ecosystems. According to these estimates more than 60% of terrestrial primary production occurs in the tropics. It is evident that tropical ecosystems generally have greater production than comparable ones at higher latitudes. However, there is much variation within and substantial overlap between categories. For example, primary production in tropical forests ranges from approximately 1000–3000 g $m^{-2}a^{-1}$, temperate forests 600–2500 g $m^{-2}a^{-1}$, and boreal forests 400–2000 g $m^{-2}a^{-1}$. In other words, some tropical forests are less productive than some of their temperate or boreal counterparts. Similar variations in plant production of savannas and grasslands are discussed in Section 8.2.

The highest rate of primary production occurs in tropical rain forests, and yet the average for these ecosystems, 2300 g $m^{-2}a^{-1}$, represents conversion of less than 0.001% of the total solar radiation reaching 1 m^2 of the planet. Agricultural experiments have yielded a maximum efficiency of only 0.2%. Why is so little of the sun's energy converted to plant production? The following analysis, based largely on Monteith (1972), accounts for these energy losses in a variety of ways, which are summarized in Figure 2-3.

2.2.1   Energy Losses Between the Outer Atmosphere and the Plant Canopy

The supply of energy from the sun is very close to a constant of 42.9 x 10^6 kJ a^{-1} at a surface of 1 m^2 held perpendicular to the sun's rays. After its passage through the atmosphere to a similar surface above the plant canopy, only 15% and 8% of this energy remains on average in tropical and temperate regions, respectively. The reasons for these losses are the earth's geometry and rotation and restrictions on transmission through the atmosphere (Figure 2-4). Higher latitudes also exhibit greater seasonal variation in solar radiation with an annual range of 210–1800 J $cm^{-2}day^{-1}$ at ground level in Berlin (53°N) compared with 1670–2010 J $cm^{-2}day^{-1}$ in Singapore (1°N). While production is potentially continuous in tropical areas provided that other factors are limiting, low light intensity and low temperatures reduce production to negligible levels during the temperate winter.

2.2.2   Interception of Light by Plants

Light energy used in photosynthesis is absorbed mainly by the pigment chlorophyll. All wavelengths are not absorbed uniformly, since chlorophyll responds only to the visible spectrum, which comprises around 50% of the total solar radiation. Within the visible spectrum, some light (particularly in the green wavelengths) is reflected. Overall a little more

Table 2-2 Estimates of standing crop and annual net primary production (dry weight) for terrestrial vegetation types. Approximate mean values are given, but ranges are large (see text). Also given is percent of land surface covered by each and its percent contribution to total terrestrial primary production. Data from Ajtay et al. (1979).

TYPE OF ECOSYSTEM			STANDING CROP ($g\ m^{-2}$)	NET PRIMARY PRODUCTION ($g\ m^{-2}$)	PERCENT OF TOTAL LAND AREA*	PERCENT OF TOTAL PRIMARY PRODUCTION
NATURAL ECOSYSTEMS						
Forests	Tropical:	humid	42,000	2300	6.7	17.3
		seasonal	25,000	1600	3.0	5.4
		mangrove	30,000	1000	0.2	0.2
	Temperate:	deciduous/mixed	28,000	1300	2.0	2.9
		evergreen/coniferous	30,000	1500	2.0	3.4
	Boreal:	closed canopy	25,000	850	4.4	4.2
		open canopy	17,000	650	1.7	1.2
Woodlands and shrublands	Tropical:	wooded savanna	9000	1500	11.0	19.2
	Temperate:	woodland	18,000	1500	1.3	2.3
		scrubland	7000	800	1.7	1.5
Grasslands	Tropical:	grass savanna	2200	2300	4.0	10.4
	Temperate:	moist grassland	2100	1200	3.3	4.5
		dry grassland	1300	500	5.0	2.8
Deserts (mainly tropical)	Desert and semidesert scrub		1100	200	6.0	1.4
	Extreme desert:	hot and dry	60	10	5.4	negligible
		cold and dry	300	50	0.7	negligible

*10% of the land surface is perpetual ice with a production of zero.

Table 2-2 (Continued)

TYPE OF ECOSYSTEM			STANDING CROP ($g\,m^{-2}$)	NET PRIMARY PRODUCTION ($g\,m^{-2}$)	PERCENT OF TOTAL LAND AREA*	PERCENT OF TOTAL PRIMARY PRODUCTION
Tundra	Low arctic/alpine		2300	350	2.9	1.2
	High arctic/alpine		750	150	2.4	0.4
	Polar desert		150	25	1.0	negligible
Swamps	Tropical		15,000	4000	1.0	4.5
	Temperate		7500	2500	0.3	0.9
Peatlands and bogs (mainly temperate)			5000	1000	1.0	1.1
MANAGED ECOSYSTEMS						
Forest plantation			20,000	1750	1.0	2.0
Agricultural	Tropical:	annuals†	60	700	6.0	4.7
		perennials	6000	1600	0.3	0.3
	Temperate:	annuals†	100	1200	4.0	5.4
		perennials	5000	1500	0.3	0.6
Desertified areas			500	100	8.0	0.9
Human area			4000	500	1.3	0.3

* 10% of the land surface is perpetual ice with a production of zero.

† Averaged over the whole year, including times when no crop is being grown.

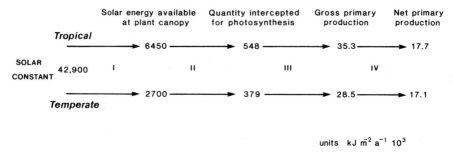

units   kJ m̄^{2} a^{-1} 10^3

Figure 2-3   Summary of the energy losses in primary production between solar radiation and net primary production. The figures compare tropical and temperate intensive agriculture from the analysis of Monteith (1972). Losses are due to: (I) geometry and rotation of the earth and transmission through the atmosphere; (II) energy not absorbed by chlorophyll; (III) rate of carbon dioxide diffusions and photochemical efficiency; and (IV) plant respiration.

than 40% of total solar radiation is photosynthetically active, irrespective of latitude.

All of the sun's rays reaching the ground are not intercepted by plants. There are some permanently barren areas such as ice caps and urban areas and many seasonally barren areas, particularly in agricultural systems that leave bare ground between one harvest and sowing of the next crop. Different plant communities intercept different amounts of light, depending on the density of the canopy. Leaf-area index, the area of leaf (m^2) arranged vertically above each square meter of ground, is

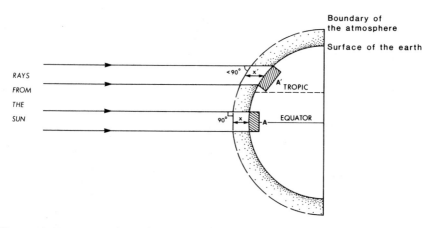

Figure 2-4   More solar radiation reaches a unit area of the earth's surface in the tropics than at higher latitudes. Note the difference in the angle of incidence, the spread of rays on the ground (area A < A$'$), and distance traveled through the atmosphere (length x < x$'$).

a useful measure of canopy density. Annual average leaf-area indices are at their highest in evergreen forests (10 or more) and their lowest in arid deserts (0.5 or less). In seasonal environments, variation through the year may be tenfold or more. For example, in an Indian grassland, Misra and Misra (1981) found an index of 0.5 in the dry season when the plants had died back and 5.0 in the wet season when they were actively growing. A green plant can grow in its particular light environment only if photosynthesis exceeds respiration. In the absence of other limitations (such as soil moisture) this fact limits the leaf area index of a plant community. Individual plants that are too shaded to obtain a net energy gain die. Because of the greater intensity of radiation received in the tropics, leaf area is sometimes greater in tropical eco-systems than in comparable temperate ecosystems.

No overall estimate can be given for interception efficiency. Defining it as the fraction of incident light intercepted, Monteith (1972) suggests annual averages of 20% for tropical intensive agriculture, 5% for tropical subsistence agriculture, and 33% for temperate intensive agriculture. This greater interception efficiency of temperate agriculture leads to tropical production being only slightly greater than the temperate estimate given in Figure 2-3. This difference is much smaller than in most natural ecosystems (Table 2-2) and illustrates that intensive tropical agriculture often does not realize as much of its potential as temperate agriculture.

2.2.3   The Effects of Plant Biochemistry

Solar energy absorbed by chlorophyll is not all converted to plant growth. The theoretical maximum photochemical efficiency of photosynthetic biochemistry is 21.5% (Monteith 1972). Of greater interest to ecologists is the efficiency of **gross primary production**, the rate of energy fixation taking into account all the ecological factors that depress this theoretical maximum. Environmental factors such as shortages of water or nutrients are obviously significant. More subtle is the slow rate at which carbon dioxide diffuses from the atmosphere to the chloroplasts. This problem is at its most acute in high light intensities (such as in the tropics) and limits the rate at which photosynthesis can proceed. Plants also respire some of their fixed energy, and this affects the amount of energy available to higher trophic levels. The amount of energy that is realized as plant growth is **net primary production** such that net primary production = gross primary production – plant respiration. It is this net primary production that is available to consumers (including humans). On average, only about 30% of gross primary production is available as net primary production in tropical rain forests compared with 40% in temperate forests (Kira 1975, Unesco 1978). This difference is due in part to the fact that respiration is proportionately faster at high temperatures than is photosynthesis. Respiration is also greater in plant communities with a high ratio of standing crop to production. In forests the photosynthetic leaves support a large nonphotosynthetic biomass of wood while in herbaceous vegetation most of the above-ground tissues are photosynthetic (Figure 2-5).

Another physiological factor of significance when comparing production at low and high latitudes is the predominance of different biochemical pathways in different environments. The Calvin cycle is the classic description of photosynthesis. Carbon dioxide is fixed initially as a three-carbon compound, phosphoglyceric acid, a reaction catalyzed by the enzyme ribulose biphosphate carboxylase. Plants with this pathway are called C_3 plants. In contrast, C_4 plants have a more recently discovered pathway that has a four-carbon compound as the first product. C_4 carbon fixation is catalyzed by phosphoenol pyruvate carboxylate to produce oxaloacetic acid. The C_4 pathway is found only in flowering plants and in less than 20 of the 300 families. It has evolved from the C_3 pathway independently in many species with fossil C_4 plants dating back more than 2.5 million years (Nambudiri et al. 1978)

A third biochemical pathway, crassulacean acid metabolism (CAM), is found in a few specialized plants that undergo extreme water stress. To conserve water the stomata remain closed during the day. At night carbon fixation occurs by the C_4 route to produce malic acid, which is stored in the cell vacuoles. During daylight the malic acid is decarboxylated and carbon refixed by the Calvin cycle. One species of fern, one gymnosperm, and species from 23 families of angiosperms exhibit this pathway. It is obligatory in some species but only occurs at times of water stress in others. CAM has evolved many times from the C_3 system and independently of the C_4 pathway.

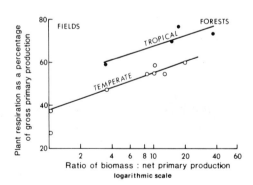

Figure 2-5   Respiration of plant communities in relation to the ratio of biomass to net production in each. Note that forests have a greater respiratory demand than fields, and tropical communities have a higher respiratory demand than those in the temperate zone. (Modified with permission of Macmillan Publishing Company from *Communities and Ecosystems* by R. H. Whittaker. Copyright © 1975 by Robert H. Whittaker.)

Table 2-3 summarizes the differences in gross physiology of the three systems of carbon fixation. The following generalizations emerge:

1. C_4 plants are the most efficient photosynthetically in optimum conditions and are adapted to hot, bright, and fairly dry climates.

2. C_3 plants are adapted to cool, moist, and shady environments.

3. CAM plants are adapted to very dry environments.

Such distributional effects do occur as illustrated by the change from 100% C_4 grasses in hot, dry lowland areas to 100% C_3 grasses in cool, moist montane environments of Kenya (Tiezen et al. 1979). However, most tropical trees have C_3 photosynthesis irrespective of macroclimate, possibly as a reflection of seedling and sapling growth in the cooler, shadier microclimate beneath the canopy. The precise conditions under which C_3 and C_4 photosynthesis are advantageous result from quite subtle interactions between light, temperature, and moisture that are not yet fully understood (see, for example, Rundell 1980). CAM plants are found only in very dry environments where their slow rate of growth does not put them at a competitive disadvantage. Desert succulents such as cacti and euphorbias fit into this category as do some orchids and bromeliads, which are canopy epiphytes in tropical rain forests. These latter plants apparently experience water stress because their tree bark substrate does not retain much moisture.

Although photosynthetic rate is greater in C_4 species, there is no compelling evidence that C_4-dominated natural communities are more productive than their C_3 counterparts (Caldwell 1975). Most plant communities are of such low efficiency, compared with their biochemical potential under optimum conditions, that the ecological significance

Table 2-3   Comparison of the gross physiological attributes of the C_3, C_4, and CAM photosynthetical pathways. Data from Black (1971), Mooney (1972), and Smith et al. (1979).

ATTRIBUTE	PHOTOSYNTHETIC PATHWAY		
	C_3	C_4	CAM
Optimum temperature for CO_2 fixation (°C)	15–25	30–45	approximately 35
Photosynthetic rate under optimal conditions (mg CO_2 dm^{-2} hour^{-1})	15–35	40–80	0.5–0.7
Light saturation (the point at which the rate of CO_2 diffusion limits photosynthesis) as % of full sunlight	10–25	100	—
Water use efficiency (number of grams of water to produce 1 g plant matter)	380–900	250–350	50

of the C_3 and C_4 pathways is more apparent in the distribution of species than in net primary production. However, the most productive agricultural crops such as sugar cane and maize are C_4 species. Careful management enables these plants to realize more of their potential production.

Even if light and carbon dioxide are used with maximum efficiency, plant production in many environments is constrained by the secondary limitations of water or nutrients. In humid forests, water is rarely limiting, but in seasonal or unpredictable rainfall climates, primary production may be directly correlated with rainfall. This theme is developed further in Chapter 8. During the wet season in many semiarid West African savannas primary production is reduced yet further by inadequate supplies of nitrogen and phosphorus (Penning de Vries et al. 1979).

2.2.4   Below-ground Production

Plants translocate significant amounts of net primary production to support the growth of roots and other underground organs. Very little is known about below-ground energy flow in the tropics, particularly beyond the producer trophic level. A compilation of available data suggests that tropical vegetation translocates a smaller proportion of net production below-ground than do comparable temperate plants. On average, tropical forests retain 90% and savannas 50% above-ground compared with 75% in temperate forests and 40% in temperate grasslands (see Coleman 1976, Unesco 1978, Coupland 1979, Sarmiento 1984).

To summarize, given sufficient precipitation and nutrients, tropical ecosystems have greater primary production than temperate ones but to a lesser extent than would be expected from the differences in solar radiation. This is because of lower interception efficiency and higher respiratory demand at low latitudes. Some tropical plants have compensated to some extent by acquiring the C_4 pathway, which is more efficient at high temperatures and high light intensities. Many ecosystems do not reach the primary production suggested by the light climate because of other factors limiting their growth.

2.3   Secondary Production ————————————

Secondary production, like primary production, can be described by transfer efficiencies between the various steps in energy utilization. From the energy budget of a population of animals, two key efficiencies have been isolated (Figure 2-6). These are **assimilation efficiency (A/C%)**, which is a measure of the amount of ingested energy left after waste production, and **production efficiency (P/A%)**, the proportion of assimilated energy used to increase the energy content of the population. Examples of energy budgets including these transfer efficiencies are given

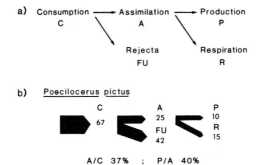

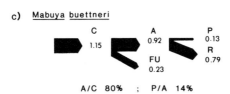

Figure 2-6   Energy budgets of animal populations. A/C and P/A are assimilation efficiency and production efficiency, respectively (see text). Units = kJ m^{-2}a^{-1}. *a*, IBP terminology. FU are initials for feces and urine. *b*, Herbivorous grasshopper population in Indian fields. (Data from Delve and Pandian [1979].) *c*, Carnivorous lizard population in the Lamto savanna of the Ivory Coast. (Data from Lamotte [1979].)

in Figure 2-6. The two animals chosen are from different taxonomic groups (invertebrate, vertebrate) and different trophic levels (herbivore, carnivore). From such data on a wide variety of organisms, the generalizations about transfer efficiencies presented in Table 2-4 have been derived. Note that in microorganisms, only production efficiency is applicable because their external digestion means that only assimilated energy enters the organism.

Production is realized either by growth in the bodies of individuals in a population (P$_g$) or by the production of new individuals through reproduction (P$_r$).The ratio between these two types of production varies considerably in different populations. P$_g$ accounts for more than 80% of total production in a tropical savanna lizard but only 9% in laboratory-reared flour beetles (Petrusewicz and Macfadyen 1970, Unesco 1979).

Waste materials other than egesta and excreta must be accounted for to obtain a balanced energy budget. Secretions, cast hair, feathers, and skin all fall into this category and are sometimes large proportions of assimilated energy. For example, earthworms in an Indian grassland secreted 1025 kJ m^{-1}a^{-1} of mucus compared with a tissue production of 678 kJ m^{-2}a^{-1} (Dash 1979).

Table 2-4   Assimilation efficiency and production efficiency as percentages for heterotrophic populations. Blank categories indicate types of population that do not exist or are of minor significance in most ecosystems. Modified from Heal and Mclean (1975).

	HERBIVORE A/C	HERBIVORE P/A	CARNIVORE A/C	CARNIVORE P/A	MICROBIVORE A/C	MICROBIVORE P/A	DECOMPOSER A/C	DECOMPOSER P/A
Microorganisms	—	—	—	—	—	—	—	40
Invertebrates	40	40	80	20*	30	40	20	30*
Vertebrates								
Homeotherms	50	2	80	2	—	—	—	—
Heterotherms	50	10	80	10	—	—	—	—

*Modified with data in Humphreys (1979).

2.3.1   Assimilation Efficiency (A/C)

The proportion of energy assimilated depends on the quality of food ingested. Biophages are noticeably more efficient than are saprophages. Animal food is usually of higher quality than plant food because of its high protein and low fiber concentrations. These differences are reflected in Table 2-4. Variations in food quality are particularly relevant to herbivores in seasonal environments. A/C in savanna-grazing mammals can be as high as 80% in the wet seasons but may drop to 20% during dry seasons. In fact, such grazing herbivores are "decomposers" during the dry season, since their intake is mainly dead grass. A similar problem of trophic categorization exists with animal decomposers such as earthworms and soil arthropods because the dead plant material they eat also contains fungi and bacteria. Most animal decomposers are, therefore, simultaneously microbivores. Termites have exceptionally high A/C compared with other decomposers, ranging from 50% to more than 90%. As a result, they have a great impact on tropical ecosystems in which they are abundant. There is no reason to suppose that A/C differs in tropical and temperate environments.

2.3.2   Production Efficiency (P/A)

A striking feature of the data in Table 2-4 is that homeothermic vertebrates have a lower P/A than other heterotrophs by an order of magnitude. An important consequence of this is that food chains including birds and mammals tend to be shorter than those involving heterotherms because more assimilated energy is dissipated as respiration at each stage. The extra metabolic cost incurred by these animals is used in part to maintain their constant body temperature, which leaves less energy for production. It is important to stress, however, that transfer efficiencies are properties of populations and relate to life history characteristics as well as the physiology of individuals. Birds and mammals often have lower rates of population growth because energy is put into large body size and parental care. Consequently, they devote less energy to maximizing production than do many heterotherms (Wieser 1984).

Detailed analysis of P/A by Humphreys (1979, 1984) shows more variation than Table 2-4 suggests, with finer subdivisions between taxonomic groups. For example, social ants and termites have a very low mean P/A (10%) compared with other insects (42%), probably because of the energy demand of social interactions. As with A/C, there is no evidence to show that there are consistent differences in P/A in tropical and temperate environments. However, Weathers (1979) has demonstrated that the metabolic rate of tropical birds is significantly lower than that of temperate birds.

Rate of production is correlated with body size in animals such that the rate of production per unit of biomass (P/B) in populations decreases as the body size of individuals increases (Figure 2-7). Rodent populations have a P/B ratio of around 3:1, while an African buffalo population has a ratio of only 0.1:1. Such relationships are well known in physiology because large animals have lower metabolic rates per unit body weight than do smaller animals.

2.3.3   Consumption Efficiency

A third transfer efficiency of importance to heterotrophs is **consumption efficiency**. It measures the amount of energy transferred from one trophic level to the next. For example, the consumption efficiency of herbivores is the total annual consumption of all herbivore populations as a percentage of annual net primary production. The food intake of

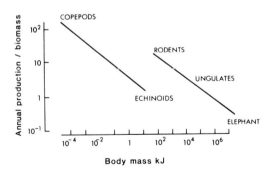

Figure 2-7   Relationships between the rate of production and body mass (energy units) for invertebrates and mammals. Per unit body mass, small animals are more productive than large ones and invertebrates less productive than mammals. The invertebrates of intermediate size between copepods and echinoids include mollusks, annelids, isopods, and insects. (Modified from Banse and Mosher [1980].)

individual animals is closely related to body size (Figure 2-8). From such information the consumption of each population in a trophic group can be estimated if its abundance is known. The consumption efficiency of the herbivore trophic level determines what proportion of energy enters

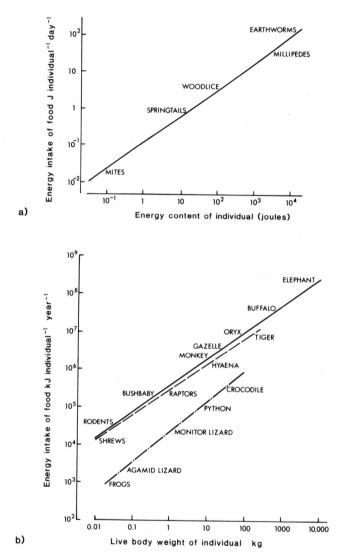

Figure 2-8   Relationships between body size and energy intake as food in various animals. *a,* Invertebrates in temperate decomposer food webs. (Modified from Reichle [1971].) *b,* Vertebrates (temperate and tropical) from different physiological and trophic categories. — = homeothermic herbivores; – – – = homeothermic carnivores; – · — · — = heterothermic carnivores. (Modified from Farlow [1976].)

the herbivore food web and what proportion passes to decomposers as shown in the following section.

2.4   Modeling of Energy Flow in Ecosystems

The study of production ecology has reached the stage at which complex energy flow parameters can be predicted from relatively simple information such as transfer efficiencies. For example, primary consumption in African savannas is estimated from data on rainfall and herbivore density in Chapter 8. While the precise nature of such relationships will be refined in the future, their value as generalizations is already such that energy flow models of the type shown in Figure 2-1 can be constructed. A basic assumption of such models is that ecosystems are in a steady state with respect to energy flow. This means that gross primary production is balanced by total community respiration (plants + consumers + decomposers). Without this balance the total amount of energy in an ecosystem would change over time, a state uncommon in natural ecosystems. In many savannas, dead organic matter does accumulate temporarily, but this is eventually decomposed by fire (see Figure 2-2).

Figure 2-9 is a model of energy flow in a hypothetical ecosystem. It is derived by setting net primary production at 1000 units, then making certain assumptions about consumption efficiencies and using the transfer efficiencies of Table 2-4. As will be seen in the next section, a fairly typical (but high) herbivore consumption efficiency of 10% is used, with half of primary consumption by invertebrates and half by homeothermic vertebrates. This model reveals important aspects of ecosystem energy flow that are not immediately apparent from Figure 2-1. Energy is recycled in the decomposer system because waste products and corpses of decomposer organisms are returned to the dead organic matter pool to be themselves decomposed.

It takes three passes through the decomposers before the energy content of the dead organic matter is less than 10% of the original quantity. Because of this conservative nature of the decomposer system and because it is composed almost entirely of heterotherms, its production is more than 98 times that of the herbivore system. Indeed, total animal production accounts for less than 6% of secondary production, the remainder being due to fungi and bacteria.

From the earliest attempts to formalize the concept of energy flow in ecosystems, it has been apparent that the pattern beyond the primary producers is Y-shaped. One arm represents the herbivore food web and the other, the decomposer food web. The two arms differ fundamentally in the way in which they can influence primary producers. Herbivores feed on living plants and, therefore, directly affect the plant populations.

What they do not eat is available, after death, to the decomposers. As a result, decomposers are not able to directly influence the rate of supply of their food.

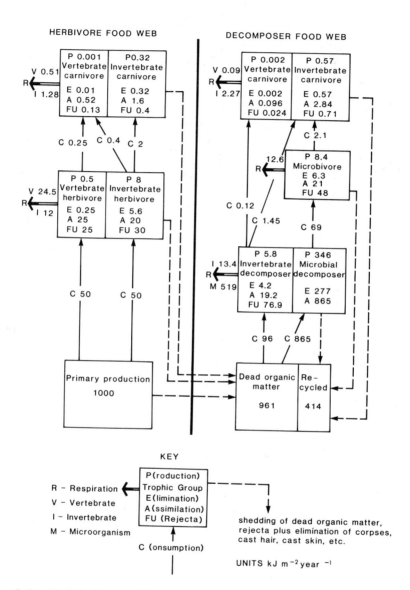

Figure 2-9   Model of energy flow in a hypothetical ecosystem with net primary production of 1000 kJ m^{-2}a^{-1}. Herbivores consume 10% of this, equally divided between invertebrates and homeothermic vertebrates. The only other assumptions necessary are the transfer efficiencies in Table 2-4 and consumption efficiencies between higher trophic levels (Heal and McLean 1975). For simplicity, some possible groups (such as heterothermic vertebrates) and some possible interactions (such as vertebrates feeding on microbivores) are omitted.

2.5   The Herbivore Food Web

Terrestrial green plants are only seriously depleted in exceptional circumstances and for short periods such as during locust plagues. The amount of net primary production consumed by herbivores usually averages less than 10% in terrestrial ecosystems, and there are no substantial differences between tropical and higher latitude environments (Table 2-5). Earlier estimates of herbivore consumption efficiencies as high as 60% in savannas appear to be exceptional and limited to small parts of ecosystems or brief periods. Why do herbivore populations not increase to consume more of the energy available to them? Answers to this question are complex and will be expanded on in later chapters. Briefly they may be given as follows:

1. Herbivore populations may be limited by predators.

2. Plants may defend themselves from herbivores by chemical and physical means.

3. Primary production may be seasonal (or variable on longer time scales) such that herbivore populations are limited by periodic scarcity of food.

Table 2-5   Percentage of above-ground net primary production consumed by herbivores in a variety of ecosystems (that is, consumption efficiency). Note that this rarely exceeds 10%.

TYPE OF ECOSYSTEM	CONSUMPTION EFFICIENCY (%)	SOURCE
FORESTS		
Tropical rain forests (various)	7–13[*]	Odum (1970), Wint (1983)
Mangrove forest (New Guinea)	5[*]–15[*]	Johnstone (1981)
Temperate deciduous forests (various)	1 11[*]	Pimentel et al. (1975), Hodkinson and Hughes (1982)
SAVANNAS AND GRASSLANDS		
Lamto savanna, Ivory Coast (large herbivores absent	1.1	Lamotte (1979)
Savannas of east and southern Africa (large hervibores only),	2–15	See section 8.2.3
North American grasslands (large herbivores absent)	1.5–5	Scott et al. (1979)
North American grasslands (large herbivores present)	5–15	Andrews et al. (1974)

[*]Consumption of leaves only.

4. Quality rather than quantity of food may be limiting for herbivores.

Although herbivores consume only a small proportion of primary production, they may destroy much more in their foraging activities. There is little information about this "wastage," but it has been estimated that cattle remove 0.75 kg of grass for each 1 kg ingested by dropping food and trampling (Anway 1978). Similarly, South African savanna grasshoppers destroy 0.4 kg for each 1 kg consumed (Gandar 1982). In tropical forests one study suggests that extensive insect damage may cause leaves to be shed. If this total leaf loss is taken into account, herbivore damage may be three to four times that reported in Table 2-5 for some tree species (Wint 1983). Primary production "wasted" in this way is passed directly to the decomposer food web.

Consumption efficiencies of predators are less well known than those of herbivores. In the Serengeti savannas of Tanzania, carnivorous mammals account for only 30% of ungulate mortality, whereas in the nearby Ngorongoro crater, hyenas kill all the ungulates that die (Kruuk 1972, Houston 1979). The Lamto savannas of the Ivory Coast have no large mammal consumers (Figure 2-2). Primary carnivores consume all their prey's production, while the secondary carnivores consume 60% of the animal production available to them (Lamotte 1979). As with herbivores, most carnivores are wasteful in their feeding and do not consume all their prey; again this wastage is available to decomposers. In addition to the more common invertebrate and microorganism decomposers, large vertebrate corpses may be eaten by more spectacular decomposers such as vultures and jackals.

2.6   The Decomposer Food Web

The dead organic matter in ecosystems comprises dead plant parts, the feces and other waste products of animals, and the corpses of dead animals and microorganisms. Dead plant remains are by far the most important quantitatively, and most of this section concentrates on the production and decomposition of dead plants. In a steady-state ecosystem, all the energy not dissipated by the respiration of the herbivore food web is inevitably decomposed, assuming that there is no net export of organic matter from the ecosystem. A broad definition of **decomposition** is used in this book to encompass all the abiotic and biotic processes involved in the breakdown of dead organic matter.

2.6.1   The Production and Decomposition of Plant Litter

Given that only a small proportion of primary production is consumed by herbivores, it follows that the primary production estimates in Table

2-2 are only slightly greater than the production of dead plant matter. The death and decomposition of below-ground plant parts are little studied, even though they may be as great quantitatively as the above-ground system. Extensive information is available about the production and decomposition of dead plant organs that fall to the ground (**litter**). Table 2-6 shows estimates of annual litterfall in forests. There is no clear distinction between litterfall in lowland and montane tropical rain forests. However, there is a marked decline from the tropics to higher latitudes.

Plant litter is not homogeneous but is composed of a mixture of plant organs. For example, in an evergreen Guatemalan forest, the litter comprised 72% leaves, 16% wood, and 7% flowers and fruit by dry weight (Kunkel-Westphal and Kunkel 1979). Neither is litterfall uniform in time. Even in continually wet forests, a peak often occurs in a wetter period of the year. In contrast, deciduous tropical trees shed their leaves in the dry season.

2.6.2   The Rate of Decomposition

Figure 2-10 summarizes information on rates of decomposition and compares tropical with temperate ecosystems. Three major factors influence this rate: (1) the quality of the material to be decomposed (for example, wood decomposes more slowly than leaves, Figure 2-10*a*); (2) the abiotic environment (Figure 2-10*b*), which operates through its effect on the third factor; and (3) the decomposer organisms, whose total activity is indicated by soil respiration (Figure 2-10*c*).

In general, decomposition is most rapid in moist tropical environments. This observation was clearly demonstrated by a pioneering study of decomposition in which the same material (alfalfa) lost 75%–95% of its weight in six months in Costa Rica, compared with 43%–73% in 12

Table 2-6   Comparison of litterfall in forests of the world. Only "small litter" (leaves, twigs, fruit, flowers, etc.) is included. Note the greater litter production in tropical forests.

FOREST TYPE (NUMBER OF SITES)	LITTERFALL ($g\ m^{-2}a^{-1}$)		
	MEAN	RANGE	SOURCE
Lowland tropical rain forests (17)	890	560–1330	Anderson and Swift (1983)
Montane tropical rain forests (3)	810	680–1010	
Temperate deciduous forests (14)	540*	340–1600*	
Temperate evergreen conifer forests (13)	440	280–610	Cole and Rapp (1981)
Boreal evergreen conifer forests (3)	32	29–53	

*The second highest litterfall was 590 $g\ m^{-2}a^{-1}$; without the exceptionally high site, the mean is 460 $g\ m^{-2}a^{-1}$.

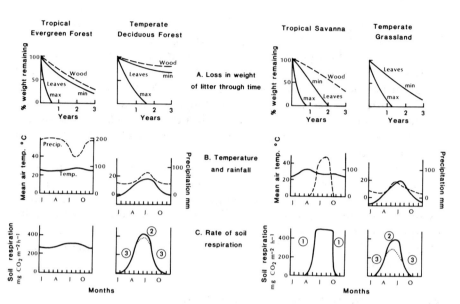

Figure 2-10 Comparative survey of decomposition processes in tropical and temperate ecosystems. In the soil respiration diagrams, ① denotes periods of moisture limitation, ② periods of probable moisture limitation and ③ periods of temperature limitation. See the discussion in text. (Modified from Swift et al. [1979].)

months in California, United States (Jenny et al. 1949). In moist tropical environments, litter decomposition continues throughout the year, whereas it virtually ceases during the temperate winter because of low temperatures. However, in seasonal tropical environments, decomposition may cease entirely during the dry season, particularly if termites are absent. Table 2-7 demonstrates just how variable decomposition rates

Table 2-7 Annual decay rates of leaf litter in a variety of environments. Estimates greater than 100% are extrapolations of rapid rates, which were measured for periods of less than one year.

TYPE OF ECOSYSTEM	LOSS OF DRY WEIGHT (% a^{-1})		
	MEAN	RANGE	SOURCE
Lowland tropical rain forest	200	35–550	Anderson and Swift (1983)
Montane tropical rain forest	55	25–100	Anderson and Swift (1983)
Temperate deciduous forest	80	20–200	Anderson and Swift (1983)
Tropical savanna grasses	65	30–95	Deshmukh (1985)
Temperate grasses in natural grasslands	30	10–60	Deshmukh (1974)

are. Complete breakdown takes from a few weeks in tropical rain forests to several years in drier savannas and temperate ecosystems.

Much less is known about the decomposition of **standing dead** remains. These are the dead parts of plants that decay slowly while still attached to the parent plant. Dead grass may remain standing for more than four months during which time it loses only 5% of its mass in Nairobi National Park in Kenya (Deshmukh 1985). Twigs in a New Guinea montane forest lost 61% of their dry weight before falling to the ground (Edwards 1977). Animals accelerate the transformation of standing dead remains to litter. Large herbivores trample dead grass to the ground, and termites feeding on standing dead grass ingested only 67% of the material they removed in a Nigerian savanna. The rest was left as litter (Ohiagu 1979).

Soil respiration is often used as an estimate of metabolism in the decomposer community. Carbon dioxide emitted from the soil is produced both by decomposers and plant roots. The latter account for between 15% and 50% of total soil respiration. Consequently, the method is an index of decomposer activity rather than a direct measure of energy release from dead organic matter. In moist environments, soil respiration declines with temperature and, therefore, with latitude (Figure 2-11). In drier tropical ecosystems, soil respiration exhibits large seasonal variations. For example, evolution of carbon dioxide in the dry season is less than 10% of that in the wet season from the soil of a South African savanna (Morris et al. 1982).

Abiotic factors not only affect decomposition by their influences on soil organisms, but they also cause litter breakdown directly. Fire has already been mentioned as a major decomposer in savannas. Wind is also important in bringing about litterfall and in transporting litter on the ground. Water from rainfall may remove soluble parts of plant tissues,

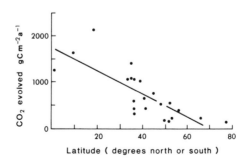

Figure 2-11   Soil respiration in forest ecosystems decreases as latitude increases (Schlesinger 1977). Temperature is the key factor. (Reproduced with permission from the *Annual Review of Ecology and Systematics,* Volume 8, © Annual Reviews Inc.)

but little is known of this process of **leaching** in tropical ecosystems. In warm, temperate *Eucalyptus* forests in Australia, almost half the weight loss from leaf litter is due to leaching. The process is variable, being related to both climate and plant species. Litter from temperate deciduous forests loses anything from 10%–75% directly as dissolved material (Swift et al. 1979). Leaching occurs throughout decomposition as new water-soluble compounds become available through the activities of the soil biota.

2.6.3   Soil Organisms and Decomposition

Soil organisms carry out two distinct processes in decomposition. **Comminution** is the reduction in size of organic particles, which occurs as a result of the feeding activities of soil animals. **Catabolism** is the biochemical breakdown of complex organic molecules, which results from the digestive processes of the soil fauna and microflora.

Table 2-8 gives estimates of population densities of soil organisms and compares tropical with temperate ecosystems. In general, tropical soils contain many fewer animals of all sizes but more bacteria. A particularly significant difference is the greater abundance of termites in the tropics. Most catabolism is performed by fungi and bacteria. However, the fauna is important qualitatively because its comminution

Table 2-8   Density of soil heterotroph populations in temperate and tropical ecosystems. Data from Swift et al. (1979).

		TYPE OF ECOSYSTEM			
	ORGANISMS	MOIST TROPICAL SAVANNAH	TEMPERATE GRASSLAND	TROPICAL RAIN FOREST	TEMPERATE DECIDUOUS FOREST
Microflora	Bacteria (cells g^{-1} of soil)	55×10^6	10×10^6	50×10^6	6×10^6
	Fungi (m of hyphae g^{-1} of soil)	?	3000	6000	3000
Fauna	Animals (m^{-2})				
Microfauna (< 0.2 mm long)	Protozoa and nematodes	30,000	500×10^6	65,000	200×10^6
Mesofauna (0.2–10 mm long)	Mites, springtails	2000	25,000	15,000	40,000
	Enchytraeid worms	0–400	10,000	0–1000	30,000
	Termites	4000	0	5000	0
	Others	500	1000	1000	500
	Total mesofauna	6900	36,000	22,000	71,500
Macrofauna (> 10 mm long)	Earthworms	1–100	750	0–250	200
	Millipedes, wood lice	<1	500	400	1000
	Mollusks, beetles, etc.	100	200	1000	200
	Total macrofauna	200	1450	1650	1400

speeds up microbial activity. Animal decomposers change the chemical environment of litter as it passes through their intestines, and they also disseminate microbial propagules. These factors cause a disproportionately large effect when compared with their catabolic activity. For example, tree leaves in a Nigerian bush-fallow lost 90% of their weight in three months when exposed to the total soil biota, but only 50% when exposed to the microflora alone (Swift et al. 1981). Because of their low assimilation efficiencies, animal decomposers ingest large quantities of litter and their stimulatory effects on microorganisms are, therefore, widespread. Earthworms are "typical" animal decomposers in this respect, as shown in the energy budget depicted in Figure 2-12a. In contrast, termites together with their symbiotic microorganisms assimilate a much greater proportion of their food. Foraging workers eat dead leaves and wood, and in many species their feces are used to construct nests. However, the mounds of *Macrotermes* are made from soil particles.

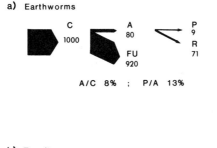

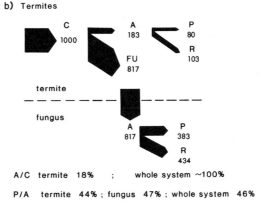

Figure 2-12   Energy budgets of two decomposer animal populations in West African savannas. Units = kJ m^{-2}a^{-1}; based on consumption of 1000 kJ. *a*, Earthworm *Millsonia anomala* population in Lamto savanna, Ivory Coast. Very low A/C results in large fecal output and low production. (Data from Lamotte [1979].) *b*, Macrotermitinae termites in Guinea savanna, Nigeria. Feces are deposited in the fungus comb parts, which are later reingested. A/C for the combined system is almost 100%, and production is 50 times that of the earthworms. (Data from Wood and Sands [1978].)

Feces are used to build fungus combs within the nest. On these combs grows the fungus *Termitomyces,* which continues the decomposition process. Older combs are in turn consumed by the termites, thereby making the termite-fungus combination approximately 100% efficient in assimilating dead plant matter (Figure 2-12*b*). Termites are also exceptional because they are active during dry periods when most other decomposers are quiescent.

Given a suitable abiotic environment, the attractiveness of plant remains to decomposers is mainly dependent on its quality, particularly in terms of fiber and nitrogen concentration. Photosynthetic tissues tend to be low in fiber and high in simple carbohydrates, which are ready sources of energy for decomposers. Consequently, leaves decompose more rapidly then do lignified stems. Nitrogen concentrations in the bodies of decomposers are up to 10 times higher than in their food. Decomposers are, therefore, attracted to litter rich in nitrogen. As a result, nitrogen deficient tissues are decomposed very slowly.

2.6.4   Dead Organic Matter Other Than That Shed by Plants

In the hypothetical ecosystem depicted in Figure 2-9, animal feces constitute 38%, animal corpses 4%, and microbe corpses 58% of the dead organic matter not shed by plants. Together these sources comprise approximately one third of total ecosystem dead organic matter. However, details of their decomposition are virtually unknown. Herbivore feces are rather similar to the plant material ingested, but the chemical and physical conditions are "improved." Particle size is reduced and moisture and nutrient content enhanced by passage through the gut in most cases. Elephant feces have a microbial respiration rate 3–40 times that of the soil beneath them in the dry season of a semiarid Kenyan savanna, mainly because of the higher moisture content of the feces. (Anderson and Coe 1974). These feces are also heavily utilized by decomposer animals—termites in the dry seasons and dung beetles during the rains.

A wide variety of fauna is associated with vertebrate corpses. In wet Costa Rican forests, more than 170 species are attracted to toad and lizard carrion. Together with the microflora they decompose all but the skeleton and skin of these corpses within one week (Cornaby 1974). Nothing is known about the decomposition of microorganism remains in the tropics. The little data available from temperate studies suggests that it is very rapid.

2.7   Heterotrophic Ecosystems _____

Some ecosystems have very few, or no, producers. The resident heterotrophs rely on the import of dead organic matter or the immigration of living organisms. Obvious examples are the ecosystems into which no light penetrates such as caves and deep water sediments. In other situations the lack of a suitable substrate precludes the growth of higher plants. For example, fast-flowing streams have few plants but abundant animal life, which is supported by dead terrestrial vegetation falling into the water. On land, the lava flows of Hawaii are colonized by the cricket *Caconemobius fori* before any plants grow there. The crickets eat dead organic matter deposited by the wind (Howarth 1979). A similar situation is found in the "bare dune ecosystem" of the Namib Desert. The windblown sandy substrate is too unstable for plants to grow, but tenebrionid beetles are able to subsist on dead grass and dying butterflies that are blown in. The beetles are eaten in turn by predatory reptiles (Walter 1979). Heterotrophic ecosystems cover only small areas on a worldwide basis but are growing due to the spread of urban environments.

Summary

Energy transformations in ecology are determined by thermodynamics and features of the biology of the organisms involved. Primary production in different ecosystems ranges over three orders of magnitude because of plant physiology and its response to environmental factors. Transfer efficiencies within and between trophic groups determine the quantitative aspects of energy and matter flows in food webs. In most terrestrial ecosystems, herbivores consume only a small proportion of primary production. As a result, most organic matter is decomposed by an interactive system involving abiotic factors, microorganisms, and soil animals. Most secondary production occurs at or below the soil surface, but the processes involved are not well quantified. Fire is an important decomposer in some ecosystems.

Study Questions

Review

1. Explain how transfer efficiencies determine the patterns of production in different trophic groups.

2. Table 2-2 shows that net primary production ranges over three orders of magnitude in different ecosystems. Explain this variation.

3. Construct a model of the herbivore food web (only) similar to Figure 2-9, but with 100% of net primary production consumed by herbivores (equally by homeothermic vertebrates and invertebrates). What is total secondary production in this herbivore food web? What is the total input of energy

to the decomposer food web? Begin with 1000 units of net primary production and use the transfer efficiencies of Table 204.

4. What factors determine the rate of decomposition in different ecosystems?

Related Topics

1. Describe the food web of a named heterotrophic ecosystem.
2. Trace the history of trophic models from Elton, through Lindeman, Odum, and Wiegert and Owen, to Heal and McLean. (In bibliography, except Lindeman, 1942, *Ecol.* 23:399–418; Wiegert and Owen, 1971, *J. Theoretical Biol.* 30:69–81.
3. Investigate the physiological and ecological significance of C_3 and C_4 photosynthesis.
4. How important are soil animals in decomposition processes?

Further Reading

Good (but rather dated) accounts of the general principles of production ecology can be found in the following:

Odum, E. P. (1971). *Fundamentals of Ecology.* 3d ed. Philadelphia: Saunders. Chapter 3.
Phillipson, J. (1966). *Ecological Energetics.* London: Arnold. (Studies in Biology No. 1).
Wiegert, R. G., editor. (1976). *Ecological Energetics.* Stroudsburg, PA: Dowden, Hutchinson & Ross. A collection of some key papers from the period 1942–1970 plus editorial commentary.

The following are more specialized:

Golley, P. M.; Golley, F. B., compilers. (1972). *Tropical Ecology With an Emphasis on Organic Production.* Athens, GA: University of Georgia. A collection of papers of variable quality, most of which describe research in India.
Swift, M. J.; Heal, O. W.; Anderson, J. M. (1979). *Decomposition in Terrestrial Environments.* Oxford: Blackwell. (Studies in Ecology, Vol 3). An outstanding account of most aspects of decomposition.

Descriptions of detailed tropical energy flow research in different ecosystems:

Lamotte, M. (1979). Structure and functioning of the savanna ecosystem of Lamto (Ivory Coast). In *Tropical Grazing Land Ecosystems.* Paris: Unesco/UNEP/FAO, pp. 511–561.

Odum, H. T; Pigeon, R. F., editors (1970). *A Tropical Rain Forest. A Study of Irradiation and Ecology at El Verde, Puerto Rico.* US Atomic Energy Commission.

The International Biological Program has published a series of methodological handbooks (Blackwell) and continues to publish syntheses of research in production ecology (mostly the Cambridge University Press).

In many tropical environments, termites are the most important decomposer animals and therefore have a large impact on nutrient cycles. Besides such effects their construction activities have a substantial influence on the structure of the environment as illustrated by this termite mound in a northern Australian savanna. (Photograph by Christopher Dunford.)

CHAPTER 3

Nutrient Cycles in Ecosystems

The bodies of organisms are made up of chemical elements or **nutrients**. These elements are obtained by plants from inorganic sources in the environment and incorporated into organic molecules using the energy provided by photosynthesis. As illustrated in Figure 1-1, nutrient cycles closely parallel the routes of energy flow within the biotic components of ecosystems, but an important distinction between the two processes is the relationship with the abiotic environment. While energy flow is profligate in the sense of being driven by an endless solar power supply, nutrient cycling is conservative, with chemical elements being drawn from finite pools and being largely retained within ecosystems. The study of nutrient cycles is also similar to that of energy flow, with emphasis on storage in living and dead organic matter and rates of flow between the various components of ecosystems.

To stress the biological, geological (in rocks, soils, and sediments), and chemical nature of the processes, they are sometimes called biogeochemical cycles. Nutrients can be subdivided into **macronutrients**, the major chemical elements used in large quantities by living organisms, and **micronutrients**, those required in much smaller quantities but nonetheless essential for life. Macronutrients include the major components of living tissues, carbon, hydrogen, and oxygen, which have cycles with an atmospheric store, and some of the nutrients obtained from the soil such as phosphorus and potassium. Micronutrients, such as copper, iron, and cobalt, also have soil-based, or edaphic, cycles. Nitrogen, another macronutrient, has inorganic pools in both the atmosphere and soil. Most plants are only able to use soil nitrogen, but some microorganisms are able to utilize atmosphere nitrogen and are, therefore, a bridge between the atmospheric and edaphic parts of the nitrogen cycle. Similar limitations apply to many edaphic nutrients. The total quantity of a particular element in rocks and soils may be very large relative to its use by the biotic community. However, large amounts are chemically bound in ways that make them unavailable to organisms. A variety of biogeochemical processes controls the links between available and unavailable forms of edaphic nutrients.

At the outset it must be stressed that the qualitative and quantitative nature of nutrient cycles in natural ecosystems is very diverse. It is not possible to describe "typical" cycles. Indeed, too few cycles have been elaborated in detail to draw more than tentative conclusions.

Human activities have increasingly disrupted local and sometimes global nutrient cycles by removal of nutrient stores in soils and vegetation and by the addition of extraneous chemicals and pollutants. Some aspects of these disruptions are discussed in Section 3.4 and others, in Chapters 9 and 11.

3.1   Atmospheric Cycles

3.1.1   The Hydrological Cycle

Strictly speaking, the hydrological cycle is not an elemental cycle because it follows the course of a compound, water. Nevertheless, the movement of water within and between ecosystems is fundamental to an understanding of nutrient cycles for several reasons. First, plants obtain their hydrogen for photosynthesis from the splitting of water molecules. Second, plants use large amounts of water to maintain their hydrostatic skeletons and to move chemicals about their bodies. Third, plants take up elements in aqueous solution from the soil; without this water flow, they would be unable to maintain the balance of minerals necessary for life.

Figure 3-1 shows the components of the global hydrological cycle and the fluxes between them. The major store is in the oceans, and the major flows are evaporation from them and precipitation upon them. However, there is a net flow of water vapor, driven by winds, from the oceans to land where it falls as rain, hail, or snow. The balance of the cycle is maintained by water flowing from the land as surface runoff or movement of groundwater into rivers and back to the oceans. A large quantity of water is locked up in the earth's crust, but this is only released in small quantities during volcanic eruptions. Similarly, the large store in polar ice caps has little effect on the hydrological cycle in the short-term because of negligible evaporation from them.

The hydrological cycle is driven by the evaporative power of solar radiation and requires 8.2×10^{20}kJ a^{-1}, which is approximately

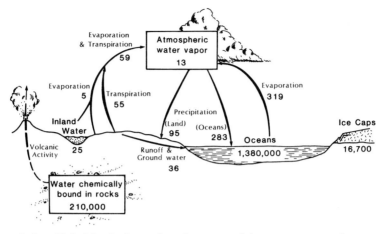

Figure 3-1   Global hydrological cycle. Most of the storage is in the oceans, but most of the flux is to and from the atmosphere. Units: stores, 10^{18} g; fluxes, 10^{18} g a^{-1}. (Modified from Ricklefs [1979] with additional data from Odum [1971].)

15% of the total radiation reaching the outer atmosphere. This proportion compares with the 0.2% used in gross primary production worldwide.

3.1.2   The Carbon Cycle

The biospheric carbon cycle is primarily concerned with the atmospheric gas carbon dioxide, its incorporation into organic matter by photosynthesis, and its subsequent release by the respiration of all the biota (Figure 3-2). Carbon also occurs in the earth's rocks predominantly as calcium and magnesium carbonates. These compounds are largely organic in origin, being the mineralized remains of the skeletons of marine organisms. By subsequent geological uplifting, carbonate rocks also occur on land and add to soil nutrients and plant nutrition by their subsequent weathering. Such transfers are small on a global scale compared with the exchanges between the biota and the atmosphere.

When net primary production exceeds community respiration, carbon-rich organic matter accumulates in ecosystems. At times in the past, such accumulations have led to the deposition of the fossil fuels coal and oil. This process occurs to a minor extent today in peatlands.

3.2   Edaphic Nutrient Cycles _____

Terrestrial plants obtain nutrients, other than carbon, in solution from the soil. Macronutrients, used in relatively large quantities, are nitrogen, phosphorus, potassium, calcium, magnesium, and sulfur. Trace elements, or micronutrients, are essential to healthy growth but required in minute quantities. They include copper, zinc, manganese, iron, boron,

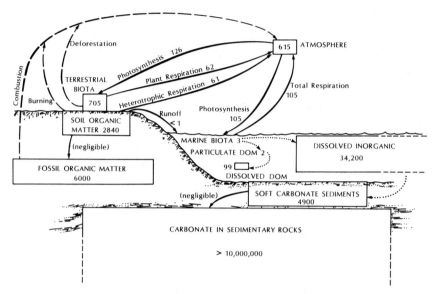

Figure 3-2   Global carbon cycle. Storage in 10^{15} g carbon; fluxes, 10^{15} g a^{-1} carbon. Solid lines indicate the cycle in balance, with dotted lines being the complex marine interactions between water, biota, and sediments, which are not well quantified. Dashed lines are fluxes resulting from human activities that have thrown the cycle out of balance (see section 3.4.1). DOM = dead organic matter. (Data from Bolin et al. [1979].)

molybdenum, and cobalt. Other elements such as silicon and aluminum are present in plant tissues but are probably not essential for growth.

One of the more complete accounts of the cycles of major elements in a tropical ecosystem is shown in Figure 3-3. Knowledge of complete cycles and details of specific interactions within the biota, and between biota and soils in the tropics, is sketchy and mostly limited to forest ecosystems.

3.2.1   The Nitrogen Cycle

As shown in Figure 3-4, nitrogen cycles are not entirely edaphic, but most have an atmospheric component linked to the soil by nitrogen fixation and denitrification. Nevertheless, plants obtain most of their nitrogen from the soil as nitrate or ammonium ions, with the former probably being more important. Worldwide, atmospheric fixation accounts for around 10 kg ha^{-1}a^{-1} of the biospheric nitrogen flow. On land, more than 60% of fixation is due to agroecosystems, and most of the remainder is fixed in forests. Other ecosystems account for about 7% of total terrestrial nitrogen fixation (Burns and Hardy 1975). Overall, atmospheric fixation represents about 2% of global nitrogen assimilation, the rest being cycled in nongaseous forms. However, most of the

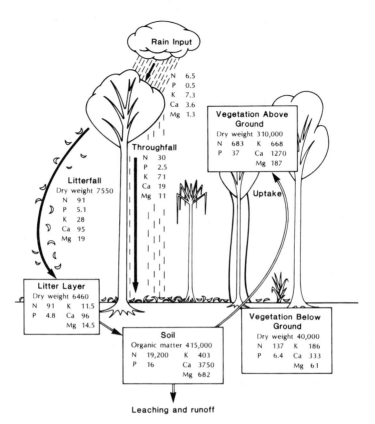

Figure 3-3 Cycles of macronutrients in a montane rainforest, Papua New Guinea. Nutrients are stored in the soil, the living biomass, and litter layers (units = kg ha^{-1}). Fluxes are uptake (unquantified in this study), litterfall, and input in rainfall (units = kg ha^{-1}a^{-1}). As rain passes through the plant canopy, it leaches nutrients from the plants, adding to those in rainfall to give the total amount in throughfall. Nutrients may be lost to the ecosystem by leaching and runoff from the soil, but these were not quantified. (From Edwards, P. J. 1982. *J. Ecol.* 70:807–827. Reproduced by permission of the British Ecological Society.)

nitrogen in soils and biota originated in the atmosphere and has accumulated over the millions of years in which nitrogen fixation has been taking place. Much of the terrestrial fixation is carried out by symbiotic bacteria living in plant roots and to a lesser extent by free-living soil bacteria. Cyanobacteria (blue-green algae) also fix nitrogen and are of major significance in aquatic systems but relatively unimportant on land. Some termites have a mutualistic gut microflora, which fixes atmospheric nitrogen. For example, arboreal termites in a Costa Rican rain forest fix between 0.25 mg and 1.0 mg per colony per hour, with the higher rates obtaining when nitrogen concentrations in food are low (Prestwich and Bentley 1981).

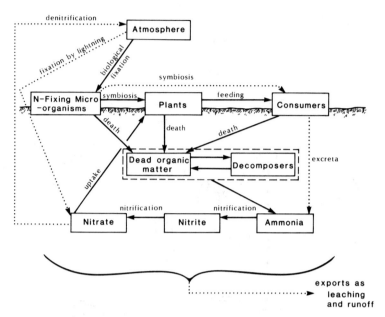

Figure 3-4   Components of the nitrogen cycle in a terrestrial ecosystem. Solid lines are major and dotted lines minor transfers and processes. The very complex processes whereby the interactions between dead organic matter and decomposers mineralize nitrogen (within the dashed box) are not shown.

Specialized but ubiquitous soil bacteria play vital roles in **nitrification,** the oxidation of the ammonium compounds produced by decomposition to nitrate. Some bacteria oxidize ammonium ions ($-NH_4$) to nitrite ($-NO_2$), and others complete the process by converting nitrite to nitrate ($-NO_3$). These chemoautotrophic bacteria obtain energy from such transformations (276 kJ and 73 kJ, respectively, for each mole of precursor). Denitrifying bacteria reduce nitrate to gaseous nitrogen in anaerobic conditions, returning a small proportion of that circulating in the biosphere to the atmosphere.

3.2.2   Cycles of Mineral Elements

Strictly speaking, mineral nutrients are those derived in the long-term from geological sources. However, in most ecosystems, mineral ions are obtained from the soil as a result of the breakdown of dead organic matter. Nitrogen is not really a mineral element in terms of its origin, but its cycling has many features in common with other edaphic nutrients and it will, therefore, be discussed further in this section. Figure 3-5 is a generalized scheme for edaphic cycles, various aspects of which are considered below.

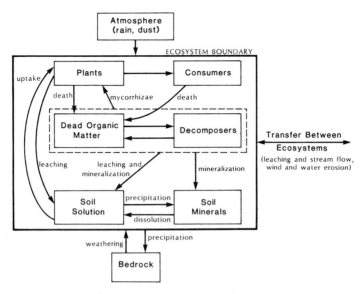

Figure 3-5   Generalized scheme of mineral cycles in terrestrial ecosystems. Most transfers are within and between the biotic and inorganic soil components. Additional inputs come to the ecosystem as rainfall (or dust) and by weathering of bedrock, and losses occur through leaching from and loss of soil and precipitation of soil minerals as unavailable forms.

Uptake and Accumulation in Vegetation.   Symbiotic fungi are often intimately associated with plant roots, and they play an important role in uptake of nutrients from the soil solution, receiving in return photosynthetic products from the host plant. These specialized **mycorrhizal** fungi exist in two forms. The sheathing (**ectotrophic**) mycorrhizae are mainly Basidiomycetes and coat the outer surface of the roots. They are typically associated with temperate trees and are less common with tropical trees, although the important dipterocarps of southeast Asia have ectotrophic mycorrhizae. Vesicular-arbuscular (**endotrophic**) mycorrhizae are phycomycetes of the family Endogonaceae. They live inside root tissues, within and between unspecialized parenchyma cells and are commonly associated with tropical rain forest trees (Redhead 1980). Both types have extensive and intimate connections with the soil and improve survival prospects for their hosts as well as facilitating nutrient uptake (Janos 1983). Some associations seem to be obligatory, with growth and survival of plants improbable without mycorrhizae, while in other cases, plants may only develop such associations on nutrient poor soils.

Many herbaceous plants including all crop species investigated can also develop mycorrhizae. For example, the African savanna grass

Panicum coloratum is more frequently associated with mycorrhizae where grazing by large animals is intense. The fungi seem to minimize the effects of grazing by promoting prostrate growth and root proliferation and probably improve the nitrogen nutrition of the plants (Wallace 1981).

Rain forest trees also take up nutrients in the canopy by forming roots in the branches beneath decomposing epiphyte litter. Larger epiphytes, such as mosses and ferns, also obtain their nutrients from this "aerial soil."

An example of the distribution of minerals in rain forest vegetation is shown in Figure 3-6. However, the concentrations vary considerably in different forests, often as a reflection of soil conditions (see Section 3.3). Elemental concentrations (mg g^{-1}) in tropical forest litterfall exhibit wide ranges, often spanning an order of magnitude or more: nitrogen, 6–21; phosphorus, 0.2–5; potassium, 1.3–10; calcium, 2.2–30; and magnesium, 1.3–5.9. Temperate forests are less variable, usually lying toward the lower end of these ranges (Anderson et al. 1983).

The Herbivore Food Web.   Little is known about the significance of herbivores and their predators in tropical nutrient cycles. As discussed in Chapter 2, herbivores only consume a small proportion of primary production in most ecosystems. In a Panamanian rain forest in which 10% of primary production was eaten by herbivores, the total herbivore and carnivore biomass contained only 0.1% of the total minerals in the biota (Golley et al. 1975). However, large mammals in African savannas may represent more substantial mineral accumulations. In the Serengeti ecosystem of Tanzania, the bones of these animals contain approximately 0.5 kg ha^{-1} of calcium, only one order of magnitude less than that in the living herbaceous vegetation. The slow decomposition of these bones may mean that a substantial proportion of calcium and magnesium is immobilized in living and dead skeletons.

Litter Decomposition and Soil Nutrients.   In a hypothetical steady-state ecosystem, nutrients taken up by plants are balanced by those returned to the soil in dead plant parts and washed from the plants by rain water. Through decomposition, insoluble organic compounds are broken down, releasing soluble inorganic ions. An exception to this pattern is in savannas maintained by burning, where some of the nutrients may be volatilized by the fire and lost to the atmosphere. Decomposition rates vary in different environments (Table 2-7), and different elements are also released at different rates to one another. Potassium is usually dissolved in the cell sap and so is rapidly leached out of litter, but few generalizations are possible about other elements as shown in Table 3-1. Much of the variation revealed in this table is due to local conditions, with some ecosystems (and leaves) richer in some elements than others. In the forests of Australia and Sarawak (Table

3-1), the leaves have especially low concentrations of phosphorus, which may be related to the slow release of this element from litter. Many features of leaf quality affect the rate of decomposition. Nutrient concentration and balance are two of these, but organic compounds such as inhibitory tannins and lignin are also important. Leaves rich in nutrients are generally decomposed more quickly as illustrated by the rough correlation between nitrogen concentration and speedy decomposition throughout the lowland tropical forest zone (Anderson and Swift 1983). However, the same study showed that other attributes can confound this relationship, with more highly lignified leaves decomposing slowly despite being rich in minerals.

To become available to plants, the various elements have to be mineralized to organic forms—in particular those that enter the soil solution. Otherwise the elements remain locked up in undecomposed dead organic matter or *immobilized* in the tissues of living decomposer organisms. Fire is an important decomposer and mineralizer in many savannas. It usually occurs during the dry seasons when most of the above-ground vegetation is dead. Some proportion of the elements enters the atmosphere as gases (N, S, P) or small particles (Ca, Mg, K), although these may be precipitated later by gravity or rainfall. However, much of the inorganic material is deposited in a nutrient-rich ash, which subsequently becomes incorporated into soils and new plant growth. For example, in Nairobi National Park, Kenya, a mostly dead standing

Table 3-1   Comparison of the rate of loss of selected elements (chemical symbols) and dry matter (DM) from tree leaf litter and grass litter in tropical forests and a savanna. The only consistent pattern is that potassium is lost most rapidly. In forests, nitrogen loss is generally at a similar rate to dry matter. Nitrogen loss, relative to dry matter, is slow in the savanna compared to the forests, but phosphorus loss is relatively rapid.

ECOSYSTEM	RELATIVE RATE OF LOSS		SOURCE
FOREST ENVIRONMENTS	FAST	SLOW	
Rain forest, Sarawak	K - Mg - Ca - DM - N - P		Anderson et al. (1983)
Rain forest, Australia	K - Mg - N - Ca - DM- P		Brassel and Sinclair (1983)
Rain forest, Guatemala	K - P - N - DM - Mg - Ca		Ewel (1976)
Dry forest fallow, Nigeria	K - Mg - N - DM - P - Ca		Swift et al. (1981)
SEMIARID SAVANNA (SOUTH AFRICA)			Morris et al. (1982)
Under trees			
Digitaria	K - DM- P - N		
Eragrostis	K - P - DM- N		
In open			
Digitaria	K - P - DM- N		
Eragrostis	K - P - DM- N		

crop of 6000 kg ha^{-1} yielded an ash containing 13 kg ha^{-1} of nitrogen, 7 kg ha^{-1} of phosphorus, 38 kg ha^{-1} of potassium, 95 kg ha^{-1} of calcium, and 11 kg ha^{-1} of magnesium. Subsequent grass growth also contained a greater concentration of these nutrients than that in adjacent unburned areas (Keiyoro 1982).

The maintenance of soil nutrients in ionic forms available to plants (for example, nitrogen as nitrate, or phosphorus as orthophosphate) results from the complex interactions of soil chemistry. The equilibria between the various forms of an element are controlled by plant uptake, return through decomposition, and the physical and chemical state of the soil. pH of the soil solution and whether the environment is reducing or oxidizing are particularly important. For example, less than 25% of phosphorus in a soil is available. Most is insoluble as iron, aluminum, and magnesium phosphates in acidic soils or calcium phosphate in basic soils. The greatest proportion of available phosphorus occurs in neutral or slightly acidic soils. In a well-drained soil, the environment tends to be oxidizing, but when the soil is waterlogged, reducing conditions are produced because decomposers respire the small amount of oxygen in the soil water. Iron and manganese are reduced to their ferrous and manganous forms, which are preferentially exchanged for calcium, magnesium, and potassium ions, releasing the latter group into solution. Phosphate may also be released as it is often adsorbed on to ferric hydroxide in oxidizing conditions.

Commonly used analytical techniques allow determination of either total or easily extractable elements. The determination of easily extractable elements is an approximation of the nutrients immediately available to plants, but transfers from unavailable to available forms can occur as physical conditions change (see above) and equilibria are reestablished following uptake or leaching. The data presented here on soil nutrients are those most often determined—total nitrogen and extractable phosphorus, potassium, calcium, and magnesium—unless otherwise stated.

Decomposition is not the only source of soluble nutrients to the soil. Rainfall contains significant quantities of mineral elements. Rain also leaches further nutrients from the plant canopy to give the total **throughfall** reaching the ground (see Figure 3-3 and Table 3-2). In arid areas, additional inputs or outputs may occur as a result of wind erosion. For example, the harmattan wind in West Africa blows from the arid northern areas during the early part of each year. The dust it carries is mainly silica and alumina, but it also contains measurable amounts of organic matter, calcium, magnesium, and potassium, which are deposited in the savannas and forests to the south (Whalley and Smith 1981).

The Rate of Nutrient Cycling.   In forests the dominant plants are mature trees, which live much longer than their decomposers or

Table 3-2   Proportional inputs of nutrients to the soil from rainfall, leaching from the tree canopy (throughfall minus rainfall), and litter fall in a range of tropical rainforests in West Africa, Oceania, and Latin America. The pattern is quite consistent for most elements, with inputs increasing in the following order: rainfall, canopy leaching, litterfall. The exception is potassium, most of which is leached from the canopy. Data from Figure 3-3, Unesco (1978), and Brassel and Sinclair (1983).

ELEMENTS (AND NUMBER OF FORESTS)	PERCENT OF ABOVE-GROUND NUTRIENT FALL (MEAN AND RANGE IN DIFFERENT FORESTS)		
	RAINFALL	CANOPY LEACHING	LITTERFALL
N (6)	7 (4–13)	18 (4–30)	74 (66–90)
P (6)	5 (2–9)	27 (6–57)	58 (38–85)
K (8)	5 (2–7)	60 (26–69)	35 (25–69)
Ca (8)	8 (1–16)	17 (9–24)	76 (60–88)
Mg (8)	9 (4–15)	35 (23–49)	56 (43–63)

consumers. However, not all the organs of trees are long-lived, with flowers lasting a few days, fruits a few weeks, and leaves several months. In terms of quantitative significance, leaves and stems are the most important (Figure 3-6), and they represent short- and long-nutrient retention, respectively. The proportion of total nutrient capital that is cycled rapidly through leaves varies with the element and the type of forest. This proportion is usually 5%–25% of the total for a particular element in the vegetation, with calcium at the lower end of this range, but other elements are variable when different forests are compared. In lowland forests, leaf litter decomposes rapidly, often in a few months, but almost invariably in less than one year. Decomposition is, therefore, unlikely to noticeably slow down nutrient acquisition by plants. Only one clear exception to this has been reported in more than 20 studies: an Amazonian forest of exceptionally low nutrient content where leaves take several years to decay (Herrera et al. 1981). In montane forests slow mineralization of litter and humus may be an important contributory cause of reduced plant production (Grubb 1977).

The **turnover rate** of nutrients in the vegetation can be measured as uptake relative to the quantity contained in the living biomass. A comparison of a tropical humid forest and a temperate deciduous forest indicates that elements are "turned over" more quickly in the tropical forest (Table 3-3). In terms of nutrient economies, these two forests lie toward the middle of the ranges of those that have been studied in each climatic zone. However, there is much variation and some overlap when different forests and different elements are compared. Table 3-3 also reflects one aspect of the relative **mobility** of elements; potassium is the most mobile (circulating most quickly), followed by magnesium. These

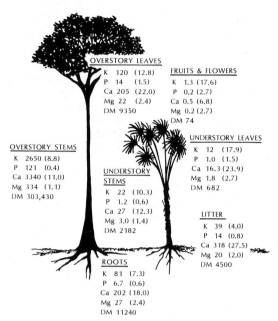

Figure 3-6   Total standing crop as dry matter (DM) and standing crop of some macronutrients in a Panamanian lowland rainforest. Units = kg ha^{-1}. Concentrations of minerals are given in brackets (milligram of element per gram DM of tissue). Note that (1) the highest concentrations of most elements are in leaves, fruit, and flowers, but the greatest store is in stems, and (2) understory organs have higher concentrations than equivalent overstory organs but have much smaller standing crops. (Data from Golley et al. [1975]; means of two forests in different seasons.)

two elements also are generally the most mobile in throughfall and release from litter. Calcium is usually the least mobile element (Tables 3-1 and 3-2).

Vitousek (1984) has assessed the efficiency of nutrient cycles in terms of the amount of organic matter produced per unit of nutrient uptake. His analysis of litterfall in 62 tropical forests shows a wide range of efficiencies, both in comparisons of a single element in different forests and when different elements are compared. Lowland tropical forests do not cycle nitrogen as efficiently as montane tropical forests or temperate forests because leaf litter has a high concentration of this element. In contrast, both types of tropical forest use phosphorus efficiently because the concentration of this element in litterfall is generally lower than in temperate forests. Vitousek concludes that the quantity of litterfall per year (an approximate measure of leaf production) is limited by phosphorus availability. Phosphorus-rich ecosystems produce more litter.

Table 3-3   Comparison of turnover time (years) of elements in a lowland, humid, tropical forest and a deciduous temperate forest. In this case, turnover is faster in the tropical forest. It is probable that such a conclusion is valid when comparing an "average" lowland tropical with an "average" temperate forest, but when the array of forests in each biome is considered, there is overlap. Turnover time is estimated as the standing crop of an element in the total forest biomass divided by the annual uptake by plants.

| | TURNOVER TIME (YEARS) | | | | | |
LOCATION	N	P	K	Ca	Mg	SOURCE
Kade, Ghana	7	7	3	7	5	Nye and Greenland (1960)
Oak Ridge, Tennessee (United States)	8	15	10	12	5	Cole and Rapp (1981)

Comparable data for savannas are sparse. Where herbaceous vegetation predominates, most above-ground tissues live only a few weeks or months, which suggests a rapid turnover of nutrients. However, decomposition rates are sometimes very low, particularly where termites are absent, and litter may accumulate over several years. Burning reactivates nutrient circulation in such situations. Concentrations of nitrogen and calcium in neotropical savanna grasses are generally lower than those in rain forest foliage. Other major nutrients seem to attain similar concentrations in the two types of ecosystem (Sarmiento 1984). In many African savannas, long-lived animals are important consumers, and they may immobilize some elements for decades in skeletal tissues. Many savanna plants adapt to defoliation by fire or herbivores by developing extensive below-ground systems of roots and shoots. In such cases most of the mineral cycling is within the soil. Almost nothing is known about nutrient cycling in these subterranean organs.

3.2.3   Cycles of Micronutrients

Little is known about the circulation of micronutrients in tropical ecosystems. The only comprehensive study is that of Panamanian rain forests by Golley et al. (1975). On average, more than 80% of the total store of these elements is found in the soil, with progressively less in vegetation, litter, and the fauna. However, the concentration of these trace elements increases significantly along food chains in the sequences soil–plants–herbivores–carnivores and litter–detritivores. For example, the concentrations of iron in the Panamanian forest were 6 ppm in the soil, 218 ppm in the vegetation, 3054 ppm in the herbivores, and 3387 ppm in the carnivores.

Many of the soil micronutrients are toxic to plants at high concentrations and a specialized flora has evolved where such soils exist.

In the Shaba province of Zaire, soil copper and cobalt concentrations reach 30,000 ppm and 5000 ppm, respectively, in local outcrops, but the normal forest vegetation of the region is absent when soils have more than 300 ppm copper and 100 ppm of cobalt. Plants growing on the high-concentration soils either resist uptake of these elements or have tissues tolerant of high concentrations, which are lethal to unspecialized plants (Malaisse et al. 1979).

3.3   Soil Types and Nutrient Cycles: Tropical/Temperate Comparisons_____

The classic picture of tropical forest nutrient cycles is that living vegetation stores a much greater proportion of ecosystem mineral capital relative to the forest floor than at higher latitudes. Most tropical soils are millions of years old, and, in the humid zone, it is generalized that they have a very low nutrient stock because of leaching and negligible input of new minerals from weathering of bedrock. As a consequence, nutrient cycles are closed, with ions taken up by plants as soon as they become available. In contrast, most temperate soils are relatively young and fertile, having developed over only a few tens of thousands of years since the last glaciation. Such soils have significant inputs of nutrients from weathering of bedrock and have not been subjected to such lengthy leaching. Nutrient cycles are more open with longer residence periods in the soil and with greater export in runoff. For any ecosystem to remain in a steady state with respect to nutrients, export in runoff must be balanced by new inputs, which are the sum of elements in rainfall and weathering of rock (see Figure 3-5). Where weathering of rock is negligible, as often supposed in most tropical soils, it is argued that nutrient exports will also be reduced.

Is this a true picture of differences between tropical and temperate forest nutrient cycles? There is now a large body of data on temperate forests, but only about 10 studies that consider most aspects of tropical mineral cycles. Of these, only five or six have information that allows direct comparisons of soil nutrients. However, such comparisons do not support the classical scheme entirely but reveal a more diverse situation. Figure 3-7 compares soil nutrients with the proportion contained in the vegetation in a range of forests from different latitudes. Of the two tropical forests, one grows on a nutrient-poor soil in the lowlands of the Ivory Coast and the other, on a richer montane soil in Papua New Guinea. These are compared with a temperate deciduous forest and a boreal conifer forest. In these examples, the tropical vegetation does hold a higher proportion of the nutrients than the temperate vegetation when all nutrients are averaged, but there is little difference between the montane tropical forest and the temperate and boreal forests. A more detailed examination reveals that the differences are even less clear-cut.

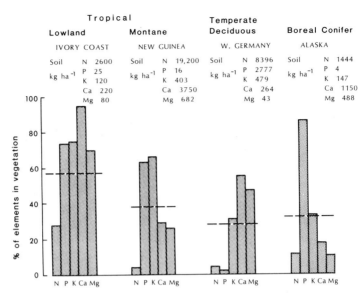

Figure 3-7   Mineral elements in the above-ground tree biomass as a percentage of the total in the vegetation plus that in the soil in various forest ecosystems. Dashed lines are the mean percentage for the five elements. See text for discussion. Tropical forests are on a relatively nutrient-poor soil (Ivory Coast) and a relatively rich soil (Papua New Guinea) (data from Unesco [1978] and Edwards [1982], respectively.) Temperate forest data are the means of three sites at Solling, West Germany. Boreal forest in Alaska is the mean of two sites (data from Cole and Rapp [1981].)

Elements with low concentrations in the soil are the ones accumulated by the vegetation in all types of forest (for example, phosphorus in the tropical and boreal forests, calcium in the acidic soils of the Ivory Coast and West Germany). Both tropical soils have low concentrations of phosphorus, a widespread phenomenon in the humid tropics. For example, Sanchez et al. (1982) estimate that 90% of Amazonian soils have too little phosphorus for agriculture. In contrast, only 1 of 14 temperate deciduous forests studied during the IBP had a soil with a comparably low level of this element. However, some temperate coniferous forest soils from that study and the boreal soils (as in Figure 3-7) have low concentrations of soil phosphorus and consequently a high accumulation in the vegetation (Cole and Rapp 1981).

Forests growing on the poorest soils exhibit the most extreme mechanisms for nutrient conservation. A dense root mat is present immediately beneath the litter, which (assisted by its mycorrhizae) absorbs nutrients as, or possibly before, they are released to the soil. In a Venezuelan rain forest, experiments showed that less than 0.1% of calcium and phosphorus reaching the upper layer of litter leaches through to the soil humus layer (Stark and Jordan 1978). Some of these elements are probably immobilized temporarily in decomposer tissues, but they are likely to become available to the roots eventually and before penetrating to deeper soil layers. In such forests, leaves are long-lived, and some nutrients may be translocated back to living tissues before they are shed.

Comparisons of input to ecosystems of elements in rainfall to losses in runoff do not support the idea that nutrient cycles in all tropical forests are tighter than those of temperate forests. Jordan (1982) has compiled the relevant data for 7 tropical and 11 temperate forests with respect to calcium and phosphorus. All the forests were effective at retaining phosphorus, but most had net losses of calcium ranging from 0.7–134 kg ha^{-1}a^{-1}. Such losses can only be made up from weathering of bedrock, which is said to be negligible in tropical forest soils according to the classical scheme outlined earlier. Detailed soil surveys in many tropical regions have not been made, but it does seem clear that vast areas of humid tropical forests are growing on highly leached, acidic soils. Nevertheless, there is a diverse array of soil types, many of which are likely to add to the soil nutrient pool by weathering. For example, a recent study indicates that weathering of certain types of forest soil is common in Southeast Asia, West Africa, and the Amazon Basin. Small but significant quantities of available nutrients are probably added to these soils (Baillie and Ashton 1983). In southwestern Amazonia, it seems that a large area (in excess of 10,000 km^2) is covered with alluvial soils that were deposited by massive seasonal flooding 7000–13,000 years ago. The flooding was caused by the recession of Andean ice caps during that period. The young soils, so deposited, will be subjected to considerable weathering in the future (Campbell and Frailey 1984).

Little is known about the influence of savanna soils on nutrient cycles. There is a range of fertility from generally poor leached soils in moist areas to richer unleached soils in drier areas. Exceptions occur where soils may have been exposed to wetter climates in the past. For this reason, some Sahelian soils have low concentrations of nitrogen and phosphorus. Nitrogen shortage seems to be a critical factor in savannas, but cycling of this element is poorly understood (see Huntley and Walker 1982). Because of its smaller biomass, the vegetation cannot act as a major store of nutrients, most of which are found in the soil. For example, over a range from moist to semiarid savannas in India, less than 5% of ecosystem nitrogen and phosphorus was found in the living and dead vegetation combined (Singh et al. 1979). In tropical America, savanna soils usually have lower concentrations of most nutrients than nearby forest soils of the same type (Sarmiento 1984).

An important feature of soil nutrient distribution, which has been clearly illustrated in savannas, is variation with local topography. Even on gentle slopes, characteristic associations between soils and vegetation develop in sequences called **catenas** (Figure 3-8). The causes of such patterns are complex and include soil processes such as weathering,

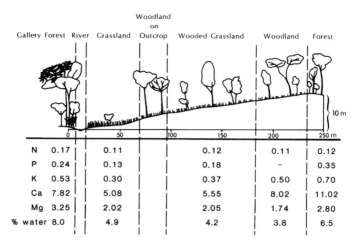

Woodland
on
| Gallery Forest | River | Grassland | Outcrop | Wooded Grassland | Woodland | Forest |

N	0.17	0.11		0.12	0.11	0.12
P	0.24	0.13		0.18	–	0.35
K	0.53	0.30		0.37	0.50	0.70
Ca	7.82	5.08		5.55	8.02	11.02
Mg	3.25	2.02		2.05	1.74	2.80
% water	8.0	4.9		4.2	3.8	6.5

Figure 3-8   Soil and vegetation catena in moist savannas in Ghana. The complex interactions between topography, geology, and vegetation produce adjacent soils of different nutrient contents. Total nitrogen and moisture = % by weight; total phosphorus = mg g^{-1}; exchangeable potassium, calcium, and magnesium = mEq 100g^{-1}. Samples taken during dry season. (Modified from Markham and Babbedge [1979].)

movement of dissolved minerals and smaller particles downslope, the deposition of alluvium by rivers, and interactions between the vegetation and the soil. In the example shown, nitrogen and magnesium show little variation, but midslopes have significantly less phosphorus, potassium, calcium, and moisture. Catenas are found in all areas, but their effects are much less distinct in the highly leached soils of wetter areas.

Soils in savannas and elsewhere may be locally enriched around some trees and termite mounds, but the methods of accumulation are not always the most obvious. In a neotropical savanna in Belize, Kellman (1979) found that surface soils had two to six times the concentrations of available nutrients around *Clethra hondurensis* trees compared with adjacent soils under herbaceous vegetation. This enrichment is due to more effective capture of nutrients in rainfall by the trees rather than the bigger plants drawing nutrients from deeper soil layers and depositing them as surface litter. In the Serengeti savannas of Tanzania, termite mounds have two to five times the concentrations of sodium and approximately five times the calcium of surrounding soils. This is only partly due to selection of building materials by the insects; another major mechanism is deposition by evaporation of soil water in the airflow created by the layout of galleries within the mound (Weir 1973). Termitaria in Amazonian forests accumulate minerals to an even greater degree, although in this case the mechanism is direct deposition of enriched fecal matter in most species. The mounds have concentrations of nitrogen 3–5 times, total phosphorus 4–10 times, and exchangeable cations 5–20 times as great as adjacent surface soils (Salick et al. 1983).

3.4   Disruption of Nutrient Cycles _____

From the setting of the first grass fires, through the development of agroecosystems, to urbanization and industrialization, human activities have increasingly disrupted the balance of nutrient cycles. Some of these alterations produce global effects, such as increasing atmospheric carbon dioxide; others produce regional effects, such as pollution of lakes. Many more lead to local changes, including the open nutrient flows of modern agroecosystems and the changes in the balance of elemental cycles induced by pollutants. Agroecosystems are dealt with in Chapter 9. A selection of other topics is discussed next.

3.4.1   Deforestation in the Tropics

The Carbon Dioxide Problem.   The concentration of carbon dioxide in the atmosphere has been increasing since people ceased to act as foragers in natural ecosystems more than 10,000 years ago. The change over the last 20 years is well documented as shown in Figure 3-9. The dashed lines in Figure 3-2 indicate the sources of this increase. Combustion of fossil fuels adds about 5×10^{15} g a^{-1} of carbon

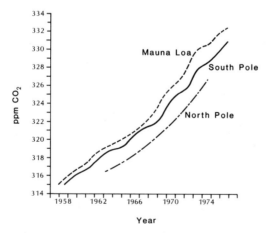

Figure 3-9   Increase in global atmospheric carbon dioxide concentrations illustrated by samples from the Arctic and Antarctic and Mauna Loa, an Hawaiian Island in the tropical Pacific. Lines are smoothed to eliminate seasonal fluctuations. (From Bolin, B.; et al. 1979. *The Global Carbon Cycle*. SCOPE Report 13. Reproduced by Permission of John Wiley & Sons.)

(80% from developed countries), and deforestation adds approximately 4×10^{15} g a^{-1} (80% in the tropics), the latter coming from loss of carbon in vegetation and soil. Some of this is absorbed by revegetation through succession, afforestation, or agriculture and some by the complex oceanic interactions between the biota and inorganic forms of carbon. Nevertheless, there is an overall increase of about 2.5×10^{15} g a^{-1} of carbon as atmospheric carbon dioxide.

The estimates just given are from Bolin et al. (1979). They fit better with recent revisions of tropical forest biomass by Brown and Lugo (1984) than do some other models. However, the widely used factor of 50% loss of soil carbon following deforestation is probably an overestimation. For example, only a 20% reduction was observed in a wet Costa Rican forest environment (Raich 1983). Whatever the precise figures, tropical deforestation, which is rapidly accelerating, is playing a significant and increasing role in carbon dioxide increase (see Chapter 11).

What are the effects of rising atmospheric carbon dioxide concentrations? One possibility is an increase in plant production, since low carbon dioxide pressure can be a limiting factor on the rate of photosynthesis. Increased production is unlikely in those natural ecosystems where water or minerals limit plant growth, but it could occur in industrial agroecosystems. Another possibility is that the concentration of carbon relative to other elements may increase even if biomass remains constant.

Much more frequently discussed is the "greenhouse effect" in which carbon dioxide in the atmosphere absorbs heat radiated from the ground and reradiates it back. As carbon dioxide concentration increases, so does this effect, suggesting a global warming of $2°-3°C$ by 2025 over the temperatures prior to widespread human activity. It is not possible to determine unequivocally whether global temperatures are rising, since any such pattern is obscured by short-term fluctuations in climate. However, most climatologists agree that such an increase will change global atmospheric circulation and alter precipitation regimes. The exact nature of such changes with respect to latitude is not clear, but most models predict a decrease of rainfall in major food-growing areas. Some models suggest that rainfall will increase in the tropics and subtropics north of the equator but decrease in the southern tropics and northern temperate zone (Revelle 1982). A minority of climatologists claim that carbon dioxide and temperature changes will be slight and may even be beneficial. The mere prospect of increased human suffering as a result of a decrease in food production suggests that caution is the best attitude and that changes in carbon dioxide be minimized. Reduction in fossil fuel use and deforestation in the tropics are the two main ways of achieving this end. Tropical deforestation is particularly significant because much of the destruction is avoidable and produces little benefit for local populations (Chapter 11).

Effects on Hydrological Cycles.   Changes in vegetation cover alter the rate, quantities, and distribution of water cycling. The nature and effects of such changes are not easy to predict, with different models producing different results. A total deforestation in the equatorial region leads to a drop in rainfall of 230 mm a^{-1} and lower temperatures in the band from 5°N to 5°S according to the model of Potter et al. (1975). A recent assessment of hydrological and elemental cycles in the Amazon Basin suggests that an equilibrium obtains overall (Salati and Vose 1984). However, continued large-scale deforestation is predicted to produce widespread erosion and flooding, reduced evapotranspiration, and ultimately reduced precipitation. In addition, temperature regimes may become more "continental" with greater seasonal variations and more frequent frosts in some agricultural areas. Such changes could seriously damage cash crops such as coffee, sugar cane, and oranges.

Effects on Mineral Cycles.   Felling of tropical forests leads to release of the large store of nutrients immobilized in the biomass. Much of this is lost to the ecosystem by removal of timber or by decomposition, including burning, followed by leaching and runoff. In effect, deforestation breaks nutrient cycles, at least temporarily. With a brief and small-scale disturbance, subsequent recovery of vegetation, biomass, and nutrients is usually quite rapid. Large-scale deforestation may cause substantial losses of the store of soil nutrients through erosion and upset the prior balance of nutrient transfers between soil and vegetation and between available and unavailable forms in the soil.

Replacement of natural forests with plantations can significantly alter nutrient cycles, particularly when conifers are planted in lowland forest environments as a study by Brassel and Sinclair (1983) showed. The amount of nutrients reaching the soil is greatly reduced because of lower concentrations in litter and throughfall and because of slower decomposition (Table 3-4). While the rate of cycling was reduced, there was no noticeable effect on cation availability in the soil in this study. However, shortages of soil nitrogen and phosphorus are suggested by similar work in Nigeria, where Central American pines are grown in the dry forest zone. Mineralization of litter takes more than three years, and consequently litter accumulates at the soil surface. Ten-year-old plantations have almost 20,000 kg ha^{-1} of litter, equivalent to 15% of nitrogen and 8% of phosphorus in total organic matter. In contrast, litter in natural forests in the same climate decomposes in a few months (Egunjobi and Bada 1979, Egunjobi and Onweluzo 1979). In some situations, return of soil fertility is delayed and reduced compared with natural regeneration. For example, concentration of soil nitrogen, potassium, and magnesium were less than 20% of those in adjacent rain forest soils after six years of growth of pines in Trinidad. Cations showed signs of leveling off at these low concentrations (Cornforth 1970). This is in marked

Table 3-4   Comparison of some aspects of nutrient cycles in a rain forest and a conifer (*Araucaria*) plantation in Queensland, northern Australia. Litterfall as dry matter is similar in both, but the quantity of nutrients reaching the ground in litter and throughfall is much less in the plantation. The rate of litter decomposition and nutrient release is also much lower in the plantation. Data from Brassel and Sinclair (1983).

	RAIN FOREST	PLANTATION
INPUTS TO SOIL (kg ha^{-1}a^{-1})		
Litterfall, dry matter	8690	8150
Potassium		
Litterfall	64	46
Throughfall	125	54
Calcium		
Litterfall	211	181
Throughfall	50	16
Magnesium		
Litterfall	29	21
Throughfall	25	6
RATE OF LITTER DECOMPOSITION (% a^{-1})		
Dry matter	121	68
Nitrogen	170	74
Phosphorus	114	68
Potassium	514	260
Calcium	132	83
Magnesium	168	92

contrast to the reestablishment of soil nutrients under natural vegetation (compare Figure 9-4*b*).

3.4.2   Chemical Pollution

Pollution is the addition of extraneous materials or effects to ecosystems as a result of human activities. Chemical pollutants are often processed in a similar way to nutrients, indeed are sometimes the same substances but at higher concentrations. In the 1960s, developed countries began to realize that many of these chemicals not only disrupt natural ecosystems but also threaten human health. These countries have massive and pervasive problems because of the use and subsequent discarding of pollutants in manufacturing industries and agroecosystems. In contrast, most tropical countries are not as industrialized or urbanized, and most pollution problems are relatively small in geographical scale.

Paradoxically, some pollutants actually enrich soils and aquatic systems with macronutrients and may increase production and the rate of mineral cycling. Inorganic fertilizers are an obvious example. They are applied to agricultural land but may spread to other ecosystems and increase their levels of soil nutrients. In a range of Sri Lankan grasslands, dry matter production was doubled by the experimental addition of nutrient solutions, with nitrogen and phosphorus being especially important. Species composition was also altered with a reduction in diversity because of a few species becoming dominant (Pemadasa 1981 a, 1983). Species composition was altered more drastically by fertilization of humid forest soils in Costa Rica after tree removal. Unfertilized plots followed the normal pattern of succession with shrubs and trees predominating after one year. However, herbaceous plants continued to dominate in the fertilized plots because their dense growth prevented the establishment of the seedlings of woody plants. Despite its vigorous growth, the herbaceous vegetation had only two-third the biomass and less than 90% of the nutrients contained in the vegetation of the unfertilized plots (Harcombe 1977).

Micronutrients have rather different effects because many of them, particularly heavy metals, are toxic to the biota at greater than trace concentrations. Waste materials from mines often have high heavy metal content and have similar effects to natural occurrences of these substances (see Section 3.3.2). For example, carelessly dumped spoil from the Kilembe copper mines in Uganda has increased the concentrations of copper, nickel, cobalt, and iron by factors of 250–3600 times in the most heavily polluted soils. No plants were able to grow on such soils or on those nearby with concentrations of copper above 600 ppm (unpolluted soils had only 10 ppm). The effects spread for hundreds of meters from the dumping ground and even further when transported from the mine by a stream (Edroma 1974). That these toxic wastes were spreading into a national park is indicative of the need for active measures to monitor and control pollution.

The concentration effect along food chains exhibited by micronutrients (Section 3.2.4) is also seen in some other toxic substances, notably pesticides. The insecticide DDT and its metabolites behave in this way. It is known to cause reproductive failure in birds and is toxic to mammals at high concentrations. Although now banned in many developed countries, its use in the tropics to control crop pests and disease vectors is profligate. In Lake Baringo, Kenya, aquatic vegetation contains only a trace of DDT; herbivorous fish, 0.09 ppm; the muscles of fish-eating birds, 0.5 ppm; and their eggs, 0.8 ppm (Lincer et al. 1981). Elsewhere in Africa, concentrations in eggs of fish-eating birds of up to 35 ppm have been recorded. A concentration of 30 ppm has been circumstantially linked to hatching failure in a terrestrial bird predator in Zimbabwe (Whitwell et al. 1974). In general, it seems that DDT and its metabolic by-products are at lower concentration in tropical ecosystems than in temperate ones, even where pesticide use is heavy. It is possible that high solar radiation and temperatures lead to more rapid breakdown of these substances and reduces the rate of accumulation. However, there is little cause for complacency as residues of these toxic pesticides have been detected in human tissues all over the world, even in areas where pesticides are not used.

Developing countries have opportunities to avoid the pollution problems that have developed elsewhere. Unfortunately, control measures are usually expensive, but the long-term benefits will often outweigh short-term economic gains.

Summary

Nutrient cycles describe the route of chemical elements around ecosystems. Plants obtain these elements in inorganic forms from the atmosphere or from the soil in solution. Animals and microorganisms consume the elements in their organic food and, through respiration and decomposition, eventually return the nutrients to the atmosphere or the soil. Because of their key role in nonatmospheric cycles, soils and their biota are important determinants of nutrient cycling. The vegetation is an important store of nutrients in forests, especially in the tropics. Human activities, particularly deforestation, are disrupting many nutrient cycles. Extraneous pollutants may enter the cycles and have detrimental effects throughout the biosphere.

Study Questions

Review

1. Compare the rate of mineral cycling in tropical and temperate forests.
2. Contrast the roles of plants and bacteria in the carbon and nitrogen cycles.
3. Discuss the importance of soils in nutrient cycles.
4. Compare the flow of nutrients and pollutants in ecosystems.

Related Topics

1. Discuss the interactions between mycorrhizae and their host plants.
2. Investigate the catena concept and assess its importance in plant–nutrient interactions.
3. Assess the effects (including possible advantages) of increasing atmospheric carbon dioxide concentrations.
4. Describe the effects of a specific pollutant on a tropical ecosystem.

Further Reading _____

Bolin, B.; Cook, R. B. editors. (1983). *The Major Biogeochemical Cycles and Their Interactions*. Chichester, England: Wiley. An account of global cycles of macronutrients, interactions between these cycles, and the disruption caused by human activities.

Golley, F. B.; et al. (1975). *Mineral Cycling in a Tropical Moist Forest Ecosystem*. Athens, GA: University of Georgia Press. Detailed report on research in Panama on several forests. Much detail, but limited comparative analysis.

Sarmiento, G. (1984). *The Ecology of Neotropical Savannas*. Cambridge, MA: Harvard University Press. Chapter 6 deals with the sparse information available on nutrient cycles in moist savannas.

The following two papers give comparative reviews of tropical forest nutrient cycles:

Edwards, P. J.; Grubb, P. J. (1982). Studies of mineral cycling in a montane rain forest in New Guinea. IV. Soil characteristics and the division of mineral elements between the vegetation and soil. *J. Ecol.* 70:649–666.

Vitousek, P. M. (1984). Litterfall, nutrient cycles and nutrient limitation in tropical forests. *Ecology,* 65:285–298.

The classification of soils and details of soil chemistry are not included in this book. For information on these aspects of nutrient cycles, see the following:

Kalpagé, F. S. C. P. (1974). *Tropical Soils. Classification, Fertility and Management*. New York: Macmillan.

Russell, E. W. (1973). *Soil Conditions and Plant Growth*. 10th ed. London: Longman.

Termite populations are a common feature of tropical ecosystems. These Nasutitermes *are from Panama. (Photograph by Eldridge S. Adams.)*

CHAPTER 4

Population Ecology I: Populations of a Single Species

A population was defined in Chapter 1 as a group of individuals of the same species in a given area. In this chapter, properties of populations are considered as though largely isolated from populations of other species. While this treatment may seem artificial, it lays the groundwork for subsequent chapters. The first part of the chapter defines population characteristics and describes some of their numerical patterns. A simple exposition of mathematical theories, which attempt to explain population growth and regulation, follows.

The foundations of population ecology were laid by animal ecologists in the first half of this century. It is only relatively recently that detailed studies of plant populations have been made. As a result, little information is available about plant populations in the tropics. Where possible, this data is integrated into the text, but some peculiarities of plant population dynamics are described in section 4.3.5.

Some authors have suggested that the characteristics of tropical populations differ from those at higher latitudes in systematic ways. Such generalizations usually depend on assumptions about the constant favorableness of tropical climates compared with the harsh winters experienced in temperate and boreal latitudes. The theoretical basis for such speculations is discussed in Chapter 6, but when these generalizations relate to specific population parameters, they are dealt with in this chapter and Chapter 5.

4.1   Describing a Population _____

4.1.1.   Size and Density

Total **size** is simply the number of individuals in a population (but remember the problem of boundaries from section 1.1.2). For example, it has been estimated that there are 430 and 3000 hyenas in the Ngorongoro and Serengeti savannas of Tanzania, respectively (Kruuk 1972). More informative are estimates of **density**, the number per unit area (or volume) of environment. Since the area of Ngorongoro crater is 250 km^2 and that of Serengeti, 25,000 km^2, the densities of hyenas are 1.7 and 0.12 km^{-2}, respectively. Table 4-1 gives the density of various populations in both conventional units and the same units (per square meter) for direct comparison. The choice of units is merely for convenience, since it is easier to visualize (and write about) a population of 8.5 sloths ha^{-1}, than 0.00085 sloths m^{-2}. Density is often related to size, with larger organisms being of low density, but it is also related to the type of environment as illustrated by differing densities of savanna trees and African buffaloes at different sites in Table 4-1.

As well as varying from place to place, population density also varies in time. Populations may remain constant, they may fluctuate, or they may steadily increase or decrease (Figure 4-1). Such changes are the main focus of population ecology. They are brought about by the interplay between four factors. **Natality** (production of offspring) and **immigration** (individuals entering a population from elsewhere) lead to an increase in density, while **mortality** (death of individuals) and **emigration** (individuals moving out of the population) lead to a decrease in density.

4.1.2   Dispersion

Dispersion is the spatial pattern of individuals in a population relative to one another. Figure 4-2 shows the three basic patterns that occur. **Regular dispersion,** with individuals more or less equidistant from one another, is rare in nature but is common in managed systems where food or tree crops are deliberately planted in this manner. Populations of animals that exhibit territorial behavior tend toward regular dispersion. For example, such spacing was observed by Rockwood (1973a) in colonies of ants in a dry Costa Rican forest as a result of aggressive interactions between members of neighboring colonies. **Random dispersion,** where the position of one individual is unrelated to the positions of its neighbors (Figure 4-2*b*) is also relatively rare in nature. Most populations exhibit a **clumped dispersion** to some extent, with individuals aggregated into patches interspersed with no or few individuals (Figure 4-2*c*). Such aggregations may result from social interactions, such as family groups, or may be due to certain patches of the environment being more favorable for the population concerned.

Table 4-1 Population densities of various organisms in their conventional units and in comparable units of individuals per square meter

ORGANISM	ENVIRONMENT	NORMAL UNITS	m^{-2}	SOURCE
Grass (*Cynodon dactylon*)	Moist grassland (India)	200–700 m^{-2}	200–700	Singh and Joshi (1979)
Tree (*Delonix elata*)	Dry savanna (Kenya; two transects)	150 km^{-2} 2 km^{-2}	0.00015 0.00002	Leuthold (1977)
Tree (*Sabal alleni*)	Rain forest (Panama)	96 ha^{-1}	0.0096	Golley et al. (1975)
Earthworm (*Stuhlmania porifera*)	Moist savanna (Ivory Coast)	112 m^{-2}	112	Lamotte (1979)
Termite (*Macrotermes malaccensis*)	Rain forest (Malaysia)	650 m^{-2}	650	Abe and Matsumoto (1979)
Lizard (*Anolis bonairensis*)	Thronscrub (Caribbean)	1318 ha^{-1}	0.1318	Bennet and Gorman (1979)
Three-toed sloth (*Bradypus infuscatus*)	Rain forest (Panama)	8.5 ha^{-1}	0.00085	Montgomery and Sunquist (1975)
African buffalo (*Syncerus caffer*)	Grass savanna (Zaire) Wooded savanna (Kenya)	14.6 km^{-2} 0.9 km^{-2}	0.0000146 0.0000009	Delaney and Happold (1979)

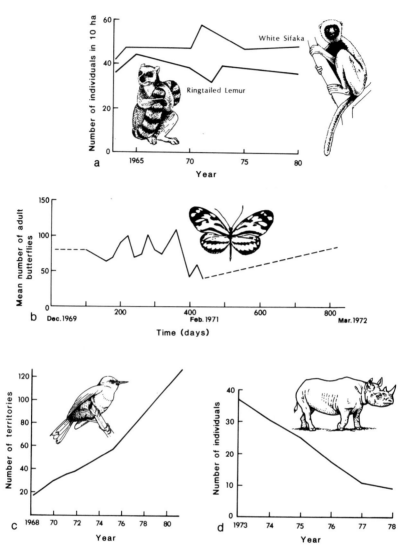

Figure 4-1   Density changes through time in all natural populations. It may fluctuate (*a* and *b*) or, more unusually, show a consistent increase or decrease over a long period (*c* and *d*). *a*, Densities of two species of Malagasy lemurs have remained remarkably constant over more than 15 years. (Data from Jolly et al. [1982a, b].) *b*, Fluctuations in the number of adult *Heliconius ethilla* butterflies in Trinidad (dashed line; infrequent censuses). (Modified from Ehrlich and Gilbert [1973].) *c*, The population (estimated as number of territories) of the Seychelles brush warbler *(Acrocephalus sechellensis)* on Cousin Island in the Indian Ocean has increased consistently for more than 12 years. (Data from Diamond [1980, personal communication].) *d*, The number of black rhinoceros *(Diceros bicornis)* in the Amboseli region of Kenya has declined markedly recently, causing concern about its possible extinction. (Data from Western [1982].)

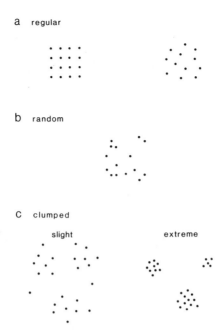

Figure 4-2   Contrasting types of dispersion pattern found in populations.
a, **Regular**, where individuals are more or less equidistant from one another.
b, **Random**, where there is no discernible relationship between the position
of individuals. *c,* **clumped**, where individuals are aggregated to a greater or
lesser extent. The clumps may themselves be regularly or randomly dispersed.

The relative frequency of these three patterns is well illustrated by trees
in a semideciduous forest in the Virgin Islands of the Caribbean. Of
sixteen species, 12 were clumped, 3 were random, and 1 was regular
in dispersion (Forman and Hahn 1980).

4.1.3   Age Structure

In most populations, individuals are of different ages. The proportion
of individuals in each age group is called the **age structure** of that
population. An understory palm tree population *(Astrocayum mexicanum)* in an evergreen forest of Mexico had 50% of individuals as seedlings
(less than two years old), 19% as saplings (eight years old), 5% as 30
year-old-adults, and so on until 70-year-old trees made up less than 2%
of the population (Sarukhán 1978).

A major problem in determining age structure of field populations
lies in determining how old each individual is. Occasionally individuals
are recognizable from when they are born, but more often indices of
age are used when some measure of size can be related to age. Trees
usually increase in circumference as they get older. In environments that

are seasonal, growth rings in the wood are formed as a result of rapid growth in wet (or warm) seasons and slow or no growth in dry (or cold) seasons. A count of these rings gives an estimate of the age. However, such rings are more difficult to interpret in tropical than temperate environments if they are formed at all. Similar types of methods can be applied to animal populations; two of them are illustrated in Figure 4-3.

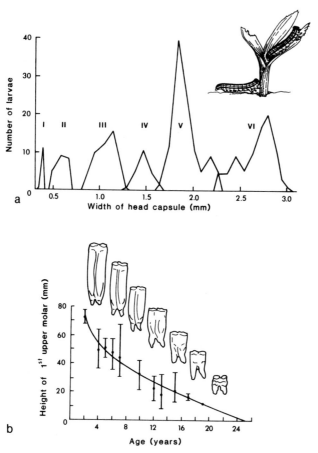

Figure 4-3   Two examples of ways in which measurements of body parts in a sample of a population may be used to predict the age of other individuals. *a,* African armyworm larvae *(Spodoptera exempta)*; each instar (roman numerals) has a characteristic range of widths of the head capsule, with only minor overlap between instars. (Data from Rose [1975].) *b,* Plains zebra *(Equus burchelli)* wear down their teeth as they get older so that tooth height may be used to estimate age. Vertical bars are the range of observed heights of first molar teeth from zebra of known age. (From Spinage, C. A. 1972. *E. Afr. Wildl. J.* 10:273–279.)

4.1.4   Production of Offspring: Natality

Natality is a general term used here to include asexual budding and fission, production of spores and seeds, laying of eggs, and the birth of live young. **Natality rate** is the number of offspring produced per female per unit time. It may represent **fecundity rate**, the maximum rate at which offspring can be produced physiologically, or **fertility rate**, the observed rate of natality in a population. The difference can be illustrated in humans for whom the fecundity rate is one child each 9–11 months, but the fertility rate may be only one child in 10 years.

Three population characteristics determine the rate at which females produce offspring:

1. **Clutch size** or the number of young produced on each occasion

2. The time between one reproductive event and the next

3. The age of first reproduction

The interplay of these factors is clearly shown by a comparison of two African rodents. The mole rat *(Tachyoryctes splendens)* produces only one young with a minimum time between births of 75 days and takes 120 days to reach maturity. The multimammate rat *(Praomys natalensis)* produces an average of 12 offspring every 23 days and takes 53 days to reach maturity. Assuming maximum fecundity and no mortality, the mole rat will leave less than 10 descendants in one year, while the multimammate rat will leave 72,000.

Natality rate usually increases during the period of maturity and then falls again as the organism gets older (Figure 4-4). Some trees seem

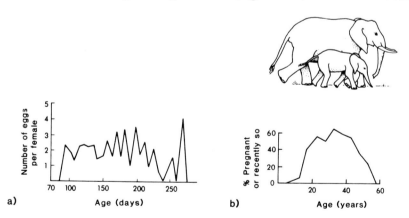

Figure 4-4   Fecundity of female animals is usually low at the onset of sexual maturity, then reaches a lengthy plateau, followed by a decline in old age. *a,* Milkweed bug *Oncopeltus unifasciatellus)* from rough pasture in Colombia and cultured in the laboratory. (Modified from Landahl and Root [1969].) *b,* African elephants *(Loxodonta africana)* in the Kabalega National Park, Uganda. (Modified from Laws et al. [1975].)

to continue to increase fruit production as they get older, but this may merely reflect a lack of data on really old trees.

Do Natality Patterns Differ in Tropical and Temperate Populations?

Is Breeding Aseasonal in the Tropics? Because of the continuously favorable temperatures, it is sometimes assumed that organisms in the tropics do not have a distinct breeding season comparable with the spring-to-summer breeding peak in most temperate organisms. As with so many generalizations about the tropics, a distinction must be made between the wet tropics and regions with seasonal rainfall. Figure 4-5 illustrates this divergence for rodent populations, which breed through-out the year in a rain forest but have two short breeding seasons in a savanna (coinciding with wet seasons) and a single prolonged breeding season in temperate areas. In fact, rodents of dry savannas may have a shorter total breeding period than their temperate relatives. The punctuated grass mouse *(Lemniscomys striatus)* in Uganda produces young during three months of the year only (Delany and Happold 1979) compared with six months or more in many temperate species.

Even in the humid tropics, natality may be highly seasonal. For instance, the breeding season of most species of passerine birds in rain forests of Sarawak (Malaysia) is hardly less distinct than that of British

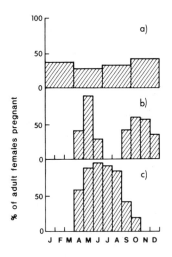

Figure 4-5   Breeding seasonality in rodents is related to the type of environment. *a,* Whitehead's rat *(Rattus whiteheadi)* in Malaysian rain forests breed throughout the year with little variation. *b,* Puntuated grass rats *(Lemniscomys striatus)* from Ugandan savannas breed during the two wet seasons that they experience. *c,* Bank voles *(Clethrionomys glareolus)* in cool temperate Britain have one long breeding season in the spring and summer. (Modified from Delany [1974].)

birds (Fogden 1972). Similarly, trees in humid tropical forests flower and fruit seasonally, although the seasons are often less pronounced than in temperate forests (see Section 8.1.3).

To conclude, in tropical areas that experience dry periods, breeding is at least as seasonal as in temperate regions. In continually humid areas, breeding may occur throughout the year, but many populations exhibit a seasonal peak. Natality is usually timed to coincide with favorable feeding conditions for the newborn, even in continuously moist environments. Exceptions to this general rule do occur, as in the case of the gelada baboons in the Simien Mountains of Ethiopia. Unlike most large herbivores in lowland savannas, these grass-eating monkeys breed during the dry season when food quality is low. The high probability of death for newborn in the wet season seems to be a more important limitation on survival than does food quality. During the rains, storms occur every day, temperatures rarely exceed $5°C$, and the young animals are susceptible to death by exposure (Dunbar 1980).

Are Clutch Sizes of Tropical Organisms Smaller Than Those of Their Temperate Equivalents?   This question is posed because theories of life history evolution suggest that clutch size in equable tropical environments will be smaller than those in the temperate zone (see Chapter 6). Where adequate information is available, the answer to this question is a qualified "yes." Many groups of birds, some plants, some insects, and some small mammals clearly exhibit this trend (Table 4-2). Among the larger mammals, most ungulates normally produce one offspring at a time. However, the smallest temperate deer may produce more (twins in roe deer and five to six in a litter of Chinese water deer) compared with only one young in the smallest African antelopes. Comprehensive information on most other groups is not available, although two studies illustrate that more offspring may be produced throughout a female life in temperate insects than in closely related tropical populations. In congeneric species of milkweed bugs from tropical and temperate America, clutch size was the same, but the average per capita egg production in the tropical species was only 60% of that of the temperate species because reproduction began later in life (Landahl and Root 1969). Similarly, in laboratory experiments using closely related *Drosophila* fruit flies in eastern Australia and the islands of New Guinea and New Britain, the tropical populations laid fewer eggs throughout their life than their temperate Australian relatives (Birch et al. 1963). Such trends are not universal. In the centipede *Scolopendra amazonica* from the Guinea savanna of Nigeria, maturity is reached in less than one year and two generations are produced each year. This species has a higher fecundity than any other centipede, including those from the temperate zone (Lewis 1970). Such exceptions to broad generalizations are to be expected because evolutionary pressures are complex and produce alternative adaptations in different populations.

Table 4-2   Comparison of tropical and temperate clutch sizes. Clutch size of tropical organisms tends to be smaller than in comparable temperate organisms. In passerine birds, this difference holds even for those in climatically unpredictable savannas and deserts.

TAXONOMIC GROUP	MEAN CLUTCH SIZE		SOURCE
	TROPICAL	TEMPERATE	
Compositae (tribe Helianthae; ovules per flower head) Type of life history:			
Annual herbs	40	62	Levin and Turner
Perennial herbs	61	99	(1977)
Perennial woody	50	69	
Aculeate wasps	5–13	15–25	Jayasingh and Freeman (1980)
Fruit flies (*Drosophilia melanogaster*); ovariole number	39	48	David and Bocquet (1975)
Passerine birds (old world)			
Forests	2.3 ⎫		
Savannas	2.7 ⎬	5.6	Southwood (1981)
Deserts	3.9 ⎭		
Tree squirrels	1.6	3.4	Emmons (1979)
Rats and mice	3.2*	4.2	Delany and Happold (1979), Fleming (1971) Southern (1964)

*Excludes the multimammate rat that has clutch sizes up to 19.

4.1.5   Mortality

A member of the reproductive caste of a termite population may live for 30 years or more if it helps to found a colony, but almost all the millions of reproductives that fly from the parent nest die within a few hours. A successful rain forest palm may live for more than 100 years, but 50% of seedlings die within a few weeks of germination. These two examples illustrate the difference between physiological potential for longevity and the actual likelihood of any one individual surviving that long. Physiological longevity is related to body size in many organisms as illustrated by most of the mammals in Figure 4-6. Irrespective of size, most organisms die when young. Mortality rate, the number of organisms in a population dying per unit time, varies considerably in different age groups. The pattern of mortality with age is best illustrated by **survivorship curves**, which plot the numbers surviving to a particular age. Such curves are usually standardized to follow the survival of a group of 1000 offspring, all produced at the same time (a **cohort**). Data from real populations are adjusted from the actual number of individuals

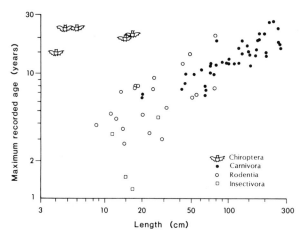

Figure 4-6   In general, larger organisms have a greater maximum life span than smaller ones as shown in several mammalian orders here. There are exceptions as exemplified by bats. (From Hutchinson, G. E. 1078. *An Introduction to Population Ecology.* New Haven, CT: Yale University Press.)

studied and for populations in which all the offspring are not produced simultaneously.

Three contrasting patterns of survivorship are shown in Figure 4-7. In the convex curve (I), probability of death is low throughout life; most individuals reach their physiological maximum age and then die. The diagonal curve (II) shows a constant proportion of organisms dying per unit time. The concave curve (III) describes a population in which most individuals die in the early stages of life but, once a certain age is reached, survival prospects are good. Survivorship of a variety of tropical populations shows that few match exactly the idealized curves. A convex pattern is found in some large mammals, particularly primates and including humans (Figure 4-8*a;* see also Figure 10-5*b*). Approximately diagonal curves are found in some rodents and birds (Figure 4-8*b, c*), while many plants and insects tend to concave patterns (Figure 4-8*d, e*). The termite curve (Figure 4-8*d*) is rather different than the others, since it shows survivorship of whole colonies rather than of individuals within a population.

Survivorship may vary in the same species in different environments. Seedlings of the plant *Eupatorium* (which has been introduced to India from tropical America) survive to maturity very well in fallow fields where few other plants are growing, but they do not live long enough to reproduce where other species of plants are well established (Kushawa et al. 1981). Such behavior is a common characteristic of weeds that establish very quickly in disturbed sites (such as agricultural fields) but are intolerant of shading by other plants.

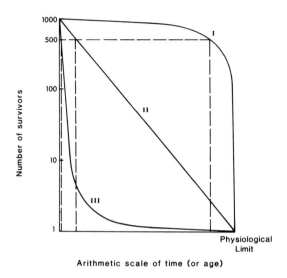

Arithmetic scale of time (or age)

Figure 4-7 Three contrasting patterns of survivorship in hypothetical populations (see text). Number of survivors is on a logarithmic scale to show the pattern in young organisms more clearly and so that a diagonal indicates that a constant proportion of individuals dies in each age group (for example, if 500 survive one year, 250 will survive two years; 125, three years, and so on). Dashed lines indicate the relative age at which half the cohort have died. Note that most die young in curves II and III.

Do Patterns of Survivorship Differ in Tropical and Temperate Populations?   It has been suggested that convex survivorship curves are more probable in fairly constant climates such as those found in the moist tropics (for example, see Margalef 1968), but evidence supporting this idea is limited at present. Indeed, convex curves sometimes come from populations in highly variable environments. As described in section 4.1.4, gelada baboons, which exhibit a convex curve (Figure 4-8 *a*), live in harsh and strongly seasonal alpine savanna. Nevertheless, some tropical populations seem to have enhanced survivorship compared with their temperate relatives. Fleming (1971) writes that survivorship is similar in rodents from both climatic zones. However, his data show that in only 1 of 18 species in the humid tropics did all individuals die within one year, compared with 15 from 30 species of temperate rodents. Evidence is also accumulating that tropical birds live longer than their temperate relatives. Individuals of tropical passerine species frequently reach ages of 7–14 years, while temperate passerines do so only rarely. Some of these tropical birds may even show a decreasing probability of death as they get old (such as the bulbul in Figure 4-8*c*), whereas survivorship remains constant in temperate birds (Fry 1980). In contrast, populations of cottontail rabbits in the tropics show lower rates of survival than those in the temperate zone (Ojeda and Keith 1982).

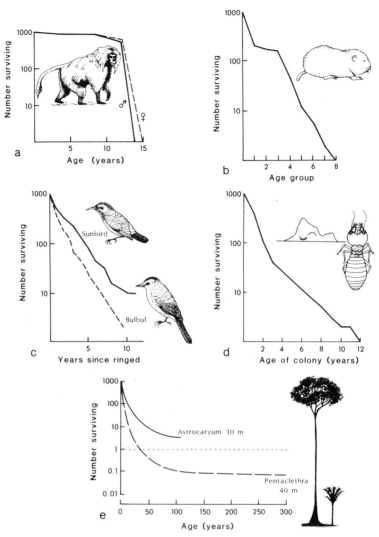

Figure 4-8   Survivorship curves of real populations rarely fit the hypothetical types of Figure 4-7 precisely. *a,* Convex curve in gelada baboons *(Theropithecus gelada)* from montane Ethiopia. (Data from Dunbar [1980].) *b* and *c* are approximately diagonal curves. *b,* Mole rat *(Tachyoryctes splendens)* from Nairobi, Kenya. (Modified from Jarvis [1973].) *c,* Brown-throated sunbird *(Anthreptes malchensis)* (solid line) and yellow-vented bulbul *(Pycnonotus goiaver)* (broken line) from forests of southeast Asia. (Modified from Fry [1980].) *d* and *e* concave curves. *d,* Whole termite colonies *(Macrotermes bellicosus)* in a Nigerian savanna. (Data from Collins [1981].) *e,* Two rainforest trees from the neotropics; solid line represents an understory palm *(Astrocaryum mexicanum)* in Mexico; broken line represents a canopy tree *(Pentaclethra macroloba)* in Costa Rica. Note the comparable abscissa with other curves is denoted by a dotted line so that survival of older *Pentaclethra* can be shown. (Modified from Sarukhán [1980].)

Similarly, populations of millipedes in Nigerian savannas have a shorter adult life than most temperate millipedes because they risk desiccation during the dry seasons (Lewis 1974).

Comparisons of survivorship between different climatic zones are even more problematic than similar comparisons of natality. Constraints on the latter are often intrinsic (genetic and physiological), whereas causes of mortality are more often environmental and therefore less predictable. The environmental determination of mortality is shown by the differential survival of many species in different environments. The weed *Eupatorium* referred to earlier is a good example, as are the human populations discussed in Chapter 10.

4.1.6   Natality and Mortality Combined: Life Tables

Information on natality and mortality in different ages and sexes can be compiled in the form of **life tables**. From these, it is possible to estimate the growth or decline of a population. As with survivorship curves, life tables are standardized to follow the progress of a cohort. Table 4-3 is a life table for a blue wildebeest population that inhabits the savanna woodlands of the Umfolozi and Hyuhluwe Game Reserves of Zululand, South Africa.

The table is subdivided vertically into three sections to emphasize the different ways in which the information is used in each. In section A, the x column is the age of individuals, the l_x column is the number surviving to each age, and the d_x column is the number dying in each age group. The q_x column depicts the proportion dying from the previous age category such that the first entry, $q_0 = (l_0 - l_2/l_0) = 0.3$. The advantage of the q_x column over the d_x column is that it gives the mortality rate in each age group. For example, between ages two to four years, 130 animals die compared with only 40 between the ages of 10–12 years. However, the rate of mortality in the second period ($q_x = 0.44$) is more than twice that in the earlier period ($q_x = 0.19$) because many fewer of the cohort are still alive. The m_x column represents fertility rate as the average number of offspring produced by each individual in the age group. In the fifth column ($l_x m_x$), the number of offspring produced by each age group is found, and the sum of this column gives the total offspring produced by the cohort throughout its life ($\Sigma l_x m_x$). Overall, therefore, the group of 1000 wildebeest produced 1325 young. The ratio of these figures is called the **net reproductive rate** (R_0) of the population; that is, $R_0 = \dfrac{1325}{1000} = 1.325$. In other words, each individual leaves, on average, 1,325 offspring, and therefore the population is increasing. A later cohort of the same population had $R_0 = 0,733$, which means that the population was declining because the initial 1000 animals left only 773 young (Attwell 1982).

The collection of such detailed data has practical as well as scientific significance. Management of these wildebeest requires knowledge of whether and why the population is changing in size. For example, hunting (for food or sport) is best done in a sustainable manner. To maximize sustainable yield, life table data are required to estimate population growth rates as shown in section 5.6.

Section B of the life table represents mortality in a logarithmic form with columns of log l_x and k (k is for killing factor and should not be confused with K, carrying capacity as used in section 4.2). **k-values** are calculated in the following way: k_0 = log l_0 - log l_2 = 3.00 - 2.84 = 0.16; k_1 = log l_2 - log l_4, etc. An advantage of this logarithmic form is that the k-values for a number of successive age groups can be added together to give total mortality. For instance, within the life table, the sum of all k-values is 3.000. The antilogarithm of 3 is 1000, meaning that all the wildebeest in the cohort have died. The k column also indicates the comparative intensity of mortality at different ages in a similar way to q_x (although summation of the latter is meaningless). k-values, therefore, combine the advantage of additive mortality of the d_x column with the intensity of mortality of the q_x column. A use of k-factors is illustrated in Figure 4-9 where the various causes of mortality in a population of Nigerian grasshoppers at different stages of its life history is given.

Section C of Table 4-3 is used in the calculation of the rate of population growth as outlined in Appendix A.

4.2   Population Dynamics

The ability to describe populations allows for changes in population parameters to be recognized. The remainder of the chapter is an account of the theoretical bases and their supporting evidence that attempt to account for such changes.

4.2.1   Approaches to the Study of Population Dynamics

Mathematical Models. Analysis of numerical changes inevitably involves the use of mathematics. Two types of mathematical model are used in population ecology. They are philosophically distinct but overlap considerably in practice. Theoretical models involve assigning algebraic notation to population phenomena and deriving equations that might describe population changes. The behavior of such models can then be tested by comparison with real populations. In contrast, simulation models use detailed data from real populations and try to predict future behavior of these populations under specified environmental conditions.

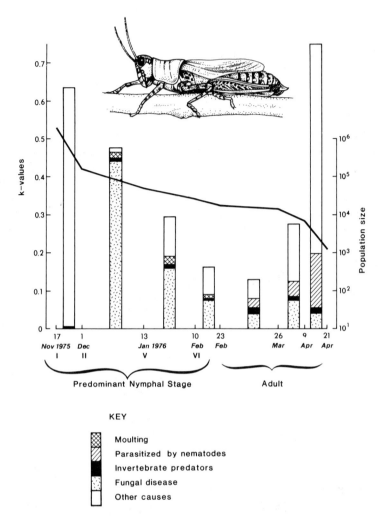

Figure 4-9   Analysis of mortality factors (k factors; see text) during the life history of the grasshopper *Zonoceros variegatus* in fields near Ibadan, Nigeria. As the insects get older, the numbers decline (line graph), but the causes of mortality are proportionately different at different stages (bar chart). Fungal disease predominates between instars II and VI, parasitization by nematodes becomes increasingly important in older adults, but insect predators are of minor importance throughout. Other causes include emigration, reproductive stress, and abiotic factors. (Data from Chapman and Page [1979].)

(Figure 2-9 is a simulation model used, in that case, to predict ecosystem energy flow). Mathematical modelling of population dynamics may seem esoteric but in fact can give deep insight into the reasons for population changes. It also forms the basis of most applications of population ecology.

Table 4-3 Life table of the blue wildebeest *Connochaetes taurinus taurinus* in Umfolozi Game Reserve, South Africa. Data from Attwell (1982).

AGE YEARS (x)	NUMBER SURVIVING (l_x)	NUMBER DYING (d_x)	SECTION A PROPORTION DYING (q_x)	FERTILITY (m_x)	NUMBER OF YOUNG BORN (l_x m_x)	SECTION B* log l_x	SECTION B* k	SECTION C† x l_x m_x
0	1000	300	0.30	0.00	0	3.000	0.16	0
2	700	130	0.19	0.12	84	2.84	0.08	168
4	570	260	0.46	0.92	524	2.76	0.27	2098
6	310	140	0.45	0.92	285	2.49	0.26	1711
8	170	30	0.18	0.92	156	2.23	0.08	1251
10	140	50	0.36	0.92	129	2.15	0.20	1288
12	90	40	0.44	0.92	83	1.95	0.25	994
14	50	30	0.60	0.92	46	1.70	0.40	708
16	20	20	1.00	0.92	18	1.30	1.30	294
18	10	—	—	0.00	0	—	—	—
		$\Sigma d_x = 1000$			$\Sigma l_x m_x = 1325$		$\Sigma k = 3.00$	$\Sigma \times l_x m_x = 8512$

*See text.

†See Appendix A.

The basic theoretical models of population growth are presented in algebraic form to give insight into how such models are constructed. Their predictions are also presented graphically for nonmathematically inclined readers, but all are urged to follow the simple algebraic arguments. These simple models have been refined by the addition of various factors to the basic equations to represent more accurately known biological phenomena. The reasons for these refinements are presented verbally and their predictions are given graphically, but the additional mathematics is beyond the scope of this text.

Laboratory Studies.   Models are of use only if they illuminate facets of the dynamics of real populations. Their predictions are often tested experimentally using laboratory populations. Such tests have two major advantages over the use of field populations. First, environmental conditions in the laboratory can be controlled and are often maintained as constant, whereas in the field, environmental fluctuations may mask changes that are inherent properties of populations. Second, taking a census of populations in the laboratory is relatively easy, but in the field accurate determination of density is often extremely difficult. Disadvantages are the inevitable artificiality of laboratory conditions and the types of organisms that can be used. Species that can be effectively cultured are not a representative sample of natural populations and may therefore bias conclusions about the general relevance of a particular model. Nevertheless, the interactions between theoretical modeling and the study of laboratory populations has produced a rich theory of population dynamics, which is sampled in the following sections.

Field Studies.   Mathematical and laboratory studies are only of value if they help to explain the behavior of natural populations. Inevitably, the field situation is much more complex, with many factors interacting to produce observed changes in populations. However, models and laboratory studies greatly assist in understanding such changes and can predict future changes with reasonable accuracy in many cases. If a field population strongly deviates from all model predictions, information gleaned from it can be used to refine subsequent models.

4.2.2   The Theory of Population Growth

Charles Darwin (1859) eloquently described the result of population growth as follows: "The elephant is reckoned to be the slowest breeder of all known animals . . . it breeds when thirty years old, and goes on breeding till ninety years old, bringing forth three pairs of young in this interval; if this be so, at the end of the fifth century there would be alive fifteen million elephants descended from the first pair."

The model is one of geometric increase and assumes that there is no environmental constraint on population growth. A geometrical series is one with a constant multiplier. The simplest biological example is binary fission in Protozoa when one individual divides to give two each time reproduction occurs. Starting with one individual (generation 0), there will be 2, 4, 8, 16, 32 . . . in succeeding generations if fission is synchronous. To determine the number (N_t) present in any generation (t), the following equation is used:

$$N_t = N_0 (2^t)$$

where N_0 is the number present in generation 0. For example, the number in generation 3 will be:

$$N_3 = 1 (2^3) = 8$$

This growth equation only applies to populations that double in number each generation. A more general equation that allows for other rates of growth is:

$$N_t = N_0 (e^{rt})$$

This is the **exponential growth** equation. The only change from the earlier equation is the substitution of e^r for 2; e is a universal constant, the base of natural logarithms, and has a value of 2.718 . . . ; r is a constant for a particular population under specific environmental conditions and is called the intrinsic rate of natural increase. The significance of logarithms in exponential growth is illustrated in Figure 4-10, which compares the predictions of the binary fission equation on arithmetic and logarithmic scales. The logarithmic plot is a straight line, meaning

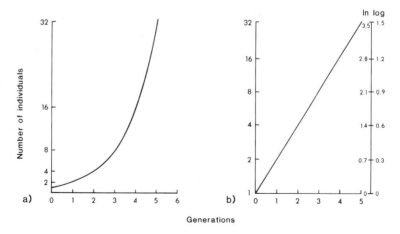

Figure 4-10 The exponential growth curve corresponding to synchronous binary fission, beginning with one individual (see text). *a,* Arithmetic scale for population size. *b,* Logarithmic scales for population size; both common logarithms (log) and natural logarithms (ln) are shown.

that the number added in any time period is proportional to the number present at the begining of the period (compare with survivorship curve II in Figure 4-7, which is an exponential decline).

The **intrinsic rate of natural increase (r)** is easy to calculate if natality and mortality rates are known. For example, for each of the 14 million people present in Sri Lanka in 1976, 0.028 were born and 0.008 died during that year; r = natality rate – mortality rate, or 0.028 – 0.008 per year. Substituting in the exponential growth equation, the number present in 1986 is predicted as:

$$N_{10} = 14 \text{ million} \times 2.718^{0.02 \times 10}$$
$$= 17 \text{ million}$$

(Note this is only strictly correct if the population is growing exponentially with constant birth and death rates and age structure, and there is no immigration or emigration.)

Since the exponential equation describes continuous growth, it can also be presented in differential calculus such that:

$$\frac{dN}{dt} \quad = \quad r \quad \times \quad N$$

| Instantaneous increase in population (change in number at any instant) | = | Intrinsic rate of natural increase | × | Number present at that instant |

When a detailed life table is available, r can be calculated using the information in section C of Table 4-3. The method is given as Appendix A. For the wildebeest population in Table 4-3, r = +0.44; for the same population in decline at a later date, r = -0.038 (see section 4.1.6 and Attwell 1982). That is, when r is positive, a population is increasing, and when r is negative, the population is declining.

A little thought reveals that most populations cannot possibly increase exponentially because their environment prevents this. Most ecosystems contain many species and present a variety of abiotic conditions such that a given population will have an upper size limit that cannot be exceeded permanently. Such a limit is called the **carrying capacity** of the environment for that population. Carrying capacity is determined by the availability of resources such as food and space.

The **logistic equation** was derived by Verhulst in 1838 to describe population growth with an upper limit. Suppose that the environment has a carrying capacity (K) for a particular population. Then the logistic equation assumes that the intrinsic rate of natural increase (r) is progressively reduced as population size increases toward that carrying capacity. If there are 10 individuals present (N = 10) and K = 100 individuals, the resources remaining can support a further K – N = 90 individuals and the proportion of resources remaining is $\frac{K - N}{K} = 0.9$.

The logistic equation simply assumes that r is reduced in relation to the proportion of resources remaining. This equation is derived from the exponential such that:

$$N_t = N_o \ e^{r \left(\frac{K-N}{K} \right)} t \ \ldots \text{ the logistic growth equation}$$

Suppose that $r = 1$; then when $N = 10$, the growth rate actually achieved, $r\left(\dfrac{K-N}{K}\right)$, is 1×0.9. When $N = 90$, $\left(\dfrac{K-N}{K}\right) = 0.1$ and the growth rate is 1×0.1, and so on, giving the progressive decline in realized growth rate. The logistic curve is S-shaped on arithmetic co-ordinates and is always less steep than its exponential equivalent (Figure 4-11a). A comparison of the same curve on semilogarithmic coordinates is given in Figure 4-11b. Using differential calculus, the logistic equation becomes:

$$\frac{dN}{dt} = rN \left(\frac{K-N}{K} \right)$$

4.2.3   Growth of Laboratory and Field Populations

From the previous discussion of carrying capacity, one would expect to find exponential growth only in populations under circumstances of no environmental constraint. Such growth is possible in laboratory cultures if excess organisms are removed so that the culture does not become overcrowded. When natural populations undergo exponential increase, they usually experience subsequent catastrophic decline. In the example shown in Figure 4-12, the insect pest population increased when

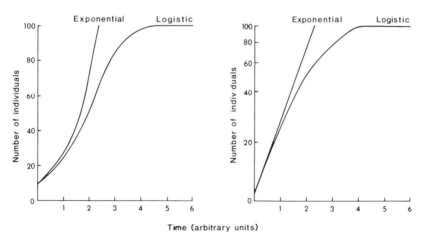

Figure 4-11 Comparisons of exponential and logistic growth curves predicted from the respective equations, beginning from the same population size, $N = 10, r = 1$, for the logistic $K = 100$. a, Arithmetic scale for population size. b, Logarithmic scale for population size.

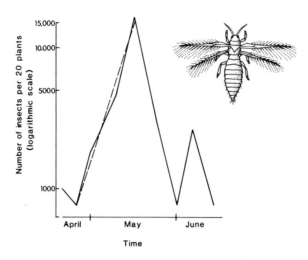

Figure 4-12   The population of thrips *(Scirtothrips maniboti)* living on cassava grows approximately exponentially in Brazilian fields (note log scale). Solid line is observed density; dashed line is an exponential prediction. Exponential growth is usually followed by catastrophic decline as in this example. (Data from Samways [1979].)

climate and food supply were favorable and then declined when these conditions became unfavorable. An unusual case of exponential growth is found in the number of insects in colonies of the termite *Macrotermes bellicosus* in Nigeria. Starting from one queen and king, the colony population increases exponentially to 800,000 in about five years after which the number declines (Collins 1981). During the growth phase the colony produces workers and soldiers who build and protect a mound that may be 5 m high and establish the fungus combs. The number of these castes stabilizes during the next five years or so, but reproductive alates, which disperse from the nest, are produced. Colonies 10–20 years old decline in size due to decreasing fecundity of the queen. The fungus combs also become less efficient in recycling termite feces, possibly because of the accumulation of toxins (see Section 2.6.3).

Most ecologists expect logistic growth to be a more common model of growth in real populations. In the laboratory, it has been observed in bacteria, yeasts, algae, and Protozoa. Figure 4-13*a* shows results of experiments by Gause (1934) using cultures of the ciliate *Paramecium*. In Metazoan laboratory cultures, a much less precise fit to the logistic is usually found, but populations do often fluctuate around a carrying capacity (such as that shown in Figure 4-13*b*).

Why do laboratory populations level off or fluctuate around a certain density? What is the biological mechanism? As density increases, so does competition between individuals (competition is defined and

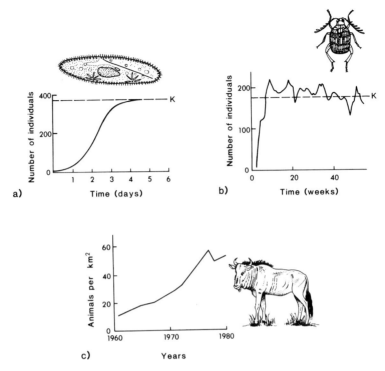

Figure 4-13   Approximations to logistic growth observed in real populations. *a,* The classical experiment of Gause using the protozoan *Paramecium caudatum* in laboratory culture. (Modified from Gause [1934].) *b,* The azuki bean weevil *(Callosobruchus chinensis)* in laboratory culture; as with most Metazoa, the population fluctuates about K. (Modified from Utida [1941].) *c,* The wildebeest population *(Connochaetes taurinus)* of the Serengeti ecosystem in Tanzania and Kenya seems to be leveling off at K determined by food availability following its low density earlier in the century, which was caused by rinderpest. (Modified from Sinclair and Norton-Griffiths [1982].)

covered in more detail in section 5.2). Following is a description of increasing competition in one population, but the exact mechanism varies in different species. The azuki bean weevil feeds on stored legume seeds. Population growth is shown in Figure 4-13*b.* The numerous laboratory experiments performed on this beetle are summarized in Utida (1943). Figure 4-14 shows results from a series of experiments that were initiated with different numbers of adult weevils. At high density, female fecundity is reduced because overcrowding leads to fewer successful matings due to mechanical interference between individuals. These collisions are the main factor causing the population to level off at high density. Subsidiary effects are increased egg mortality and

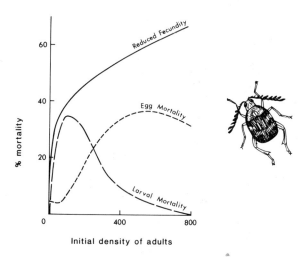

Figure 4-14   Intraspecific interactions between azuki bean weevils (*Callosbruchus chinensis*) change as population density is increased in laboratory cultures. Fecundity is reduced (regarded here as equivalent to an increase in mortality), and mortality among younger stages shows more complex and contrasting patterns (see text). (Modified from Utida [1941].)

increased larval competition for food at high density. The egg mortality is due to adults knocking eggs from the sites at which they were laid. In Figure 4-14, the leveling off of egg mortality and the reduction in larval mortality at high adult density result from reduced fecundity and increased egg mortality, respectively. The complexity and differential timing of these factors lead to the oscillations in density seen in Figure 4-13*b*.

In the field, direct observations of the initiation and subsequent growth and limitation of a population are rare. A well-documented example that approximates to the logistic model is that of the wildebeest in the Serengeti-Mara ecosystem in northern Tanzania and southern Kenya (Figure 4-13*c*). However, only an approximate correspondence with the logistic model should be expected as carrying capacity is rarely constant from year to year in savanna ecosystems (see section 8.2.3). An alternative test of the logistic theory in field populations is to determine whether the populations remain constant at their carrying capacities. Caution is needed in drawing such conclusions, since a steady population density does not always mean resource limitation. In some cases, predation may cause a constant population held below its carrying capacity. Many populations fluctuate around a level rather than remaining constant. In some cases the fluctuations are very large (such as that shown in Figure 4-12); in others the population changes only slightly (see Figure 4-1*a*). Can such fluctuations be accounted for and yet retain the essence of logistic theory? The next section discusses refinements of the logistic model that include more realistic assumptions about organisms and their environments.

4.2.4   Modifications of Logistic Theory

Implicit in the logistic equation are the following assumptions:

1. The environment is constant such that r and K do not change.

2. There are no time lags in the response of the population to changes in its density.

3. Individuals have identical ecological properties in terms of their responses to density, regardless of age or sex.

4. Growth is continuous.

5. The age structure does not change in successive generations.

These assumptions are unlikely to be met even in laboratory populations and will never obtain in the field. For example, larval and adult insects can be totally disimilar in feeding habits and abiotic environmental requirements. Many species have aquatic larvae and terrestrial flying adults. Modifications to the logistic model have been proposed to make the assumptions more biologically reasonable. Some of these are briefly discussed next.

Inclusion of a Time Lag.   That individuals in laboratory populations of Metazoa do not respond instantaneously to changing density explains a large component of fluctuations of the type seen in Figure 4-13b. Eggs take time to mature in the female body, but the number laid will often respond to the food available when the eggs were initiated rather than when they hatch. Such time lags allow K to be exceeded, temporarily. High mortality results because resources are then inadequate, followed by a drop below K. Once below K the population begins to rise again. The result is a series of oscillations around the carrying capacity. Incorporation of a time lag into the logistic equation produces predictions of the sort shown in Figure 4-15a. Different curves are produced by differences in r and the length of the time lag.

Fluctuations in Carrying Capacity.   The carrying capacity for populations in the field is rarely constant for long. Figure 4-15b shows predictions when carrying capacity shows regular oscillations as it might in a seasonal environment. When the response time of the population to changes in K is long compared to the periodicity of the fluctuations in K, the population is approximately stable. When response time is short, population density follows the changes in K.

Chance Events.   Whether an individual survives and reproduces is subject to chance to some extent. One or a few individuals may happen

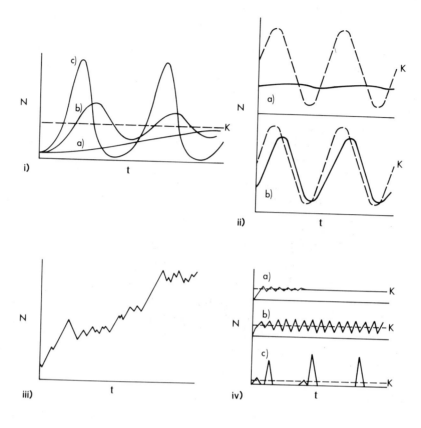

Figure 4-15 Some modifications to population growth equations include more sophisticated biological assumptions. All can produce fluctuations in population size, a condition found in most metazoan populations. *a,* Time lag in the response of the population causing a delayed response to K; curves (i) to (iii) have increasing time lags but the same r. See Hutchinson (1978) for further details. (Modified Cunningham [1954].) *b,* When K fluctuates, the predicted response of the population varies, depending on the relationship between r and the time period of the fluctuations; (i) small r population is almost constant, and (ii) larger r population tracks fluctuations in K but with time lag. See May (1981a) for further details. (From May, R. M. 1981. *Theoretical Ecology.* Oxford: Blackwell.) *c,* An example of predictions from a model that includes chance events. This model separates the effects of changes in density upon birth and death rates and determines the likelihood of a resulting increase or decrease in population size in a probabilistic manner. See Poole (1974) for further details. (From Poole, R. W. 1974. *An Introduction to Quantitative Ecology.* New York: McGraw-Hill, p. 70.) *d,* Population growth predicted from a simple difference equation (see text). Depending on the value of r, the equation predicts (i) a constant population (r = 1.8), (ii) stable oscillations about K (r = 2.3), and (iii) chaotic fluctuations that vastly exceed K and cause periodic extinction (r = 5.0). Population size is measured as N/K. (Modified from May [1981a] where further details are given.)

upon a particularly rich supply of food and therefore reproduce more quickly. Others may be killed by a landslide. In arid areas, rainfall at a particular site and time has a large random component, but rainfall determines the outcome of many population processes in such environments. Figure 4-15c is one example of the way in which chance events can be incorporated in the logistic type of model.

Discontinuous Growth. Many populations grow in discrete steps, with little or no overlap between successive generations. A complete separation of generations occurs in annual plants and in many insects that have a dormant egg or pupal stage. Growth in such populations cannot be described accurately by continuous equations such as the exponential and logistic equations. Instead, they are represented by difference equations of the following general form:

$$N_{t+1} = F(N_t)$$

which means that the population in any generation (N_{t+1}) is some function (F) of the population in the preceding generation (N_t). F can be assigned various formulas of varying complexity that include values for r and K (see May 1981a). Such equations produce a suite of predictions from gradually decreasing oscillations to chaotic variation (Figure 4-5d). The nondifferential forms of the logistic and exponential equations are simply special cases of the general difference equation.

4.2.5   Conclusions About Mathematical Models of Population Growth

The mathematical derivation of the logistic equation is readily understood and predicts adequately the behavior of many unicellular organisms in the laboratory. Populations of Metazoa do not obey the logistic model strictly but tend to fluctuate in density around the carrying capacity. Similar fluctuations are predicted to some extent by incorporating more realistic assumptions into the logistic equation. Nevertheless, such modifications are always gross simplifications of the multitude of influences on field populations.

Are, then, mathematical models of any use? The answer is an unequivocal "yes" for the following reasons:

1. As theoretical abstractions, they set the crude limits to possible patterns of population growth (such as, is growth exponential or limited?). Observations on real populations can be assessed within these limits.

2. As models become more refined, they more closely fit the behavior of real populations, allowing for a greater understanding of the factors that affect population growth. Increasingly accurate predictions of population behavior can then be made.

3. From Nos. 1 and 2, it follows that population models can be applied to practical problems such as human population growth and the management of plant and animal populations (see section 5.6 and Chapters 9 to 11).

4.3   Regulation of Population Density

The logistic model and its derivatives assume that a population will level off at its carrying capacity, that there is an upper limit to population density set by the environment, and that the population is regulated at or around that level. The significance of regulation in natural populations has been a topic of controversy among ecologists for 50 years. The historical development of the arguments is reviewed by Krebs (1985) and will not be dealt with here except to clarify particular points.

4.3.1   What Types of Factors can Influence Population Density?

As previously established, population density can only be increased by natality or immigration and decreased by mortality or emigration. These parameters may be density dependent or density independent in their effects. **Density-dependent** factors increase in their proportional effect as a population increases. A classic example is the influence of carrying capacity in the logistic equation where the rate of population growth is increasingly depressed by intraspecific competition as density is increased. Density-dependent factors are often biotic environmental factors such as competition and predation. Figure 4-16 shows the density-dependent mortality of legume seeds due to predation, with a greater proportion of seeds lost at higher densities. **Inverse density**

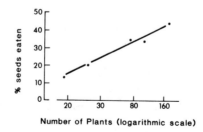

Number of Plants (logarithmic scale)

Figure 4-16   There is a density-dependent relationship between the number of plants in stands of the neotropical legume *Bauhinia ungulata* and the proportion of their seeds, which are destroyed by herbivores. (Modified from Heithaus et al. [1982].)

dependence also occurs if the intensity of the factor decreases proportionately as density increases (an example is larval mortality above 200 adults in Figure 4-14). **Density-independent** factors do not vary systematically in their effects as density changes. Abiotic environmental factors are often density independent but not always so. Floods or fires may destroy all populations regardless of their densities. Even in the tropics, cold weather may cause large-scale mortality. For example, many monarch butterflies *(Danaus plexippus)*, which overwinter in the montane areas of Mexico, die as a result of damage by storms and freezing temperatures (Calvert et al. 1983). Much of this mortality is density independent. Natality rate may also respond in a density-independent fashion. For example, total egg production of the Jamaican mud wasp *(Sceliphron assimile)* is greater in favorable weather whatever the density of adult wasps (Freeman 1980).

Both density-dependent and density-independent factors interact in most populations to determine observed densities. Figure 4-17 shows the interplay of both types of factor in a Nigerian tsetse fly population. Density-independent mortality due to climate is mainly caused by low humidity, which is at the same level in both January and November. However, the tsetse fly population is reduced only 12% in January compared with a 32% reduction in November, during which density-dependent factors are at their most intense.

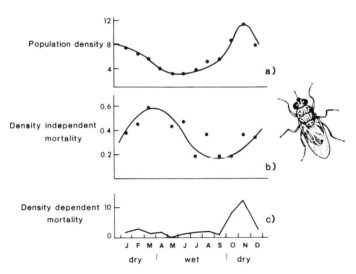

Figure 4-17   Population density *(a)* of tsetse flies *(Glossina morsitans)* in Yankari Game Reserve, Nigeria, is affected by both density-independent *(b)* and density-dependent *(c)* factors, which have different intensities at different times of the year. Population density is the number of females trapped; methods of determining density-dependent and density- independent effects are complex and are given by Rogers (1979), upon which this figure is based.

That density-dependent and density-independent and biotic and abiotic factors must be taken into account in population dynamics is an important conclusion because of the controversy between ecologists as to which type of factor is the more important in determining the size of natural populations. Put very simply, the "climatic school" of thought maintains that natural populations rarely attain densities high enough for density-dependent factors to have noticeable influences. Instead, climatic factors, by their effects on mortality and natality, determine the level that populations can reach during favorable periods. The opposing "biotic school" claims that average density in most populations is regulated by density-dependent factors such as competition and predation and that density-independent factors serve only to cause fluctuations in population size. Since the word **regulation** implies a tendency to return to some equilibrium, only density-dependent factors can regulate a population. The dispute can, therefore, be condensed to the question, "Are populations regulated or not?" It seems likely that relatively stable populations are regulated (see Figures 4-1a, b, and 4-13). Those undergoing violent fluctuations (see Figure 4-12) are probably not regulated, at least for certain periods, and the densities observed are likely to be determined by density-independent factors. Clearly the best approach is to analyze the factors affecting population size and to determine which are important, rather than to deny the significance of any factor as an article of faith.

4.3.2   Key Factor Analysis

This technique enables analysis of mortality factors to determine which are the most important quantitatively and which may be regulatory. The k-values for each mortality factor (section 4.1.6), together with total k (the sum of all such factors), are plotted for several successive generations (Figure 4-18). The k-factor that most closely follows the pattern of k_{total} is called the **key factor**. In the insect population shown in Figure 4-18a, loss of adults through migration or death is the key factor. Notice that the key factor is not necessarily the biggest killing agent (which is larval parasitism) nor is it necessarily density dependent. Indeed none of the k-factors affecting this insect were demonstrably density dependent (Banerjee 1979). Juvenile mortality (k_1)is the key factor in the African buffalo population (Figure 4-18b). The young are more susceptible to diseases and predators than are adults. However, only adult mortality (k_2)was shown to be density dependent and, therefore, potentially regulatory (Sinclair 1973).

When two k-factors show an inverse relationship in successive generations, the two factors may tend to cancel one another out and stabilize density. Such a pattern occurs in the failure of females to lay eggs and death of female pupae in the tea moth (k_3 and k_4 in Figure 4-18a). Banerjee (1979) suggests that adult density may be stabilized by immigration from outside the study area when adult density is low.

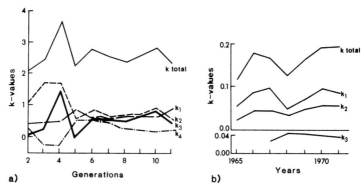

Figure 4-18 k-factor analysis of two animal populations; key factors are identified by an asterisk. See text. *a,* The moth *Andraca bipunctata,* whose larvae are pests in Indian tea plantations. k_1, failure of eggs to hatch; k_2, parasitic mortality of larvae; k_3*, failure of females to lay eggs (includes mortality and emigration); k_4, mortality of female pupae. (Modified from Banerjee [1979].) *b,* African buffalo *(Syncerus caffer)* in Serengeti National Park, Tanzania. k_1*, juvenile mortality; k_2, adult mortality; k_3, reduced fertility. (Modified from Sinclair [1973].)

4.3.3   Self-regulation of Populations

Intraspecific density-dependent interactions regulate many populations in the laboratory, and in that sense these populations are self-regulating. Mechanical interference and starvation have been mentioned, but ultimately these are caused by the external constraints imposed by the environment. More intrinsic forms of regulation occur in some cultures, such as flour beetles, in which the effects of accumulating waste products depress population growth in a density-dependent manner.

Social behavior may also be regulatory, particularly when the number of organisms able to breed is affected. Many species have individuals (or groups of individuals) that defend food or breeding partners in a **territory**. When food is included, a territory will normally contain a sufficient amount to support a breeding group and their young offspring. In most cases the adult males aggressively defend the territory. Other members of the population are excluded and may starve or fail to mate if they are unable to establish territories of their own. Such deaths or failures to mate are density dependent and can, therefore, regulate the population density.

Territory size varies from year to year and place to place, depending on the resources available per unit area, as shown by the golden-winged sunbird (*Nectarinia reichenowi*) in Kenyan savannas. Territory area ranges from 2–50 m², but each contains approximately 1500 *Leonotis* flowers. The nectar from these flowers is the major food supply of the birds. By defending this number of blooms, the sunbirds are able to meet their daily energy need without having to forage further afield (Gill and Wolf 1975a). The effects of territorial behavior on population density and the associated behaviors have been studied in many species of vertebrates and insects (see Davies and Houston 1984, Price 1975).

Social behavior that regulates population density may result in an approximate balance between resources and population density. However, it is important to realize that such mechanisms evolve because they benefit the individuals that survive and reproduce and not because the whole population avoids overexploitation of its resources (see section 6.1.6).

4.3.4   Immigration, Emigration, and Population Dynamics

Of the four factors that affect population density, natality and mortality have been discussed at length, but immigration and emigration have only been mentioned in passing. Most models take no account of such movements; in most laboratory studies, these factors are precluded by the experimental containers; and in field studies, they are assumed to be of minor importance or to balance one another. Immigration and emigration are features of **dispersal**, whereby individuals (singly or in groups) move from one population and die if no suitable environment is found, establish a new population, or join an existing one at a new locale. Beyond the immediate concerns of population dynamics, dispersal has important consequences for the geographical range of a species and for evolution as shown in other chapters. Little detailed work has been done on such topics as differences in dispersal by different age groups. Without recognition of specific individuals, it is often impossible to separate the various factors involved in population changes. In some cases an apparent high survival rate may actually be a function of immigration rather than reduced mortality and apparent low survival rate may be explained by emigration. Members of some populations rarely remain in one place throughout their lives. For example, female mud wasps in Jamaica rarely lay their eggs at the site where they emerge. At the edge of their distribution, population density is low and is maintained by immigration from denser areas, but emigration from these denser areas is not density dependent. Dispersal in these insects seems to be genetically determined and unrelated to resource availability (Freeman 1980, 1981).

Recently Taylor and Taylor (1979) and their colleagues have stressed the importance of the **spatial dynamics** of populations. Empirical evidence shows that variability in density of a population from one place to another is directly correlated with the mean density in a logarithmic manner (Figure 4-19). In other words, as the mean population density increases so does the variability between samples. These authors argue

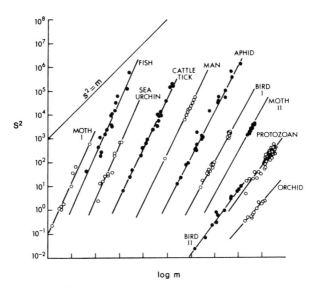

Figure 4-19   The spatial variability in density, estimated as the statistical variance (s^2) of a population, is closely related to the mean density (m) when plotted on logarithmic coordinates. This relationship holds for all organisms studied from Protozoa to orchids to humans, although the slope is rarely unity ($s^2 = m$). (From Taylor, R. A. J.; and Taylor L. R. 1979. In: Anderson, R. M.; Turner, B. D.; Taylor, L. R. editors. *Population Dynamics*. Oxford: Blackwell.)

that this pattern results from the behavior of individuals rather than of whole populations. Organisms within a population may congregate in areas of high environmental quality (leading to high variance over the range as a whole) or spread out when environmental quality is low or when regions of high quality are not located (leading to lower variance). The pattern overall is one of changing dispersion in which individuals are usually on the move but never able to deplete resources on a wide scale. Population parameters such as density and dispersion reflect this continual search for resources. According to this hypothesis, the factor limiting populations is the ability (or otherwise) to find adequate resources. Observations on year-to-year changes in the distribution and density of moths and aphids in Great Britain support these conclusions. Whether similar patterns occur in other organisms (such as large perennial plants and vetebrates) and other regions is unknown. The idea that resources are abundant overall is supported in herbivores at least, since they consume only a small proportion of primary production (section 2.5). However, food may be superabundant in quantity but of inadequate quality to support a larger population, as discussed in sections 6.3 and 8.2.3.

4.3.5   Special Features of Plant Population Dynamics

Plant populations behave in similar ways to animal populations in many respects. However, there are some characteristics peculiar to plants that need mention. Most higher plants are **modular** organisms, developing from a single zygote but producing an indeterminate number of repetitive structure (modules) vegetatively. A clump of herbs, grasses, or trees may be the product of one zygote proliferating vegetatively or the product of many zygotes germinating close together. In plants there are two levels of population structure: the individual produced from a single zygote, termed a **genet** (equivalent to N in the population equations), and the vegetative offshoots with various names such as ramet or tiller depending on the growth form (Harper 1977). The relative proportions of genets and vegetative individuals are an important feature of plant populations but are often difficult to determine. In the palm *Podococcus barteri,* which grows in Cameroonian forests, asexual stolons are produced from all sizes of tree, but sexual branches only appear in trees taller than 1.4 m. At least 59% and possibly 96% of a study population had originated asexually from stolons (Bullock 1982).

Another distinctive feature of higher plants is that they cannot move to mate or disperse. As a result, they have evolved to use gravity, wind, water flow, or animals to move pollen, seeds, or vegetative propagules. Plants are also ill equipped to escape from competitors and predators in space but may do so in time. The consequences of and evolutionary adaptations to this sedentary habit are discussed further in the next two chapters.

Many aspects of the growth and development of individuals and populations of plants are density related. For example, size and seed production of individuals often declines at high density, while total biomass of the population increases initially (Figure 4-20). Two generalizations have been demonstrated repeatedly in temperate plant populations and presumably apply to tropical ones also. First, yield of plants in dry weight per unit area is independent of the initial density of genets sown. This is true of the poppies in Figure 4-20b above a density of 600 plants m^{-2} . The tropical grass *Cynodon dactylon* shows a similar response, with a constant yield of 300 g m^{-2} at densities ranging from 400–3200 genets m^{-2}. Second is the **3/2 thinning law**. As plants grow and compete, some individuals usually die. As the remaining individuals grow, the increase in weight of these survivors more than compensates for decreasing density such that the total weight of the population increases. It has been demonstrated that the line relating weight of each individual to density has a slope of -1.5 (or $-3/2$). The slope would be -1 if increasing density was exactly compensated by reduction in weight of individuals. Thinning is normally inversely density dependent as illustrated in Figure 4-21, but does not always occur if the growth

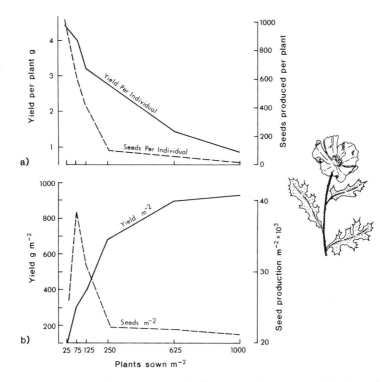

Figure 4-20   Many characteristics of plant populations are related to the density at which seeds are sown as shown in this example of experiments in India on the neotropical poppy *Argemone mexicanum*. (Data from Ramakrishnana and Jeet [1972].) *a,* Yield of biomass and seeds from each individual decreases as sowing density increases. *b,* Total biomass production of the population increases as sowing density increases at lower densities, reaching a plateau at higher densities, while total seed production reaches a maximum at 75 m⁻² plants sown and is much lower at higher densities.

of plants is extremely plastic. No mortality was found in the experiments with *Cynodon* just described. The thinning law has been verified many times for a wide variety of temperate plants from mosses to trees. A series of experiments with two species of tropical poppies approximate to it, although the slope is slightly less steep (Figure 4-21). It is probably safe to assume that the 3/2 law is universal, although the exact reason for its occurence is not known.

4.3.6   Do the Dynamics of Tropical and Temperate Populations Differ?

Earlier in the chapter it was concluded that there is limited evidence suggestive of lower rates of natality and mortality in some organisms

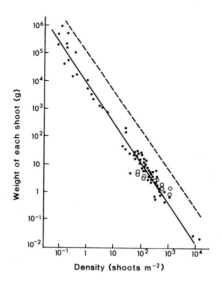

Figure 4-21   The 3/2 thinning law: the relationship between dry weight and density of shoots in populations of 29 species of temperate plants. Broken line is the predicted relationship if increasing density was exactly compensated by reduced weight of each individual. (Modified from Gorham [1979], with added data from tropical poppy populations [open circles] from Ramakrishnan and Jeet [1972].)

and some environments in the tropics compared with similar organisms from the temperate zone (sections 4.1.4 and 4.1.5). Where such differences occur, they obviously affect population growth and regulation. A comparison of the intrinsic rate of natural increase (r) in closely related *Drosophila* populations revealed a faster potential rate of growth at higher latitudes (Birch et al. 1963). Such differences could easily affect the stability of these populations, since smaller values of r lead to smaller fluctuations in density in several modifications of the logistic model (Figure 4-15). Is it possible to generalize and state that tropical populations are more stable than temperate populations?

Some tropical populations are remarkably constant from year to year (see Figure 4.1*a*, *b*). Available data suggests that rodent populations in moderate temperate environments do fluctuate more widely than those from tropical forests. Using amplitude of fluctuations as an inverse measure of stability, 12 species of tropical rodents have an average amplitude of 5.0 compared with 6.6 for 14 species from temperate ecosystems (data compiled from Fleming 1971, 1974, Happold 1977). However, fluctuations in tropical savannas with regular burning and in arid areas may well be as great or even greater than the famous eruptions of rodent populations in northern temperate and arctic areas (see

Happold 1977). Large herbivorous mammals exhibit fluctuations with amplitudes of 2–5, with no obvious differences between tropical and temperate areas (Leigh 1975).

Elton (1958, 1973) and others suggest that insects of tropical forests show only minor variations in density (as in Figure 4-1b). However, in a series of investigations, Wolda (1983) has shown that large fluctuations in abundance of many insect taxa occur in seasonal and aseasonal forests of Central America. These fluctuations are of similar magnitude to those observed in temperate species. Locusts are another insect group that undergoes massive fluctuations in density in tropical environments.

Because of the quality of most of the data and the short duration of most studies relative to the life span of the organisms involved, no broad generalizations about the stability of tropical populations are possible at present.

Summary

Parameters that affect population density are natality, mortality, immigration, and emigration. Measurement of these factors allows for the construction of life tables and the analysis of changes in population size. Mathematical models of population growth make predictions that can be tested in laboratory and field studies. The growth of real populations is more complex than the predictions of simple models because of features inherent in the biology of the organisms and because of environmental factors. Populations may be regulated by density-dependent factors but may also change in an unregulated manner in response to other factors. Tropical species seem to have lower natality rates than closely related temperate populations.

Study Questions

Review

1. Discuss how natality, mortality, immigration, and emigration interact to determine population density.
2. How well are the predictions of population growth models borne out by observations on natural populations?
3. Write an essay on population regulation.
4. Describe the steps needed to perform a key-factor analysis.

Related Topics

1. Review the biotic and climatic schools of population control. Why did the ecologists involved come to their divergent conclusions?

2. Contrast population growth in modular organisms with that of other organisms.

3. Review the Taylors' ideas of spatial dynamics. How can they be accommodated by more conventional models?

4. Investigate intraspecific variation in survivorship curves (that is, between cohorts, sexes, and environments).

Further Reading

See "Further Reading" at the end of Chapter 5 for literature on population ecology.

One extreme of prey (plant)-predator (giraffe) interaction is shown here in Maasai Mara, Kenya.

Population Ecology II: Interaction Between Populations of Different Species

In Chapter 4, populations were discussed as though they existed independently of other species. When other species were mentioned, it was only with regard to their effects on the population under consideration. Most ecosystems contain populations of many species that interact in vital ways such that changes in one population will have effects on many others. Three distinct classes of interaction were introduced in Chapter 1, and these are the subject of this chapter. Competition is usually between members of the same trophic level, and predation is always and mutualism is often between populations on adjacent trophic levels.

The affects of one population on the growth of another (indicated by + and − signs) can be categorized as follows:

Competition: Population 1 $\xrightleftharpoons[\quad -\quad]{\quad -\quad}$ Population 2

Predation: Population 1 (prey) $\xrightleftharpoons[\quad -\quad]{\quad +\quad}$ Population 2 (predator)

Mutualism: Population 1 $\xrightleftharpoons[\quad +\quad]{\quad +\quad}$ Population 2

A pair of species does not necessarily have a constant type of relationship. For example, in the Serengeti savannas of Tanzania, large herbivorous mammals are superficially visualized as competitors. Wildebeest probably do compete with Thomson's gazelles, taking the whole year together. However, early in the dry season, the large migratory population of wildebeest actually increase the grass production in the areas they pass through. This extra food is often eaten by the Thomson's gazelles (McNaughton 1979). The wildebeest population has a positive short-term effect on the gazelle population, which can be characterized as:

Commensalism: Wildebeest $\underset{0}{\overset{+}{\rightleftarrows}}$ Gazelles.

A fifth type of interaction, **amensalism** ($\underset{-}{\overset{0}{\rightleftarrows}}$), is also possible. Allelopathy between species of plants, in which one species produces a chemical that inhibits a second species but with no reciprocal interaction, is a form of amensalism (see section 5.1.3).

In this text, mathematical models of interacting populations are dealt with graphically and verbally. Examples of their relevant predictions are given in the appropriate sections, and a brief account of their algebraic structure is included as Appendix B. For details of these and other models, including the assumptions involved, see the references listed in "Further Reading" at the end of the chapter.

5.1   Interspecific Competition

Intraspecific competition, between members of the same population, was introduced in Chapter 4 and should be contrasted with interspecific competition between individuals of different species. Both may be defined in the following way. **Competition** occurs when individuals attempt to obtain a resource that is inadequate to support all the individuals seeking it, or, even if the resource is adequate, individuals harm one another in trying to obtain it.

The first category, often called **scramble** competition, will occupy most of this section. It is this type of competition (intraspecifically) that was mentioned as an important density-dependent factor regulating populations in the previous chapter. The wildebeest population of Figure 4-13 is thought to be regulated by intraspecific scramble competition for a limited supply of grass of adequate quality in the dry season. This example illustrates that competition does not have to occur all the time (grass is plentiful in the wet season) for it to be an important factor. The second category, called a **contest** (or interference competition), is less common but does occur. For example, lions often take kills from other large predators such as hyenas and cheetahs even though prey may be abundant. The lions do not physically harm their competitors. However, the energy expended by the hyenas or cheetahs in catching their prey is wasted, and survival prospects may be reduced as a result.

The resources competed for can be divided into two types:

1. Raw materials such as light, inorganic nutrients, and water in autotrophs and organic food and water in heterotrophs.

2. Space to grow, nest, hide from predators, etc. In higher plants this is manifested in spatial patterns; in animals, by spatial patterns or movements.

The classical mathematical model of competition between two species is called the Lotka-Volterra competition equations in honor of two of the cofounders. Based on the logistic model, these equations predict the following three types of outcome, depending on the characteristics of the competing populations (specific predictions are shown in Figure 5-1).

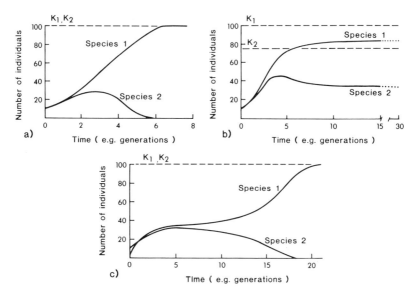

Figure 5-1   Examples of the predictions of the Lotka-Volterra competition equations (see Appendix C). *a,* Species I has a greater negative effect on the growth of species 2 than vice versa and drives it to extinction. In this example, values of r and K are the same for both species and, as in Figure 4-11, depicting logistic growth. Note that species 1 takes much longer to reach K in the presence of species 2 than it does when alone in Figure 4-11. *b,* Both species coexist indefinitely at levels below their designated carrying capacities. In this example, both species have the same r and interspecific effects on each other; K_1 and K_2 are set differently so that the two growth curves do not coincide. *c,* Only the species with the higher initial density survives. In this example, all parameters are set the same for both populations and the same as for species 1 in *a.* Note that species 1 here takes much longer to reach K and species 2 much longer to be eliminated than in *a.*

1. One species only survives, it being the one with the greater negative effect on its competitor (Figure 5-1*a*). Growth of the surviving population to its carrying capacity is slower than if the second population had been absent.

2. Both species coexist indefinitely. This occurs when interspecific competition is less intense than intraspecific competition in both species (Figure 5-1*b*). Neither population reaches the carrying capacity it would have in the absence of the other species.

3. The species beginning at higher density persists, and the other is eliminated. This is a special case where the populations have equally negative effects on the growth of each other, but interspecific competition is stronger than intraspecific competition.

Tilman (1982) has recently developed a series of graphical models based on the interactions between the growth of competing populations and their shared resources. When two or more species compete for a single limiting resource, only the one with the lowest requirement for that resource survives. (Requirement here is defined as the amount of resource needed to maintain an equilibrium population size.) This conclusion is qualitatively similar to the first prediction of the Lotka-Volterra equations just given. However, when two populations compete for two different resources—a situation addressed by classical competition theory—several outcomes are possible. If one species has lower requirements for both resources, it alone survives. When the two species each have a lower requirement than the other for one of the resources, both species may coexist indefinitely. This outcome happens when each population consumes proportionately more of the resource that has a greater limiting effect on its own growth. To date, supporting evidence comes mainly from planktonic algae, but this approach has potentially important implications for the structure of terrestrial communities as well (see section 7.4.1).

5.1.1   Laboratory Experiments and Interspecific Competition

Does the Lotka-Volterra model adequately predict the behavior of real populations? Simple laboratory experiments lead to the elimination of one species or the other. Gause (1934) performed experiments using pairs of species of ciliate Protozoa. Typical results for *Paramecium caudatum* and *P. Aurelia* are shown in Figure 5-2. In all cases, one species was eliminated. The agreement with the equations was not precise because after the elimination of *P. caudatum, P. aurelia* did not attain the carrying capacity that it achieved when cultured alone. Also *P. caudatum* had a higher growth rate initially in the mixed culture than in the monospecific culture. These observations serve to emphasize that

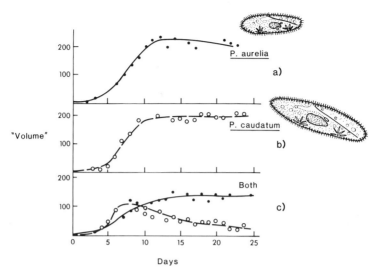

Figure 5-2 Laboratory experiments on population growth and competition between two species of ciliate Protozoa. *a* and *b,* The two species when cultured separately approximate to logistic growth. *c,* In mixed culture, *Paramecium aurelia* survives and *P. caudatum* dies out. (Modified from Gause [1934]; Volume of the animals was used rather than numbers to adjust for the different size of the two species.

simple mathematical models can only approximately predict the complex behavior of living populations.

Comparable experiments using insects fit the model less closely, which is not surprising given their crude approximation to logistic growth. Nevertheless, the prediction that two species cannot coexist indefinitely is met in simple experiments using beetle pests of stored flour (Figure 5-3). Predicting which species will survive is less straightforward than with Protozoa because changes in the abiotic conditions change the result of competition between these beetles. *Tribolium castaneum* is the sole survivor in hot and moist environments, while *T. confusum* "wins" in cool, dry cultures. With intermediate combinations of temperature and moisture, either species may win but with differing probabilities (Table 5-1). The outcome of competition can also be reversed in some cases by addition of the sporozoan parasite *Adelina* to the experimental cultures. *T. confusum* survives in 67% of experiments, whereas it wins only 10% of the time in the absence of the parasite. Such probabilistic outcomes seem to result from chance interactions between the beetles and their environments (Mertz et al. 1976), supporting the need to include random effects in the construction of population models (see Figure 4-15c).

The Lotka-Volterra model also predicts the possibility of coexistence, but such an outcome has not been observed in the simplest of

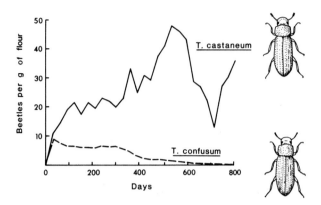

Figure 5-3 Laboratory experiment on competition between two species of flour beetles; *Tribolium castaneum* survives, and *T. confusum* dies out. (Data from Park [1948].)

experiments using species with near-identical resource use and population growth. However, in experiments with a heterogeneous environment or with greater interspecific differences, coexistence may be found. In finely ground wheat flour, *Tribolium* always displaces another flour beetle *Oryzaephilus,* due largely to predation on the pupae of the second species by *Tribolium.* When cracked whole wheat grains are used, the species coexist indefinitely because *Oryzaephilus* larvae pupate within the grains, thereby escaping from *Tribolium.* That the effect is largely one of physical protection for the pupae was demonstrated by further experiments using ground flour to which capillary tubes were added. The *Oryzaephilus* pupated inside the tubes and both populations persisted (Crombie 1946).

Table 5-1 Percentage of experiments in which each of two flour beetle species was the sole survivor under different abiotic conditions. The line separates a greater proportion of "wins" for *Tribolium castaneum* (hotter and moister) and *T. confusum* (cooler and drier). Data from Park (1948).

	HOT (34°C)	WARM (29°C)	COOL (24°C)
Moist (relative humidity 70%)			
T. confusum	0	14	71
T. castaneum	100	86	29
Dry (relative humidity 30%)			
T. confusum	90	87	100
T. castaneum	10	13	0

Such results led to the formulation of what is often called **Gause's hypothesis,** although it was never stated explicitly by Gause. Slobodkin (1961) states the hypothesis as follows: "Given a region of physical space in which two species do persist indefinitely . . . there exists one or more properties of the environment or species, or of both, that ensures an ecological distinction between the two species. . . ." In the case of competition between *Oryzaephilus* and *Tribolium,* the ecological distinction is the ability of *Oryzaephilus* larvae to pupate in confined spaces safe from predation. In the homogenous flour, this distinction is absent and the species cannot coexist. Gause's hypothesis was restated by Hardin (1960) as the **competitive exclusion principle,** which in its simplest form, states that *complete competitors cannot coexist.* This concept is a cornerstone of ecology, particularly in the form of its corollary of **ecological segregation,** the idea that populations will evolve to use the environment in slightly different ways, thereby reducing competition and permitting coexistence (see section 5.1.3).

5.1.2   Competition in Field Populations

It is commonly assumed that most populations are regulated by competition, primarily for food. Many field studies attempt to prove this by demonstrating ecological segregation, but seldom have resources been estimated precisely enough to show that competition must be occurring. In most cases, ecological differences between similar species have been found, but when no such difference is detected, it is often assumed that further study would reveal a difference. Such conclusions are unscientific. Ideally, science proceeds by the investigation of falsifiable hypotheses. The search for ecological divergence between competing species has an element of tautology and is ultimately not falsifiable. Unfortunately, it is often difficult to conclusively demonstrate competition in the field. Among the problems are that resource shortage may occur only occasionally or that specific nutrients may be inadequate, rather than the bulk of food available. To demonstrate competition unequivocally, natural experiments or deliberate manipulations, in which changes in the density of one species affect the density of other species, are required.

Competitive displacement has occasionally been observed in the field. Successive introductions of braconid wasp parasites were made in an attempt to control a fruit fly pest in Hawaii. First, *Opius longicaudatus* was introduced, but levels of parasitization were low. *O. vandenboschi* was introduced next but with limited success. Finally, *O. oophilus* was released, attained high levels of parasitization of the fruit flies, and apparently displaced the other two wasp species (Bess et al. 1961). Such introductions by humans have become increasingly common and often result in the displacement of native species (see section 5.2.3). Another example is the tramp ant *Wasmannia auro-punctata,* introduced to Santa Cruz island in the Galápagos group. The tramp ant population is displacing all the native ants in some areas (Clark et al. 1982). In neither of these cases is there any evidence of aggression by one species against the other. Competition for resources seems to be the cause of displacement.

Direct experimental evidence of competition in the field has been demonstrated among the *Anolis* lizards of the Antilles in the Caribbean (Roughgarden et al. 1984). *A. wattsi* was introduced to a tiny island upon which only *A. gingivinus* was present. Despite surviving for up to two years, 75% of the *A. wattsi* were eliminated during the first six months, and the few remaining were unable to establish territories in the preferred habitat. The density of *A. gingivinus* remained roughly constant. Removal of 35%–50% of the resident *A. gingivinus* increased the survival of *A. wattsi* somewhat and allowed a few to establish territories in the center of the preferred habitat. It seems, therefore, that interspecific competition for territories is an important feature of survival in these lizards. Further experiments on St. Maarten island, where both species co-exist, are also indicative of competition. When alone, *A. gingivinus* grew faster, produced more eggs, had more food in their stomachs, and consumed larger prey than when present at the same density but accompanied by *A. wattsi.*

Examples of coexistence, apparently through ecological segrega-tion, abound in the literature. Only one illustration is given here, but others can be found in sections 5.1.4 and 7.4.1. Two species of *Leptomyrmex* ants occur together on the floor of montane rain forests in Papua New Guinea. They are particularly abundant as the forest regenerates on the abandoned fields of shifting cultivators. Although *L. lugubris* can displace *L. fragilis* from particular nest sites, the two species coexist. There are minor differences in activity patterns, but the main seg-regation between the species is in the selection of food items (Figure 5-4). *L. lugubris* feeds mainly on plants and *L. fragilis* mainly on other insects. The greatest overlap is in insect food, but closer study reveals that other ants are the main prey of *L. lugubris,* while *L. fragilis* captures flying insects. Also, *L. lugubris* captures larger food items than *L. fragilis* (mean weight 3.5 mg and 1.5 mg, respectively), leading to further divergence in use of food resources (Plowman 1981).

5.1.3   Interspecific Competition in Higher Plants

Competition theory and ecological segregation as applied to animal populations lead one to expect that species will often differ in their food requirements. In contrast, higher plants use the same basic "foods"—light, water, and nutrients. Ecological segregation often takes the form of

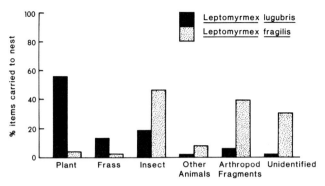

Figure 5-4 Ecological segregation between two species of rain forest ants in Papua New Guinea with respect to food items carried to the nest. The two species select different proportions of the food available. (Frass is the feces of leaf-eating insects that falls to the ground.) (Data from Plowman [1981].)

specialization in growth form or degree of tolerance of environmental stresses rather than food requirements. Furthermore, the distinction between scramble and contest competition is often obscure in plants. For example, a tall individual with a dense canopy makes light a limited resource and harms its smaller neighbors by reducing their growth.

Relatively few investigations have looked at interspecific competition between tropical plants. A problem with experimental studies is that displacement may take many years because of the slow turnover of individuals. Replacement series experiments overcome this problem by enabling the outcome of competition to be predicted from a series of experiments in which total density is kept constant, but the proportions of each species are varied. An example of this approach using montane grasses from Sri Lanka is shown in Figure 5-5. When two species are grown in the same pot, that which has a convex curve is the more successful competitor because its yield is reduced less by the interspecific interaction. In low-density experiments, *Cymbopogon* (Figure 5-5*a*) and *Pennisetum* (Figure 5-5*b*) actually yield more in mixed culture than in pure culture. At higher total density this effect is less pronounced but still significant, with the species mixtures yielding more than the monocultures. From these graphs, the grasses can be ranked in terms of their competitive abilities in the order *Pennisetum, Cymbopogon,* and *Eulalia,* and this is reflected in their distribution in the field. Where *Eulalia* and *Cymbopogon* coexist, *Eulalia* is restricted to the tops of hummocks while *Cymbopogon* occupies the more favorable hollows in between. *Pennisetum* is a weed species because it rapidly colonizes areas disturbed by human activity, and in so doing it occupies the hollows, displacing *Cymbopogon* to the hummocks (Amarsinghe and Pemadasa 1982).

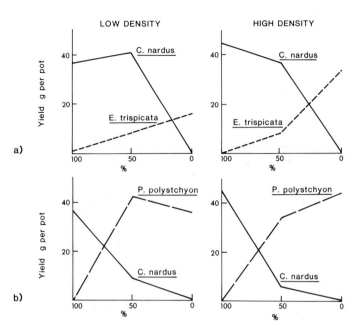

Figure 5-5   Results of replacement series experiments on montane grasses from Sri Lanka. Species are grown alone (100%) and in paired mixtures (50%:50%) at two different total densities (low six plants, and high 54 plants per 25 cm² pot). The species with a convex curve is the more successful competitor in each case. (Modified from Pemadasa and Amarsinghe [1982].) Abscissa is percent of *C. nardus* in each case. *a, Cymbopogon nardus* versus *Eulalia trispicata: Cymbopogon* "wins." *b, C. nardus* versus *Pennisetum polysrichon: Pennisetum* "wins."

In temperate forests, when a trench is dug around a plot of understory trees, these small trees grow and survive better. The understory trees no longer have to compete with the canopy trees for water and nutrients because the trench cuts any overlapping roots. A similar study in a tropical forest produced no effect, with trees in experimental and surrounding plots doing equally well (Connell 1979). Whether this is a common difference between tropical and temperate trees is unknown, and more experiments are needed.

Interspecific competition between plants may manifest itself by chemical aggression, or **allelopathy**, whereby one species produces chemicals that inhibit or kill competing plants. These amensal associations have been demonstrated in the laboratory, but their common occurrence and significance in the field is unclear (Harper 1977). Allelopathic interactions may even be intraspecific, as in the Australian rain forest tree *Grevillea*. The roots appear to produce a water-soluble substance that inhibits the establishment of adjacent seedlings of the same species (Webb et al. 1967).

5.1.4.   Competition and the Ecological Niche

Since its first use in 1910 the term niche has taken various meanings in ecology, but the most common current usage is that due to Hutchinson (1965). Any population can only survive and reproduce within certain environmental limits. For example, it will only tolerate a certain range of temperature and humidity as shown graphically in Figure 5-6*a*. These two abiotic factors are **niche dimensions**. If a third is added, say size of food items eaten, this can be represented diagrammatically as in Figure 5-6*b*. The rectangular box is the **niche space** in which the population can survive and reproduce with respect to these three important factors. Further dimensions, such as type of substrate or size

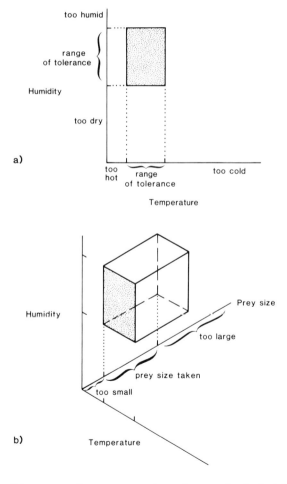

Figure 5-6   Diagrammatic representation of an ecological niche. *a,* In two dimensions (for example, temperature and humidity). *b,* In three dimensions (for example, prey size added).

of nesting sites available, could be added to produce a conceptual n-dimensional hypervolume or **niche,** although only three dimensions can be drawn on paper.

The Hutchinsonian niche just described is " . . . obtained by considering a hyperspace every co-ordinate of which corresponds to a relevant variable in the life of a species of organism. A hypervolume can therefore be constructed, every point of which corresponds to a set of values of the variables permitting the organism to exist" (Hutchinson 1965). The **fundamental niche** of a species is the hyper-volume that a population can fill in the absence of competitors. Closely related species of competitors will have similar requirements along the niche dimensions such that their niches will overlap one another. If the niche of one species completely overlaps that of another, then one of the species will be eliminated according to the competitive exclusion principle (Figure 5-7*a*). If the niches overlap partially, coexistence is

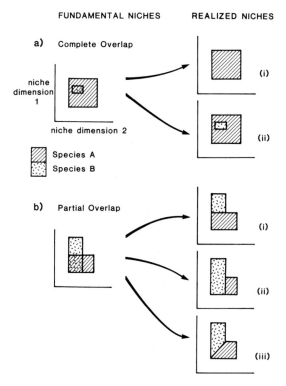

Figure 5-7   Patterns of overlap of fundamental niches of two competing species (see text). *a,* complete overlap, (i) One or other species eliminated completely or (ii) from area of overlap. *b,* Partial overlap. (i) and (ii) One species fills its fundamental niche at the expense of the other, or (iii) both species have smaller realized niches.

possible in two ways. Either one species fully occupies its own fundamental niche, excluding the second species from part of its fundamental niche and leaving it to occupy a smaller *realized niche* (Figure 5-7*b* [i, ii]). Or both species have restricted realized niches, each utilizing a smaller range of particular niche dimensions than they would in the absence of the other species (Figure 5-7*b* [iii]).

Niche theory as described leads to ecological segregation and therefore can be regarded as an extension of Gause's hypothesis. For example, the subterranean legless lizard *, Typhlosaurus lineatus* of the Kalahari Desert, Botswana, has a fundamental niche that includes eating roughly equal proportions of small- and medium-size termites. However, where the smaller *T. gariepensis,* which feeds mainly on small termites, is also present, *T. lineatus* eats more medium- and large-size termites (Pianka 1981). In other words, *T. lineatus* adjusts to a realized niche with respect to food size, thereby reducing interspecific competition.

Methods of calculating niche width and overlap along particular niche dimensions are discussed by Hutchinson (1978). Such measurements lead to questions about how narrow niches can be, how much they can overlap, and how many can be packed along any niche dimension. These questions are discussed in section 7.4.1.

5.2   Predation

The term predation is used here to cover all feeding interactions between populations on adjacent trophic levels such that plant → herbivore, animal → carnivore, host → parasite are all prey → predator relationships. Also included is the host → parasitoid interaction. **Parasitoids** are insects (mostly some wasps and true flies) that have free-living adults, but whose larvae develop parasitically within the bodies of living arthropod hosts. The host sustains the larval stages of the parasite, but is eventually almost entirely consumed and killed. This contrasts with other host-parasite systems in which the parasite consumes only a small portion of the host's production and often does not cause death. Predation is easier to study directly than competition because the act of predation is directly observable, whereas competition may occur without the organisms ever meeting. However, predation is difficult to model accurately and comprehensively because of the complex behavioral interactions involved.

The first attempt to model prey-predator systems was the independent derivation of equations by Lotka and Volterra, based on the exponential growth equation (see Appendix B). These equations predict stable oscillations with the two populations out of phase (Figure 5-8). When the prey population increases, more food is available for predators, which also begin to increase. However, more predators eventually means a decline in prey, followed inevitably by a decline in the predator population. The prey are again able to increase, leading to the cyclic pattern. The amplitude of the oscillations depends on the initial densities

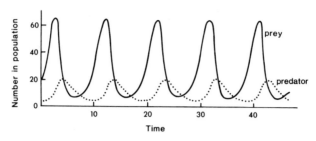

Figure 5-8   Oscillations of prey and predator populations predicted by the Lotka-Volterra predation equations beginning with 25 prey and 10 predators. (From Krebs [1978] where values of parameters are given. Reproduced by permission of C. J. Krebs.)

of the two populations. More complex models also tend to predict oscillations, which may be constant in amplitude through time; may decrease in amplitude, leading to constant populations; or may increase in amplitude, leading to the extinction of one or both populations (see May 1981b).

5.2.1   Laboratory and Field Studies of Predation

Laboratory experiments to demonstrate the Lotka-Volterra and other simple predation equations have not been successful. Gause's (1934) experiments with Protozoa convincingly demonstrated the logistic and interspecific competition models, but in his experiments with *Paramecium caudatum* and its ciliate predator *Didinium nasutum,* both populations became extinct (Figure 5-9a). The predator was so voracious that it consumed all the prey and then died out itself. In a second experiment a sediment was present in the culture vials. The prey survived because the predator could not find it in the sediment, but the predator died out as a result (Figure 5-9b). Gause was able to induce reciprocal oscillations only by periodically adding *Paramecium* to the cultures to prevent its extinction (Figure 5-9c). Although he failed to verify the predation equations, Gause's experiments did point to the importance of environmental heterogeneity in the sediment experiments and, in particular, the significance of a refuge in which the prey is protected from the predator. As Salt (1983) points out, *Didinium* does not obey the Lotka-Volterra assumption that predator growth rate must be dependent on the rate of food intake. The experiments were, therefore, invalid as a test of these specific equations, but they remain of empirical interest.

Oscillations between prey and predator have been produced in more complex experiments. Although these oscillations are not described precisely by the Lotka-Volterra equations, they do lend support to the predictions of most prey-predator models: oscillations occur in

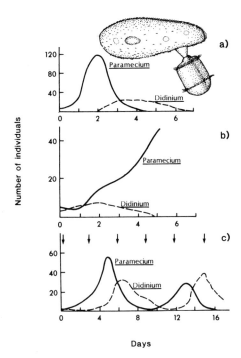

Figure 5-9   Laboratory experiments on prey *(Paramecium)* – predator *(Didinium)* relationships in ciliate populations. (Modified from Gause [1934].) *a,* In an homogeneous culture medium: *Didinium* "overeats" and both populations are eliminated. *b,* In a culture medium with a sediment: *Didinium* is unable to find prey and dies out. *Paramecium* survives. *c,* In an homogeneous culture with periodic introductions of prey (indicated by arrows): oscillations of both species obtained.

a closed system. Figure 5-10 shows the results of experiments using the Indian meal moth, a pest of stored grain, and an ichneumonid wasp parasitoid. Both populations oscillate for a year. However, to prevent extinction of the prey at high predator density, safe pupation sites were provided within the container, confirming the importance of refuges for prey survival.

In field populations, fluctuations in prey and predator density are often observed, but it is usually impossible to ascertain whether they are caused by population interactions or other environmental factors. Figure 5-11 depicts the population fluctuations of mosquitos attracted to artificial tree holes. Small sections of bamboo were half filled with water and attached to tree crops (coconut, mango, and cashew) near dwellings and in patches of rain forest in coastal Kenya. The mosquitos laid eggs in the bamboo traps, and 14 species of larvae were found associated with the large predatory larvae of *Toxorhynchites.* For over

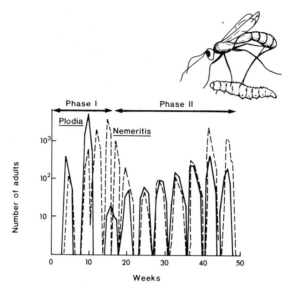

Figure 5-10   Laboratory experiments on prey-predator interactions between the Indian meal moth (*Plodia interpunctella:* solid curve) and its ichneumonid wasp parasitoid (*Nemeritis canescens:* broken curve). In carefully designed experiments, oscillations continued for a year (see text). Note that the oscillations in phase II (after the settling period, phase I) are synchronized rather than reciprocal (see also Figure 5-8). (From Podoler, H. 1974. *J. Anim. Ecol.* 43:657.)

two years the predator and mixed prey populations oscillated sometimes together and sometimes out of phase (Figure 5-11). However, rainfall seems to be a major influence on the fluctuations, with increased density during the rains when mosquito breeding sites are abundant and decreased density when dry weather returns. Although prey-predator interactions affect both populations, climate seems to be the driving force of the fluctuations in this case.

Caughley (1976) has suggested that the relationship between elephant and tree populations in the Luangwa Valley, Zambia, is a classical Lotka-Volterra type of relationship. He proposes that when tree population density is high, elephant populations increase. When elephants in turn reach high density, they destroy the trees (their food) causing their own population to decline and allowing the trees to recover. However, elephants feed on grass as well as trees, and their food supply is not inevitably decreased by tree mortality. Many alternative explanations exist for the interactions between tree and elephant populations, although none seems to be definitive (see section 11.5.3).

There are several features of plant-herbivore interactions that distinguish them from other predatory systems. First, herbivores feeding

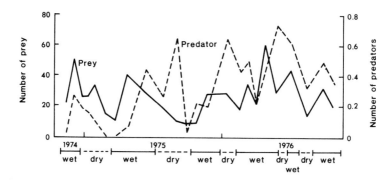

Figure 5-11   Fluctuations in the density of mosquito larvae in artificial tree-holes set in a variety of environments in coastal Kenya in relation to wet and dry seasons. The predator is the mosquito larva *Toxorhynchites brevipalpis* and the prey a variety of other mosquito larvae. (Modified from Lounibos [1979].)

on vegetative parts do not normally kill whole higher plants but consume only certain organs or fluids. Plants have well-developed powers of regeneration, and losses can often be replaced. In this respect, the interaction is similar to host-parasite systems, particularly if the herbivore is restricted to one species of plant. An exception is seed predation where a whole, as yet undeveloped, plant is destroyed. Second, plants tend to be larger than their predators, whereas most carnivores are bigger than their prey. Many savanna mammals are larger than their herbaceous plant food, but such animals only account for a small proportion of herbivory worldwide. Finally, plants cannot escape from herbivores by movement but can develop other methods of defense such as spines and chemicals that deter consumers. Another mechanism is to translocate material below-ground where it is unavailable to many herbivores.

In most ecosystems, most of the time, herbivores do not extensively deplete their food plants (section 2.5). Moderate grazing may even stimulate plant growth and asexual reproduction. Adult woody plants are rarely killed by the direct actions of small herbivores, but it can happen. For instance, the bush *Scaevola taccada,* which grows on the coast of Aldabra Island in the Indian Ocean, is killed by heavy infestations of the sap-sucking coccid scale insect *Icerya seychellarum* (Newbarry 1980). In some species, defoliation leads to a loss of reproductive potential even when the plant survives. Removal of live leaves from deciduous forest trees in Costa Rica reduced fruit production by 50%– 100% in different species (Rockwood 1973b). Much more vulnerable than the vegetative organs are the flowers, seeds, and seedlings, which determine the reproductive success of the parent. For example, the neotropical legume *Bauhinia ungulata* loses 92 of every 100 ovules between flower formation and seed germination, and 82 of these losses are due to herbivores (Heithaus et al. 1982). Indeed the intensity of seed predation in tropical forests may often approximate to 100% most of the time (Connell 1979). Seedlings are also susceptible to death if attacked by herbivores. Clipping of shoots of gum arabic *(Acacia senegal)* seedlings reduced the rate of survival from 90% to less than 40% in the most sensitive age group. These are the five-week-old seedlings, which have used up the parental food store in the seed, but are not well-enough established to withstand browsing (El-Din 1971). Since these trees live in semiarid savannas of Africa, they are likely to be browsed by goats or small antelopes.

5.2.2   Behavioral Aspects of Predation

The failure of simple equations to accurately describe the interactions between prey and predator is largely due to the complexity of behavior of the organisms involved. The Lotka-Volterra equations (and some other simple formulations) assume that as prey density increases, there is a linear increase in the rate of predation by each predator. This **functional response** of predators to prey density is rarely so simple. Logically, there must be an upper limit of prey density at which the predator is unable to further increase the rate at which it can attack and eat prey. Holling (1959) formalized this idea by proposing the functional response patterns portrayed in Figure 5-12. Two distinct factors affect functional response. First is the **rate of encounter** of prey by predators. This is assumed to

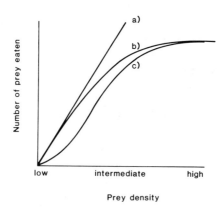

Figure 5-12   Functional responses of a single predator to changes in the density of its prey. *a*, Predator encounters and consumes prey in proportion to their density. *b*, Predation rate declines as prey density increases because handling time takes up an increasing proportion of total time from one encounter to the next. Holling type II response. *c*, At low to intermediate prey density, predation rate increases due either to decrease in handling time or increase in encounter rate as predator learns to deal with a particular type of prey. Holling type III response.

increase linearly with prey density (more prey, more encounters) in the type II response (Figure 5-12, curve b). Second is the **handling time** required to subdue and consume the prey and prepare for the next attack. Handling time becomes an increasing proportion of the total time needed from one encounter to the next as prey density increases in the type II curve. As a result, the response levels off at high density and is always less steep than the linear response (compare curves *a* and *b* in Figure 5-12). In the type III response, the rate of predation increases at intermediate densities because the predator learns to seek and handle the particular type of prey more efficiently (Figure 5-12 curve c). An example of such behavior is found in many predators that hunt by sight. They become accustomed to a particular type of prey and search for that preferentially and more efficiently than suitable alternatives that are also present.

Functional responses of types II and III have been observed in many laboratory experiments and a few field situations. These results suggest that invertebrates often have a type II response and vertebrates, a type III response, although the rule is not universal. Other types of functional response do occur as shown by that of the parasitoid wasp *Melittobia* in Jamaica. This insect parasitizes mud wasp larvae within their nests. A type II response obtains until high host densities, but the curve then declines as fewer larvae are killed at the highest density of cells in the mud wasp nests. This decline is due to the morphology of the nests and mutual interference between the parasitoid adults, which "waste time" investigating cells that have already been killed (Freeman and Ittyeipe 1976).

Laboratory experiments with a predatory mosquito larva show yet another pattern (Figure 5-13). The predator continued to eat more prey as more were offered, with no leveling off. Furthermore, the predator killed many more prey than it consumed. Prey densities were higher than those normally found in the field and the overkill reaction may result from this unnatural condition. Overkill is probably rare in nature but was observed by Kruuk (1972) when a group of hyenas killed 100 gazelles in one night but consumed only small portions of a few of them.

Optimal Foraging.   Evolution by natural selection leads to organisms becoming adapted to their environments. The theory of optimal foraging illustrates this by predicting that predators will forage efficiently. The simplest meaning of efficiency in this context is the maximization of the ratio energy gain from food to energy dissipated in obtaining that food. In terms of an individual's energy budget, this means maximizing consumption (C) relative to respiration (R) in the short-term (section 2.3). Other adaptations to improve nutritional efficiency such as physiological mechanisms to increase assimilation efficiency or production efficiency, are also important but are not considered in optimal foraging studies.

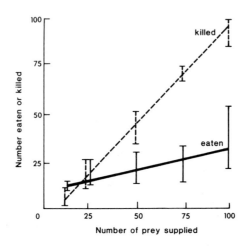

Figure 5-13   Number of larvae of the mosquito *Aedes aegypti* killed and eaten by larvae of the predatory mosquito *Toxorhynchites brevipalpis* in laboratory experiments. The number eaten increases with no maximum (in the range of experimental densities) as prey density is increased; the number killed but not eaten increases in direct proportion to the number supplied. (Modified from Lounibos [1979].)

Short-term maximization of net energy gain through foraging behavior has been demonstrated in many organisms (see Krebs, J.R. 1978). There are relatively few clear-cut tropical examples, and most of these come from nectar-feeding hummingbirds and sunbirds. For example, the sunbird *Nectarinia kilimensis* feeds on the nectar of mistletoe flowers in Kenyan savannas. There are two types of flower: those closed and not previously visited and those that have already been opened by foraging sunbirds. Closed flowers provide a greater energy gain per unit time and are, therefore, predicted to be preferred. This is indeed the case as they are visited 32%–91% more often than their abundance would suggest (Gill and Wolf 1975b). However, when food supply is unpredictable (in the same sense that an animal's feeding experience cannot forecast future availability of a particular type of food), the most profitable items may not be the ones that enhance survival to the greatest degree. Greenwood (1984) predicts that, in some circumstances, animals will adopt a low-yield, low-risk (but predictable) food rather than a high-yield but unpredictable one when the difference in yield is small compared with the difference in risk.

The basic predictions of optimal foraging theory have been extended to include greater complexities such as the dispersion of prey, the selection of prey type (including quality and familiarity), and the allocation of energy to other activities such as defense of territories and reproductive behavior. An example of these developments is **prey switching**. Theory predicts that if more than one type of prey is available, the predator will take the most profitable and switch from one to another as profitability changes. The Etolo-speaking people of Papua New Guinea obtain most of their protein by hunting or trapping mammals. In the "cloud season" of overcast skies, trapping predominates, with bandicoots, wallabies, marsupial cats, and rodents as major prey. During the clearer "pandanus" season, hunting of possums, phalangers, and bats is more important (Dwyer 1982). Traps are set close to the settlements and are used in the period of intensive gardening, during which time the catch declines. When less time is needed for gardening, the people switch to hunting, which involves parties being away for several days. During this latter season, the populations of animals that are trapped close to home have a chance to recover.

Variety of food rather than maximum energy gain is an important criterion of selection in leaf-cutting ants of dry Costa Rican forests. These ants bring leaves from greater distances than is dictated by the distribution of palatable species. Their foraging behavior is very complex and seems to relate to both secondary compounds in the leaves and to maintaining a suitable balance of moisture and nutrients for the fungal mutualists in their nests (Rockwood 1976).

Much optimal foraging theory has been developed from studies of vertebrates. Griffiths (1981) has proposed that "suboptimal" foraging may be common in invertebrates. His studies of ant lion larvae in Tanzania clearly demonstrate that the larvae attack their prey even when energy costs outweigh the gains. This response is probably due to a reduced attack threshold, which is a physiological response to hunger.

5.3   Mutualism

Mutually beneficial interspecific interactions are more common in the tropics than elsewhere. Descriptive studies of such interactions, mainly between populations of plants and herbivores, abound. Microorganismal populations are also involved in mutualisms as previously discussed in the examples of plants and mycorrhizae and termites and their fungus gardens. Mutualisms between potential competitors and animal prey and their predators are rarer, but examples include Müllerian mimicry and human–domestic animal interactions, respectively (see sections 6.3.5 and 9.2.3).

Mathematical models of mutualisms have received much less attention than those of competition and predation (see Appendix B). The simplest model, based upon the Lotka-Volterra competition equations, produces unrealistic predictions, with both populations increasing indefinitely. A degree of reality is introduced if it is assumed that the benefit to at least one of the populations has a saturation value above which no further benefits accrue. Such a model predicts that both populations attain a stable density, higher than either would attain alone.

However, these equilibria are less stable in the face of disturbances than are nonmutualistic interactions (see May 1981b). It is tempting to conclude that the large number of mutualisms found in the humid tropics is a reflection of smaller climatic disturbances than elsewhere, but there is no direct evidence to support this.

By defining persistence, rather than mathematical stability, as the criterion of continued existence, Vandermeer (1984) shows that the Lotka-Volterra models can produce useful predictions. Where both species are facultative mutualists, both may persist indefinitely; where both are obligate or one obligate and the other facultative, persistence of both or the extinction of obligate partners may occur. Extinction is most likely when interdependence is strongest. Using a different model approach, Heithaus et al. (1980) showed that the addition of a species predatory on one of the mutualists can lead to stability of both mutualistic populations at feasible population densities.

Mutualism has an important evolutionary dimension as the two (or more) populations may become progressively better adapted to each other and, as a combined system, to their environment. Some of the large body of observations on this aspect of mutualism are discussed further in Chapter 6.

5.4   Interspecific Interactions and Population Regulation

In section 4.2 it was concluded that only density-dependent factors regulate populations, although density-independent factors can be important in causing nonregulatory changes in density. Interspecific competition is usually density dependent and, therefore, a potential regulator of population size. The situation with predation is more complex. The type II functional response is inversely density dependent in the sense that the rate of predation per prey individual decreases as prey density increases. Such a predator has a destabilizing effect on prey populations. In the concave portion of the type III response, when prey density is low, predation is density dependent and can have a regulatory effect on the prey population, provided that its density does not get high enough to enter the convex part of the curve. When several species of prey are eaten, predation may be **frequency dependent**. If the predator reduces the population of one species of prey, to a level at which search cost is high, it may switch to another prey species that is more abundant or easier to find. The population of the first prey is then able to build up again. In this way, one predator may regulate the populations of several prey species. Mutualism is less well understood. Models tend to predict low levels of stability for both populations, but field evidence suggests that these interactions may persist for long periods in natural communities.

The simplest way to determine the effects of one population on another is to remove one of them and observe the effects on the other. The best evidence of this sort comes from mutualisms. For example, removal of ants from species of *Acacia*, which normally have them as mutualists, greatly reduces the survival prospects of the *Acacia* bushes (see section 6.3.4). Predation can keep prey populations at lower levels than they would otherwise attain, as clearly demonstrated by the successes of biological control of pests (see section 9.3.2). However, there is scanty evidence from the tropics to directly demonstrate the regulation of populations by interspecific interactions. Much more of the indirect evidence is presented in Chapter 7, which describes the role of these interactions in determining community structure.

5.4.1   Are Interspecific Interactions More Intense in the Tropics?

Several authors have argued that competition and predation are more intense in the tropics, particularly in the moist tropics, than in temperate environments (for example, Dobzhansky 1950, Elton 1973). The suggestion is that these biotic factors control population density in relatively constant (tropical) environments, whereas abiotic factors are more important in variable (temperate) areas. One idea, following from Gause's hypothesis, is that increased interspecific competition leads to more specialized organisms with narrower niches. This claim holds for some bird communities in terms of their use of different levels of forest canopies (MacArthur et al. 1966). Work on insect herbivores that feed exclusively on one plant species suggests that some groups of Lepidoptera and some beetles are more specialized in this way (that is, have narrower niches) than their temperate relatives (Gilbert and Smiley 1978). In contrast, Beaver (1979) found that there was less host specificity in bark and ambrosia beetles from the tropics. In West Malaysia and Fiji, 61% and 68%, respectively, of species feed on more than one species, compared with 9% and 6% in France and California, United States, respectively. Similar trends away from a high degree of specialization in the tropics have been identified in some bees, wasps, and ants (Culver, 1974, Heithaus 1979). While there clearly are latitudinal gradients in specialization and in some niche dimensions, they do not follow a consistent pattern, and generalizations are premature.

Predictions that predation rates will be at their highest in the humid tropics are supported by many reports on nest predation in birds, although qualitative data are rare. Nests of several bulbul species in the forests of Gabon experience predation rates of up to 80%, and in some cases these birds need to lay four to five clutches to fledge one young per year (Brosset 1971, 1981). Among the passerines of a Venezuelan savanna, 65% of nests are lost, mainly due to predation (Ramo and Busto 1984). In temperate regions, rates of predation on nests of birds of similar size and habits rarely exceed 50%. (Lack 1966). Experimentally it has been shown in the Americas that predatory ants remove prey more quickly in tropical forests, with 50% removal in 1 day compared with 5 days in temperate forests. There is no similar difference in comparisons between farm fields in the two climatic zones. Predation, mainly by birds, on web-spinning spiders during the daytime is at twice the rate in tropical Peruvian and Gabonese forests that it is in a temperate forest in the United States (Rypstra 1984).

Much more work is needed before general conclusions about latitudinal gradients in competition and predation can be drawn (see also section 7.4). However, most ecologists agree that mutualistic interactions are much more common in the tropics. For example, in the Americas there are no obligatory ant-plant mutualisms and no orchids with specific bee pollinators north of 24° and no pollinating or seed dispersing bats north of 33° (Farnworth and Golley 1974).

5.5   Distribution of Populations _____

There are two spatial scales at which the distribution of populations is studied. The first is local distribution, which involves defining the boundaries of a particular population. The second is the geographical range within which populations of a particular species can be found.

5.5.1   Local Distribution

Life in some form exists over a wide range of physical and chemical environments, but most populations are able to persist in only a small portion of that range. The distribution of a population is usually such that organisms are abundant around their optimum and absent outside their range of physiological tolerance for a particular abiotic factor. Between these extremes is a region that is stressful but inhabitable at intermediate abundance. The distribution of several species of Indian earthworms with respect to their tolerance of soil acidity is shown in Figure 5-14. Figure 5-15 shows how two abiotic factors combine to limit the distribution of tsetse flies throughout Africa. The line encloses predicted combinations of temperature and saturation deficit (a measure of the drying power of the air), which allow populations to persist. In general, abiotic factors set the outer boundary within which a population can survive, but other factors may cause a population to be absent from a particular site within that boundary.

Interspecific interactions affect distribution of populations by displacing or precluding competitors, predators, or mutualists. Competitive exclusion is an obvious example (see section 5.1.2). Niche theory is another way of looking at distribution. For example, in Figure 5-15, two dimensions of the fundamental niche of tsetse flies are portrayed. There are many sites at which flies could persist in terms of temperature

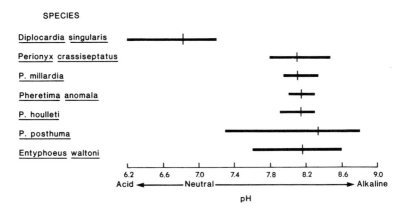

SPECIES

Figure 5-14 Physiological tolerance of Indian earthworms with respect to pH. Horizontal bars represent the range of pH in which each species can survive, and vertical bars represent the pH of the soil from which the species was taken. (Data from Varma and Chauhan [1979].)

and saturation deficit but at which none are found because they are precluded by other niche dimensions. Biotic factors of importance to these flies include the presence of suitable veretebrate hosts for their blood meals and bushes or trees on which to rest after feeding.

Prey-specific predators obviously are limited to areas in which their prey occur, as in the case of the neotropical butterfly *Heliconius sara* whose larvae feed only on the passion flower vine *Passiflora auriculata* (Gilbert 1980). A predator may exclude a population simply by eating all of it, as happened to rock iguanas on some Caribbean islands following the introduction of domestic cats and dogs (Iverson 1978).

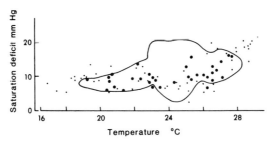

Figure 5-15 Combinations of temperature and atmospheric saturation deficit in which persistence of tsetse fly populations *(Glossina morsitans)* in Africa is predicted are surrounded by the solid line. Large dots are combinations of these variables in which populations are found and small dots locations from which they are absent. All but one population occur within the predicted limits. Suitable areas may have no flies because of other environmental factors. (Modified from Rogers [1979].)

Obligate mutualists must coincide in distribution if their populations are to persist. Fig trees and their wasp pollinators are entirely interdependent in this manner. Facultative mutualists, however, can have independent distributions. Where the two species occur together, a loose association may develop, while in other areas there may be no mutualism, or alternative species may be involved. Often these facultative mutualists have a better chance of surviving in those areas where mutualisms are formed. This is illustrated by the greater reproductive success of the herb *Costus* in the presence of ant mutualists (see Table 6-7).

5.5.2   Geographical Distribution

Given the presence of suitable environments, large-scale distribution results from a combination of powers of dispersal and historical accidents. **Cosmopolitan** species (such as our own), are widely dispersed throughout the biosphere, while others may be restricted to a single population. For example, the Sokoke Scops owl (*Otus iraneae*) is found in 220 km² of coastal forest in Kenya and has a population of 1500 pairs (Britton and Zimmerman 1979). Such species, when restricted to a particular area, are said to be **endemic** to that area.

On a geological time scale, movements of continents, changes in sea level, upthrust of mountains, and changes in drainage patterns have profoundly affected distributions, often splitting populations, which evolve into distinct taxa. Marsupial mammals, mostly extinct elsewhere, have persisted in great diversity and abundance in Australia because of the absence of the placental mammals that have displaced them on other continents. A different type of distribution is shown by another evolutionary "leftover," the cycads. These plants have a patchy and scattered (**disjunct**) distribution throughout the humid tropics. They are a primitive family of trees that were a dominant vegetation type before the evolution of the flowering plants.

Climatic change also leads to successive expansions and contractions in the distribution of species. While temperate regions were in the grip of the most recent ice age, the tropics were much drier than they are now. At the time of the glacial maximum in Europe 15,000 years ago, aridity restricted the forests of Africa to five small areas totaling approximately 1500 km² and separated from one another by 500–1500 km (Hamilton 1982). As a result, mountain gorillas are now restricted to only two forest areas of equatorial Africa, a mere remnant of their former distribution when forests were more widespread and continuous. When climate becomes more favorable to a number of species, those with more effective means of dispersal will spread more rapidly than the others. During the last 12,000 years, forests have spread from their arid-period refuges in eastern Zaire and southwest Uganda, across Uganda, to western Kenya. The plant species dispersed by flying animals have spread much more quickly than those dispersed by wind or terrestrial animals (Hamilton 1982). Powers of dispersal also affect colonization of remote areas. The cloud forest of Serranía de Macuira in Colombia is separated from the nearest similar moist area by 250 km. Although only a few thousand years old, the mountain carries a diverse forest. Of the total flora, 85% of species have small propagules dispersed by wind or flying animals (Sugden 1982). Wind seems to be a more effective dispersal agent in this exposed montane environment than in the lowland forests of Uganda.

5.5.3   Human Influences on Distribution

As competitors, predators, domesticators, pollutors, and travellers, people profoundly affect the distribution of other organisms. Humans relentlessly try to eliminate competitors for food and space, as amply demonstrated by the dramatic decline of many populations and communities. As predators, people are capable of eliminating or reducing natural populations on which they feed or hunt for other purposes. Examples include the drastic decline of many fisheries and the decline of all species of rhinoceros in Asia and Africa. Pollution creates new environments that are unsuitable for the previous occupants of an area, and widespread use of fire eliminates or restricts other species.

In contrast, the distribution of some species that are mutualistic (crops, domestic animals) or predatory (human diseases) have greatly increased. Widespread human travel between regions transports organisms to alien environments where they often die but sometimes succeed, as documented by Elton (1958). Such introductions are often deliberate for food, fiber, or ornamentation, but many are accidental. For example, the African malarial mosquito was accidentally introduced to coastal Brazil around 1930. By 1939 the insect population, together with its parasite, had spread along 400 km of coast and up to 300 km inland. Widespread, debilitating malaria resulted because the alien mosquito rests in houses and native malaria carriers do not. Fortunately a vigorous campaign costing $2 million eradicated the alien mosquito population (Elton 1958).

5.6   Applications of Population Ecology

Knowledge of population dynamics is essential to solving many practical problems. Some of these are discussed briefly in later chapters: pest control in Chapter 9; human population and disease control in Chapter 10; and management of populations for conservation in Chapter 11. A problem not dealt with elsewhere is **optimum yield**; or how to harvest a population efficiently without causing it to decline, thereby reducing future yield. This situation of humans as predators of wild populations has been applied to the management of fisheries, but it is equally applicable to the cropping of wild terrestrial animals for food or other products.

The logistic equation predicts that **maximum sustainable yield** is obtained when the population is cropped at half its carrying capacity—which is also the point at which it is growing most rapidly. Figure 5-16 shows such a relationship with K = 1000 and r = 0.6. Maximum sustainable yield is achieved by keeping the population at 500 individuals and utilizing a cropping rate of r/2 to yield rK/4, or 150, per year. Attwell (1982) applied this approach to the Zululand wildebeest population illustrated in Table 4-3. Using the 1970 census of 6363 animals as approximating to K, he calculated a maximum sustainable yield of 450 animals. Actual culling was excessive and amounted to 1500 wildebeest, which accounts for the observed negative estimate of r reported in section 4.1.6.

This method of calculating sustainable yields obviously assumes that the population is growing in a logistic manner. When this is not the case, alternative models have to be applied, such as those discussed by Krebs (1985) and Beddington (1979).

Summary

Competition, predation, and mutualism are interspecific interactions that affect characteristics of the populations involved. Mathematical models of these interactions between pairs of species predict a range of behaviors

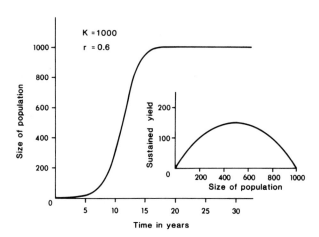

Figure 5-16   Relationship between population size and sustainable yield (inset) for the logistic curve shown. Maximum sustainable yield is obtained when the population is harvested to maintain it at half of its carrying capacity. (From Caughley, G. 1977. *Analysis of Vertebrate Populations,* p. 180. Reprinted by permission of John Wiley & Sons, Ltd.

from stable coexistence of both, through extinction of one or the other, to extinction of both. Real populations display equally diverse behaviors. Competition may lead to exclusion of one population by another or to specialization such that the species coexist but in different ecological niches. Predation in laboratory populations often leads to first prey and then predator dying out. Coexistence often requires an ability of the prey to escape temporarily from the predator. There are complex behavioral interactions between individuals of prey and predator populations. Mutualisms are more common in the tropics than elsewhere. Suggestive evidence indicates that predation on some groups may be greater in the tropics than in temperate areas. The distribution of populations is determined by local environmental factors, the history and geography of each species, and, more recently, by human influences.

Study Questions

Review

1. How well are the predictions of mathematical models of population interactions borne out by observations on laboratory populations?
2. Write an essay on ecological segregation.
3. Discuss the influence of predator behavior on prey-predator interactions.
4. How can the niche concept be applied to study the distribution of populations?

Related Topics

1. Write a crical essay on optimal foraging.
2. Investigate the development of mathematical models of mutualism.
3. Evaluate the contention that competition between field populations is very common.
4. Describe how human activities affect the distribution of populations.

Further Reading

Three excellent texts devoted to population ecology are as follows:

Begon, M.; Mortimer, M. (1981). *Population Ecology. A Unified Study of Animals and Plants.* Oxford: Blackwell. A comprehensive, up-to-date account.

Harper, J. L. (1977). *Population Biology of Plants.* London: Academic Press. A comprehensive, detailed account of plant population ecology, by the leading worker in the field.

Hutchinson, G. E. (1978). *An Introduction to Population Ecology.* New Haven, CN: Yale University Press. An eloquent personal account by one of the important thinkers in population and community ecology.

The following three pamphlets (about 70 pages each) cover the field concisely, but clearly, from somewhat different points of view:

Hassel, M. P. (1976). *The Dynamics of Competition and Predation.* London: Arnold. (Studies in Biology No. 72.) Particularly good on predation theory.

Moss, R.; Watson, A.; Ollason, J. (1982). *Animal Population Dynamics.* London: Chapman & Hall. Strong on vertebrates.

Solomon, M. E. (1976) *Population Dynamics.* 2d ed. London: Arnold. (Studies in Biology No. 18). Strong on invertebrates.

Following is the only text with a specifically tropical bias; it is rather patchy and idiosyncratic but a rich source of information:

Young, A. M. (1982). *Population Biology of Tropical Insects.* New York: Plenum.

More general ecology texts with strong sections on population ecology are as follows:

Krebs, C. J. (1985). *Ecology.* 3d ed. New York: Harper & Row. (See Chapters 2–18.) Very good on factors limiting distribution and good at explaining models.

May, R. M. editor. (1981). *Theoretical Ecology.* 2d ed. Oxford: Blackwell. (See Chapters 2–8.) Strong on mathematical models; written by leading theoreticians.

Ricklefs, R. E. (1979). *Ecology.* 2d ed. New York: Chiron Press. (See Chapters 15–16, 27, 31–35.) Great clarity and very readable. Has more tropical examples than other texts.

A good basic text on the geographical aspects of the distribution of populations:

Cox, C. B.; Healey, I. N.; Moore, P. D. (1976). *Biogeography.* 2d ed. Oxford: Blackwell.

Mutualisms between plants and animals are common in tropical environments. This pollination system between plant and butterfly was observed in Guadeloupe in the Caribbean. (Photograph by Robert K. Colwell.)

CHAPTER 6

Ecology and Evolution

6.1 The Evolutionary Synthesis

Dobzhansky's statement that "Nothing in biology makes sense except in the light of evolution" is no less true of ecology than any other branch of biology. The populations that ecologists observe are the product of past and present evolution. Failure to take account of evolutionary processes can, therefore, lead to a misinterpretation of short-term or chance ecological events. All natural historians are struck by the extent to which organisms are adapted to their environments. However, it is incorrect to assume that, say, a pollinator that appears highly specialized is adapted to a particular species of flower, unless the two populations are known to have been in contact for many generations. Another mistake is to perceive some kind of "harmony of nature" at the level of communities or ecosystems, with all populations striving to achieve an overall balance. Most of what we know of evolution suggests that any equilibria in communities are temporary and result from ecological and evolutionary tensions, rather than adaptive harmony.

The predominant mechanism of organic evolution is natural selection, a process that was described independently by two naturalists, Charles Darwin and Alfred Wallace, largely as a result of their experiences in the tropics. Darwin had the earlier insight and collated the greater mass of evidence, which he published in 1859 as *On the Origin of Species*. No modern biologist doubts that evolution has and continues to take place, and few would doubt that natural selection as originally described is by far the most important mechanism. The theoretical basis and supporting evidence have increased enormously with the development of Mendelian, population, and molecular genetics; paleontology; and ecology since the time of Darwin and Wallace. Darwin's seminal influence is recognized in the description of the modern evolutionary synthesis as neo-Darwinism.

The following sketch of these theories in sections 6.1.1 to 6.1.6 provides a background to the remainder of the chapter (and indeed the remainder of the book) and should not be regarded as a complete account. A familiarity with elementary genetics is assumed.

6.1.1   The Neo-Darwinian Synthesis

The theory of natural selection can be stated as follows:

1. Individual organisms within a population differ from one another in morphology, physiology, and behavior, and much of this variation is inherited.

2. Populations have the potential to grow exponentially, but many individuals die and populations cannot grow indefinitely.

3. Some inherited characteristics increase the probability of survival and reproduction of the individuals that bear them. As a result, their descendants (which also have these characteristics) come to make up an increasing proportion of the population as the generations go by.

The result of these processes, leading to systematic differences in survival and reproduction, is known as **natural selection**.

6.1.2   Variation

Every individual in a population is unique. In some populations a zygote may split (identical twins) or asexual reproduction may predominate (clones), producing genetically identical individuals (that is, having the same **genotype**). However, **phenotypic** differences will result from the different environments encountered as the organisms grow and develop. Because only the genotypic component of variation is inherited, it is the only part subjected to natural selection, but the amount of phenotypic plasticity of individuals may itself be genetically determined. For example, the growth form of a tree is less rigidly predetermined than that of a vertebrate, but both carry genes controlling that growth form and the extent to which it is likely to vary.

Novel genes are produced by the process of **mutation** (spontaneous changes in DNA sequences), which is random with respect to the environmental demands upon the organism. As a result, most mutations are disadvantageous. However, they are also rare, with a given mutation occurring in the range of one gamete for every 10^{-4} to 10^{-8} produced.

Mutation alone does not produce much variation as new raw material for selection. In sexually reproducing organisms the process of recombination produces several thousand times as much variation as mutation. **Recombination** involves the mixing of parental genes such that each offspring produced has, on average, half maternal and half paternal genes. The development of molecular genetics has revealed much more genetic variation at the biochemical level than was previously suspected, providing a plentiful supply of material on which selection can act.

When the effects of many genes combine to produce a phenotypic effect (a polygenic character), the product is often a **continuous variation**. A good example is adult height in human populations in which the height of each individual is controlled by many genes, but environmental factors, particularly nutrition, also affect the character (see also Figure 6-3). **Discrete variation** occurs when different genotypes produce distinct phenotypes, with no intermediate forms, to produce **polymorphisms**. Section 6.3.5 discusses a visual polymorphism in butterflies and Figure 10-10, a physiological polymorphism in humans.

6.1.3   Natural Selection

Selection takes three distinct forms. Not only can it produce directional change in a trait, but it can also lead to a divergence of characters or to a reduction in the amount of variation within a population (Figure 6-1). In natural populations the intensity of selection can be very high. Some phenotypes commonly have a 10%–40% better chance of survival than average, and advantages up to 90% have been observed. Even very small selective differences can produce dramatic changes when seen on a geological time scale. Such a change is the increase in brain size from early *Autralopithecus* to modern *Homo sapiens,* which has been much faster than most other paleontological changes. Berry (1982) calculates that a change of 0.04% per generation is sufficient.

As Darwin recognized, **sexual selection** plays a large part in the ability of sexually reproducing organisms to leave offspring. In many cases, females can choose with whom they mate. Such choices are significant evolutionarily because they affect the prospects of survival and reproduction of offspring. For example, male long-tailed widowbirds (*Coliuspasser progne*) with the longest tails mate more frequently than those with shorter tails. Andersson (1982) demonstrated this by clipping tails or adding longer tails to birds in Kenya. Because longer-tailed males mate more frequently, they are likely to leave more offspring who will also have longer tails. This preference by female birds was exercised regardless of other ecological attributes of the males such as ability to hold a territory. The logical outcome of selection over several generations would be tails of ever-increasing length. However, very long tails are likely to carry disadvantages unrelated to mating, such as reduced efficiency in flying and foraging and increased susceptibility to predation.

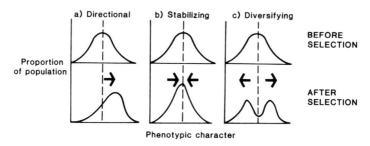

Figure 6-1   The three distinct ways in which natural selection operates to change an inherited trait. Arrows show the direction in which selection is working. *a,* **Directional selection:** phenotypes at one end of the range have a selective advantage. An example is the increase in body and bill size in *Geospiza fortis* (one of Darwin's finches) during a drought on Daphne Major Island of the Galápagos archipelago; only larger and harder seeds were available as a staple food source (Boag and Grant 1981). *b,* **Stabilizing selection:** extreme phenotypes are at a selective disadvantage compared with intermediates. For example, human babies in England range from less than 0.5 kg to more than 4.5 kg at birth, but mortality at these extreme weights is high, leading to most surviving babies being closer to the "optimum" birth weight of 3.4 kg (Karn and Penrose 1951). *c,* **Diversifying selection:** extreme phenotypes have a selective advantage over intermediates. For example, in the Ethiopian race of the swallowtail butterfly *Papilio dardanus,* long tails are favored in nonmimetic females and short tails in mimetic females of the same population (Clarke and Sheppard 1962).

The "optimal" length of tail will therefore reflect a balance between a variety of selective forces. Such "compromises" between the many agents of selection are a common feature of natural selection.

In field populations it is often impossible to measure the selective differences between genetic traits. Instead the evolutionary success of particular phenotypes can sometimes be estimated as **reproductive success**—the number of surviving offspring reared by a particular individual relative to the number reared by other members of the population. All organisms are faced with an adaptive compromise between devoting energy to producing as many offspring as possible and diverting some of that energy to enhancing the survival prospects of each offspring with the consequence of being able to produce fewer of them (see section 6.2.1).

6.1.4   Geographical Variation and Speciation

In addition to intrapopulation variation, different populations of the same species may be somewhat distinct genetically and form locally adapted **races** as exemplified by the swallowtail butterflies discussed in section 6.3.5. Such species, which include *Homo sapiens,* are said to be polytypic. When races or populations become geographically isolated from one another, they may diverge sufficiently to become distinct species whose members can no longer interbreed successfully. Ecological studies rarely encompass the time scales needed for speciation, but an understanding of the process is important to a discussion of species diversity in the tropics (see Chapter 7). For speciation to occur, many biologists believe that, in most cases, a period of complete geographical separation of populations is required with no intermigration. An exception to this generalization is found in some higher plants where new species sometimes arise very rapidly through crossbreeding between two species (hybridization) and subsequent multiplication of the chromosome complement (polyploidy). A new species formed in this way is unable to breed with either of its "parent" species.

6.1.5   Chance and Evolution

Evolution by natural selection is a probabilistic process that works on random mutations as raw material but leads to individuals that are adapted to their environments. It is not therefore, a random process taken as a whole. Nevertheless, it is possible for the process of **random genetic drift** to cause genetic changes in populations by chance alone. This process is usually envisaged in small populations as a result of "sampling errors" in mortality. Some causes of death, such as landslides, floods, and fires, may kill individuals whether or not they are adapted to more persistent features of the environment. In small populations, and where a gene is likely to increase survival and reproduction only slightly, it is statistically possible that the gene may become extinct over a number of generations by such chance effects.

Most evolutionary biologists believe that random genetic drift has only minor significance in evolution compared with selection, but many recognize that a special case of chance events, called the **founder effect,** may be more important. In populations that undergo catastrophic density-independent mortality (see section 4.3.1), the few survivors from which the population reestablishes will have only a sample of the total genetic variation of the original population. Some genes will be eliminated by chance during such population crashes. A similar situation arises when a few individuals (founders) migrate from an established population and begin a new population elsewhere. Being only a small sample they will not have the same genetic makeup as the parent population and, as a result, will probably evolve differently in the future, even if the environment is similar.

6.1.6   The Evolution of Altruism

A superficial understanding of natural selection as just formulated leads to the expectation that individuals should be "selfish" to propagate their genes at the expense of those of other individuals. Yet examples of cooperation that benefit other individuals abound in nature, both intraspecifically in social systems and interspecifically in commensal and mutualistic associations. This apparent paradox is resolved in two main ways. The first is through **kin selection**, when altruistic acts are between closely related individuals that carry many genes in common and therefore are helping to preserve copies of their own genes when they help these relatives. Kin selection may explain much of the cooperation in social units in which family ties are strong, particularly in social insects and many birds and mammals. The second set of mechanisms is where reciprocal advantages result from altruistic behavior, thereby potentially increasing the chances of survival of all participants. This process can lead to mutualistic relationships both interspecifically and intraspecifically. The important conclusion is that altruistic and cooperative behavior can evolve through individual selection and it is not necessary to invoke other mechanisms as some biologists have done. For example, **group selection** is one hypothesis put forward to explain the evolution of altruistic traits. The self-regulation of population density to ensure optimal sustained use of resources for the benefit of the whole population is a case in point (see section 4.3.3). It is suggested that populations without this trait will become extinct due to overexploitation of resources, leaving only those populations possessing the self-regulatory trait. Group selection is a superficially appealing idea that has been induced in the laboratory but seems unlikely to persist for long in nature. If one individual becomes "selfish" with respect to resources, by mutation or by migration from another population, the system breaks down because the genes conferring "selfish" behavior will spread rapidly through the population by natural selection.

6.1.7   Coevolution

Although evolution proceeds by changes in gene frequencies within a population, it must be remembered that all populations exist in communities that include populations of other species, which are also subject to evolutionary change. An evolutionary change in one population may, therefore, impose selection pressures on coexisting populations, which may themselves evolve as a result. Such mutually adaptive changes in populations of different species are called **coevolution**. The reciprocal nature of coevolution is stressed in Janzen's (1980a) definition: "an evolutionary change in a trait of the individuals of one population in response to a trait of the individuals of a second population, followed by an evolutionary response by the second population to the change in the first." Janzen felt forced to frame this rather clumsy definition because the term has become increasingly used to describe almost any evolutionary change responding to the biotic environment. Such usage is virtually synonymous with evolution itself and thereby makes coevolution redundant as a special concept.

Coevolution is often antagonistic. Reciprocal genetic changes may occur as one population evolves to become more efficient in escaping predators or competitors, or a predator becomes more efficient at capturing its prey. Coevolution can also produce mutualisms when populations evolve in ways that turn out to benefit one another. A recent review identifies the situations in which coevolution is most likely to occur. Parasites and their hosts (including monophagous herbivores and their food plants) coevolve more frequently than other prey-predator relationships. Interspecific competitors coevolve most often when they share similar traits, such as when plants compete for the same animal pollinator. In all instances, a high frequency of physical contact between the coevolving populations is required (Thompson 1982).

A coevolutionary response to interspecific competition is **character displacement** in which coexisting competitors diverge phenotypically (and presumably genetically) to increase the ecological segregation (and thereby reduce competition) between them. Darwin's finches, so called because of the description of their ecology and evolution in *On the Origin of Species,* provide one of the best examples of character displacement. These birds show a remarkable adaptive radiation on the Galápagos Islands, even though they probably all originate from a common ancestor that came from South America. There are fourteen species of which five are seed-eating ground finches. Different combinations of these species occur on different islands and these subgroups show ecological segregation with respect to feeding habits. *Geospiza fortis* and *G. fulginosa* eat a similar size range of seeds when they occur on different islands, but when they coexist on the same island, *G. fortis* usually takes larger seeds than *G. fulginosa* (Figure 6-2). Character displacement has been demonstrated convincingly only in a few cases, and some argue that it is very rare in nature (see Connell 1980).

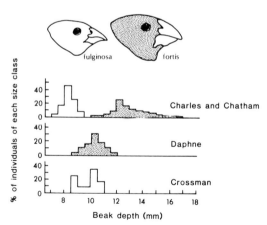

Figure 6-2   Character displacement in bill size of two species of Darwin's finches of the Galápagos Islands. When only one species is present (*Geospiza fortis* on Daphne, *G. fulginosa* on Crossman), the overlap in bill size is complete, but when the species coexist (on Charles and Chatham), there is no overlap. Bill size is a good indicator of food preference so it is reasonable to suppose that the divergence is due to interspecific competition. (Modified from Ricklefs [1979] and Lack [1947].)

6.2   Evolution and Population Ecology ⸻

Most of the characteristics studied in population ecology such as natality and mortality are merely averages hiding the genotypic variation between individuals who may be subjected to natural selection. Variation in population traits may be continuous (such as shown in Figure 6-3) or polymorphic. In the latter case, different morphs may occupy different ecological niches or have distinctly different levels of reproductive success. For example, predation rates often differ on different morphs when the predator hunts by sight (see section 6.3.5). Dispersal behavior may also be polymorphic as in a leafhopper population studied in Zimbabwe, in which some individuals were more prone to disperse than others (Rose 1972).

Most population ecologists use the mean values of population parameters to draw conclusions and to construct models, but the importance of accounting for individual differences is becoming increasingly clear (see Lomnicki 1978, Harper 1982). Indeed, it has been suggested that parts of individual plants may be important evolutionary units. While reproductive organs in animals are usually immune from the effects of mutations in other body cells, this may not be the case in higher plants. Their growing points (meristems) are undifferentiated and are presumably subject to mutations in the same way as other cells. Mutations in meristems may affect not only the "body" that grows from them (such as a branch) but also reproductive structures on that branch. Major branches of the same caper tree (*Capparis odoratissima*) in Costa Rica differ consistently in their fruit production and seed set (Gill and Halverson 1984).

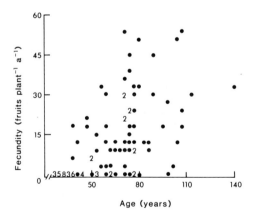

Figure 6-3   Phenotypic variation in a population parameter, the fecundity of a rain forest palm tree *(Astrocaryum mexicanum)*, in relationship to age. The digits (2 to 8) indicate more than one individual of a particular age and fecundity. The wide range of fecundities at any age results from genetic and environmental interactions. (From Piñero, D.; Sarukhán, J. 1982. *J. Ecol.* 70:461–472. Reproduced by permission of the British Ecological Society.)

6.2.1   Evolution of Life History Characteristics

Besides causing variation, evolution imposes constraints on the character-istics of organisms that are often strongly correlated with body size. On average, populations of larger organisms simply cannot grow as fast as those of smaller organisms (Figure 6-4*a*). Larger organisms have longer life spans and take longer to reach maturity (Figures 6-4*b* and 4-6). The scatter of points in Figures 4-6 and 6-4 is some measure of the extent to which evolutionary variation is possible given the constraint of body size. Evolutionary history also imposes an inertia on radical change in many cases. The fact that all African bovids (the family that includes antelopes and cattle) produce only one young at a time is due only in part to size, since bovids range from 500 kg buffalo to 5 kg dikdik. Their common ancestry is a contributory factor to litter size as shown

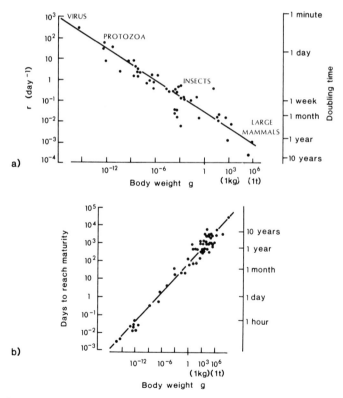

Figure 6-4   Many population parameters are related to body size (see also Figures 4-6 and 6-6). *a,* The intrinsic rate of natural increase (r) is smaller in larger organisms; note that this more complex parameter has a wider scatter of points than *b*. *b,* Larger organisms take longer to reach reproductive maturity (which is an approximation of generation time). (*a* and *b* from Blueweiss, L.; et al. 1978. *Oecologia* 37:267.)

by larger litters in other mammalian orders. Among the carnivores in the same savannas, 150 kg lions produce three or four and 20 kg wild dogs produce seven young at a time.

The extremes to which life history characteristics of particular populations may deviate from relationships such as those in Figure 6-4 were characterized by MacArthur and Wilson (1967) as being the outcome of r- or K-selection. Populations that are **r-selected** have a high intrinsic rate of growth (r) and tend to "boom" when environmental conditions are favorable and "bust" when those conditions deteriorate. As a result, they exhibit large fluctuations in density and, incidentally, have the potential for large genetic changes through the founder effect. **K-selected** populations have relatively constant density at or near the carrying capacity (K) of the environment. These contrasting patterns of population change can be visualized by referring to Figures 4-12 and 4-1a, respectively.

Many other characteristics may be associated with these tendencies, some of which are listed in Table 6-1. In natural populations, selective forces are diverse and usually result in any given population having a mixture of these traits. Table 6-2 illustrates this point by comparing two Costa Rican rodent populations. The species from an evergreen forest is mainly K-tendency compared to its relative from the more climatically variable dry forest, as predicted. Contrary to the generalization in Table 6-1, the species from the evergreen forest exhibits greater seasonal fluctuations in density and produces more offspring per year due to a

Table 6-1   Some predicted trends in r- or K-selected populations and the environments in which they are likely to occur. Data from Pianka (1970) and Southwood (1977).

	r-SELECTED	K-SELECTED
Environment	Variable and unpredictable	Constant or predictably variable
Population Characteristics	Survivorship curve concave (Figure 4-8e)	Survivorship curve convex (Figure 4-8a)
	High fecundity	Low fecundity
	Density variable usually well below carrying capacity	Density fairly constant at or near carrying capacity
	Scramble-type intraspecific competition	Contest-type intraspecific competition
Individual Characteristics	Small body size	Large body size
	Good dispersal powers	Poor dispersal powers
	Outcrossing	Parthenogenetic or vegetative reproduction
	Low level of social organization	High level of social organization.

longer breeding season. This example serves to emphasize that one should not expect to find entirely r- or K-selected populations. Indeed, the whole concept has been severely criticized in recent years as being too simplistic and yielding more exceptions to its predictions than verifications (see Boyce 1984 for a review). For example, K-selection may favor an increase in efficiency of resource use (say, taking more prey individuals or consuming a greater proportion of each), but it may also result in a reduction in the resources used per individual (perhaps by a reduction in body size). Similarly, increase in r may be brought about by greater reproductive output or by enhanced survivorship (which may involve an increase in body size). Nevertheless, the r-K dichotomy remains a useful, if artificial, guide within which to view the evolution and ecology of life histories. The predictions of Table 6-1 merely represent possible and extreme ends of continuums between which real populations lie.

Against this theoretical background, Southwood (1977) proposed a scheme to classify the types of environment in which populations with different life history characteristics can be expected to persist. Recognizing that environments vary in space and time, he identified generation time and range of movement of organisms as key features determining the ability to persist in different environments (Figure 6-5). From this

Table 6-2 Comparison of two closely related species of heteromyid rodents in Costa Rica to show characteristics related to r- and K-selection. Compare with the predictions of Table 6-1. Data from Fleming (1974).

SPECIES	Liomys salvini	Heteromys Desmorestianus
r/K tendency	r	K
Environment	Deciduous forest	Evergreen forest
Rainfall seasonality	5 months/year <100mm	All months >100mm
Body weight (g)	45	72
Number of embryos (mean)	3.8	3.1
Breeding season (months)	6	10
Mean number of young per year	6.8	9
Age at maturity of females (months)	3 or 4	8
Minimum annual probability of survival	18% (females and males)	31% females; 21% males
Mean home range area (m²)	1833	1324
Amplitude of annual density fluctuations (max – min/median)	0.67	0.88

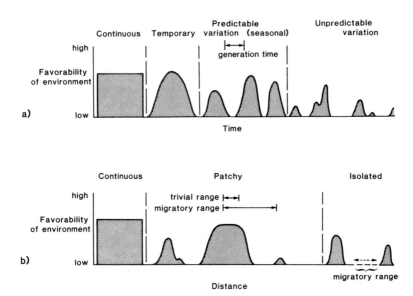

Figure 6-5   Classification of the way in which the environment for a particular organism varies in time and space (see text). (Modified from Southwood [1977].) *a,* **Time.** A population can only persist if its generation time is greater than the unfavorable periods in a variable environment. As a result, when favorability is temporally unpredictable, the environment may be uninhabited even during favorable periods. *b,* **Space.** A population can only persist in favorable areas in which its trivial range (nutrient-gathering area) is smaller than the size of favorable patches, and it may not occur in favorable areas that are outside its migratory range.

analysis it is possible to predict adaptations concerning life history and potential for escape to new environments that are likely to occur in different situations. In temporally unpredictable or temporary environments, r-tendency, short generation time, and efficient escape mechanisms are likely. Such escape may be in time (dormancy) or space (dispersal). K-tendency is most likely in environments that are constantly favorable in time and space, but such environments are rare in nature. In some environments r- or K-tendency may be favored in different populations such that large organisms (with large foraging area and long generation time) bridge the unfavorable periods or areas, while smaller organisms use favorable periods or patches temporarily, but remain dormant or disperse when conditions become unfavorable.

Southwood recognized that most organisms have two distinct ranges of movement, the **trivial range** within which nutrients are obtained, and the **migratory range** over which nonreproducing organisms can move actively (as in many animals) or passively (as in most higher plants). Like generation time, trivial range is correlated with body size in many organisms but much less tightly. Organisms of comparable size may diverge markedly if they are from different taxa or if their foraging behaviors differ (Figure 6-6). Two extreme examples illustrate this. Ant lions, which are larval insects of the order Neuroptera, bury themselves at the bottom of a pit that they excavate. Ants (or other insects) fall into the pit and are consumed by the ant lion. The trivial range of such a "sit and wait" predator is little larger than its body size. In constrast, the Serengeti wildebeest move over thousands of kilometers in their search for adequate forage. Migratory range is important in patchy or isolated environments (Figure 6-5). Areas that are otherwise suitable may remain unoccupied by a particular population simply because the individuals cannot reach it.

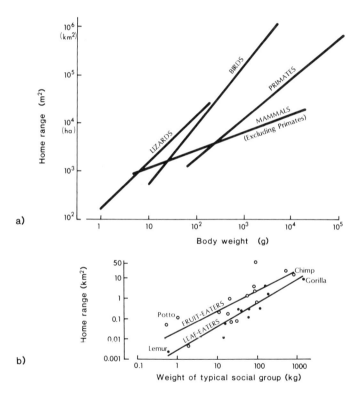

Figure 6-6 In vertebrates, the home range (similar to the trivial range in Figure 6-5b) is greater in larger animals. a, Various vertebrate groups (data from McNab [1963], Armstrong [1965], Turner et al. [1969], Milton and May [1976]). b, Primates only, to show the considerable amount of variation in analyses such as a and how different food types and foraging patterns result in different home ranges: ∘ = fruit eaters (patchy food source); • = leaf eaters (more evenly dispersed food source). Weight of whole social group is used as a measure of body size, which accounts for age and sex structure of the population. (From Clutton-Brock. T. H.; Harvey, P. H. 1977. *J. Zool. Lond.* 183:1–39. Reproduced by permission of The Zoological Society of London.)

It is important to emphasize that Southwood's scheme relates to the environment as experienced by specific populations and is not intended to classify the range of environments found in communities independently of their occupants. For example, a tropical rain forest cannot be classified as temporally constant; it may appear so for some leaf-eating insects, but it is seasonal for fruit-eaters, unpredictable for colonizers of clearings, and temporary for occupants of a decaying log.

Life history evolution is a controversial subject and the picture will inevitably change in the future. Some ecologists even suggest that the r- and K- theory be abandoned. Less drastic modifications will probably prove a more acceptable development. For example, an additional type of selection pressure, adversity selection, was proposed by Whittaker (1975b) and is discussed by Southwood (1977). This development has been extended by Greenslade (1983) who suggests **A-selection** (adversity selection) as leading to a third set of life history characteristics in addition to r- and K-selection. He proposes that in environments that are predictable but unfavorable because of abiotic factors, such selection will predominate. A-selection leads to long-lived specialized organisms held below carrying capacity by the adverse environment. Communities of A-selected populations are of low diversity and have a low level of interspecific competition. Greenslade's work on wood-decomposing beetles in rain forests of the Solomon Islands (part of Melanesia) support the existence of this extra category. In lowland forests, K-tendency predominates but in montane forests A-selection may be more important. At high altitude, fallen logs are permanently cold internally, leading to low decomposer activity (see Figure 2-10); that is, the abiotic environment is predictable, but adverse. The beetle fauna has the A-selected characteristics listed by Greenslade. A-selection may also be important in alpine savannas for similar climatic reasons (see Chapter 8). Whether other tropical terrestrial environments are A-selecting is questionable. Greenslade suggests that some organisms in hot deserts may be A-tendency. However, the abiotic environment in these deserts is generally unpredictable, and in some cases interspecific competition seems to be intense (see Chapter 8).

6.3   Adaptations to the Biotic Environment

A probable cause, and an undoubted effect, of the large number of species found in the tropics is that selection pressures are imposed by the biotic environment to a great extent. Particularly well studied are certain aspects of plant-herbivore interactions (sections 6.3.1 to 6.3.4) and predator-selected mimicry (section 6.3.5). These are by no means the only examples of biotic specialization found in the tropics.

6.3.1   Herbivores and Plant Defenses

Recently a large amount of information has accumulated about the effects of plant chemistry on animals that feed on plant parts or products. This section stresses the evolutionary response of plants to deter foliage-feeding herbivores. Adaptations to attract fruit- and nectar-feeders are discussed in the following two sections. Other types of herbivore such as sap-sucking bugs and root-eating insect larvae are also common, but little is known about them in tropical ecosystems.

Plants have various means of physical defense such as tough coverings on leaves, stems, and seeds, and some produce more specialized structures such as thorn and hairs. Many trees and bushes of African savannas have thorns, which are presumed to reduce browsing by large mammals, an idea supported by the absence of thorns in Australian *Acacia,* which have fewer large browsers. Less obvious are small hairs on leaves, which can impale and kill invertebrates.

Plant chemistry plays an important role in discouraging herbivores. Most plant tissues are low in organic and inorganic nutrients relative to herbivore tissues. A leaf-eating herbivore therefore needs to eat large amounts to secure adequate nutrition and will tend to select foods with higher than average nutrient concentrations, if available. One method of reducing herbivory is to have leaves of low nutrient value. As they mature, many plant tissues become less nutritious, particularly because of decrease in protein concentration as illustrated by the grasses in Figure 6-7. Concomitant changes are an increase in fiber and a decrease in digestibility. Grasses are of even lower food value because of the deposition of indigestible silica (Figure 6-7).

Many plants produce **secondary compounds,** which play no known role in the basic metabolism of the organism, but many of which are known to have deleterious effects on herbivores and pathogens. These protective chemicals include nitrogenous compounds (notably alkaloids, nonprotein amino acids, and cyanides) and phenolic compounds (particularly tannins and terpenoids). Their effects on herbivores range from directly poisonous (such as alkaloids), through the inhibitory tannins, to colored compounds such as carotenoids, which are often used to attract beneficial animals for pollination or seed dispersal. Alkaloids are found more frequently in the tropics than at higher latitudes (Figure 6-8), but no comparable evidence exists for other compounds. Leaf tannins, for example, showed inconsistent trends with latitude in Australia (Lowman and Box 1983). In fact, tannins occur in most flowering plants, but their effects as feeding deterrents become pronounced only at high concentrations. Tannins are not directly toxic but may cause the precipitation of proteins, thereby rendering digestion inefficient. However, Waterman (1983) suggests that the effectiveness of tannins in inhibiting digestion in herbivores has been exaggerated. He proposes alternative roles for these substances, such as discouraging

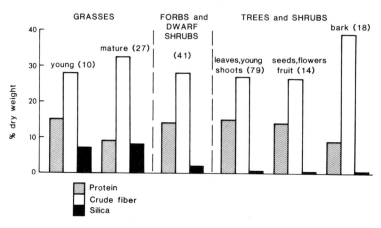

Figure 6-7   Chemical composition of plant parts eaten by large mammals in Kenyan savannas. Note in *a* the inverse relationship between protein and fiber. *b)* Reproductive structures of woody plants have similar composition to leaves. *c)* Young grasses are similar to forbs and leaves of woody plant; *d)* high levels of silica and fiber, but low levels of protein of mature grasses. Figures in brackets are numbers of species analyzed. (Data from Dougall et al. [1964].)

fungal attack and suppressing the activity of nitrifying bacteria. The latter effect may prevent loss of nitrogen from the soil to the atmosphere as litter decomposes.

Defensive chemicals such as tannins have been termed **quantitative**, since they are more effective as concentration increases, to contrast them with the **qualitative** effects of substances like alkaloids, which deter many herbivores even at low concentrations (Figure 6-9). Because all

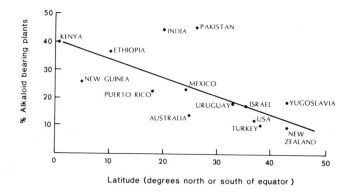

Figure 6-8   Larger proportion of tropical plant species generally produces alkaloids than those of higher latitudes. (From Levin, D. L. 1976. *Am. Nat.* 110:261–284. © By the University of Chicago; reproduced with permission.)

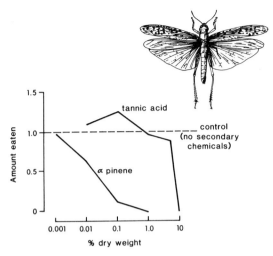

Figure 6-9   Laboratory comparison of the effects of chemical defenses on feeding activity of the desert locust, *Schistocerca gregaria*. Tannic acid deters feeding (that is, below the control level) only at high concentrations (quantitative defense) and may even be stimulatory at low concentrations. The terpene α-pinene deters the locusts even at low concentrations (qualitative defense). (From Bernays, E. A.; Chapman, R. F. 1978. In: Harborne, J. B. editor, *Biochemical Aspects of Plant and Animal Coevolution*. Orlando FL: Academic Press, p. 124.)

these secondary chemicals deflect energy from growth and reproduction, they impose a cost on the plant and therefore tend to be concentrated only in plants (or organs) that are easily found by herbivores. Feeny (1976) emphasized this point by dividing plants into those easily found (**apparent**) and those less easily found (**unapparent**) (Table 6-3). He further predicted that unapparent plants will tend to invest less energy in defense by using qualitative substances to deter polyphagous herbivores, but that such chemicals will often act as attractants to herbivores specializing on a particular species. Conversely, apparent plants will have quantitative defenses to deter all herbivores, despite the high cost. Extensive tests of these hypotheses have yet to be performed, although indirect evidence is supportive. For example, alkaloids are more frequent in woody species of disturbed environments (unapparent plants) than they are in nearby mature forests (apparent) in both Gabon and New Guinea (Levin 1976, Hladik and Hladik 1977). Unfortunately, no equivalent analyses were carried out on quantitative compounds in these studies. Also, the feeding preferences of the herbivorous monkeys *Presbytis johni* of Indian forests and the vervet monkey (*Cercopithecus pygerythrus*) of African savannas seem to be influenced by quantitative defenses. Staple foods are low in tannins, but alkaloids seem to have little effect, possibly because they are detoxified by gut microflora (Oates et al. 1980, Wrangham and Waterman 1981).

Table 6-3   Apparent and unapparent plant tissues. Some plant tissues are apparent and others unapparent to herbivores. The degree of apparency is a relative attribute referring to herbivores that are not highly specialized.

		APPARENT	UNAPPARENT
Whole plant		Perennial	Ephemeral, annual
		Climax	Pioneer
		Common	Rare
		Clumped, regular spacing	Wide, irregular spacing
Plant tissues		Mature leaves	Young leaves
		Evergreen leaves	Deciduous leaves
		Fruits	Seeds
		Animal-pollinated flowers	Wind-pollinated flowers

Losses of nutrients to herbivores are more difficult to replace for plants growing on infertile soils than for those growing on richer soils. Because of this, it has been postulated that more energy will be devoted to defense of plant tissues on poorer soils. Evidence comes from a comparison of two African forests, although they are widely separated geographically. The Douala-Edea forest in Cameroon grows on infertile white sand soils and high concentrations of quantitative (that is, energy-expensive) phenolics are common, while on the richer soils of the Kibale forest in Uganda, qualitative alkaloids predominate as secondary compounds (McKey et al. 1978). Similar reasoning may explain why plants in arid areas are often well protected, both mechanically and chemically. In this case lack of water, rather than nutrients, is the reason for slow growth and the need to protect hard-won production.

The common observation that most leaf-eaters prefer young leaves has led to suggestions that they are less well defended chemically than are older leaves (see McKey 1974a, Young 1982). However, it seems that young leaves have equivalent or even higher concentrations of secondary compounds. Preference for young leaves is more likely due to higher protein, lower fiber, and therefore greater digestibility of younger tissues (Table 6-4).

Ehrlich and Raven (1964) proposed that the antagonistic interactions between qualitative plant defenses and butterfly larvae may be strongly coevolutionary to the extent of having led to repeated bouts of speciation in both groups. They envisage several species of closely related plants with no (or similar) defensive chemicals, all of which are fed on by a suite of species of larvae. A mutation in an individual plant may produce a novel chemical protecting it from the larvae. Such a mutation will rapidly spread, by natural selection, through the population possessing it. Indeed, the greater survival and consequent reproduc-

Table 6-4   Nutritional value and chemical defenses in young and mature leaves. Most leaf-eating herbivores prefer young leaves because they are more nutritious, containing high levels of protein and low fiber levels, and are therefore more digestible, (see also Figure 6-7), and despite the fact that they are as well defended chemically as mature leaves and more difficult to find (Table 6-3). The top portion of table is for leaves from a moist tropical forest in Western Ghats (South India). Alkaloids were estimated qualitatively and were found to occur with similar frequency in young and mature leaves (Oates et al. 1980). The bottom portion of table is for leaves from a moist tropical forest on Barro Colorado Island (Panama). Qualitative tests for condensed tannins suggested similar occurrence in both types of leaf. Cell wall, as a measure of fiber, is inversely related to digestibility (Milton 1979). All measurements are given in percent of dry weight.

| | NUTRITIONAL VALUE | | CHEMICAL DEFENSES | |
	PROTEIN	DIGESTIBILITY	TOTAL PHENOLICS	CONDENSED TANNINS
Young leaves	15.1	39.8	7.2	7.6
Mature leaves	11.6	31.8	5.5	6.9

| | NUTRITIONAL VALUE | | CHEMICAL DEFENSES |
	PROTEIN	CELL WALL	TOTAL PHENOLICS
Young leaves	12.4	27.4	8.4
Mature leaves	9.3	43.1	7.6

tion of these plants may be such that populations of them spread widely and eventually produce new species, all of which are protected from herbivores. A butterfly that now has a mutation enabling it to overcome the inhibitory effects of this secondary compound will be able to feed on all these plants, free from competitors. Such a gene is likely to spread through the butterfly population, assuming that it has no deleterious effects that outweigh the benefit of plentiful food. This new herbivore genotype may in turn undergo an adaptive radiation to produce several species, each specializing in one of the new plant species. Subsequently, the plant species may evolve new defenses independently of one another, leading to a new round of speciation and specialization.

Although this scheme has attracted widespread support, there is little direct evidence to back it up. The best indirect evidence comes from the work of Gilbert and associates working in tropical America on the interactions between nymphalid butterflies of the tribe Heliconiini and their host plants, passionflower vines of the family Passifloraceae. Perhaps 50% of the heliconian larvae are monophagous, and many more feed on only two or three species of closely related vines. There have been five main radiations of heliconian species, and the most ancient species of each radiation is associated with a different species of the oldest of the Passifloraceae. Similarly, the most recent butterfly species are associated with the most recent radiation of *Passiflora* species (Benson et al. 1975). It is reasonable, therefore, to conclude that the radiations of butterfly species have followed those of their hosts, but as yet there is no direct evidence that herbivory has significantly affected the formation of new species of passionflowers. Nevertheless, coevolutionary adaptations by the plants are highly likely, given their elaborate defenses. Besides various secondary chemicals, some of the vines have structures that mimic species-specific butterfly eggs to deter the laying of real eggs, which would overload the plants with larvae. Some butterflies have evolved to overcome this defense with sophisticated behavioral repertoires to detect artificial eggs. Other vine species have deciduous stipules that fall off after eggs have been layed upon them (see Gilbert 1975 for these and other adaptations).

6.3.2   Herbivory on Fruits and Seeds

In contrast to temperate species, most woody plants in the tropics are adapted to seed dispersal by vertebrates (Table 6-5). To attract suitable animals (most importantly birds, bats, and monkeys) the seeds are surrounded by pulpy fruits, which are easily digestible energy sources. Although tropical fruits are generally richer in protein and fat than temperate fruits, they are not especially rich sources of protein compared with other plant tissues (Figure 6-7; Milton 1981).

Although fruiting in individual trees or species is seasonal even in humid tropical forests, some fruit is available at all times of the year. Specialization in fruit-eating is therefore possible, unlike the situation in the temperate zone where fruits are absent for much of each year. Full-time fruit-eating in tropical animals has led to speculation about

Table 6-5   Proportion of tree species with fleshy fruits adapted for seed dispersal by vertebrates. The proportion is greater in tropical than in temperate forests. Data from Howe and Smallwood (1982).

FORESTS	PERCENT OF SPECIES WITH FLESHY FRUIT
TEMPERATE	
Deciduous forests, United States	10–25
TROPICAL	
Neotropics, wet/humid forests	78–98
Neotropics, dry forests	51–77
Paleotropics, wet forests, Borneo (includes shrubs and lianes)	35–40
Paleotropics, moist forests, Nigeria	46–80

coevolution and mutualism. However, as Wheelwright and Orians (1982) point out, no frugivore population eats only one fruit species, and close reciprocal adaptation or mutualism is unlikely, since the plant has little to gain from the type of mutualism found in some pollination systems (see section 6.3.3). Also, many plants that produce fleshy fruits evolved before their modern frugivores, indicating that close coevolution has not taken place (Snow 1981, Howe and Smallwood 1982).

For frugivores to be efficient seed dispersers, they should digest the fruit and deposit the regurgitated or defecated seed unharmed at a site suitable for its germination. None of these conditions can be guaranteed; some seeds are dropped beneath the parent tree or at other inappropriate sites and others are digested. For example, only 6% of the seeds leaving a Costa Rican fig tree were dispersed by birds without damage. The remainder were destroyed by invertebrates or consumed by parrots, which break or digest the seeds (Jordano 1983). Many seeds are adapted to passage through the vertebrate gut, and for the seeds that survive, germination may even be enhanced. *Acacia tortilis* seeds in Africa are more likely to germinate after passage through the intestines of savanna ungulates (particularly impala) if they are not digested. Subsequent damage by bruchid beetles seems to be reduced in those seeds that have had this treatment (Lamprey et al. 1974).

In contrast to fruit-eaters, some tropical seed-eaters are highly specialized, with many species being monophagous. A high degree of specificity is found in bruchid beetles as exemplified by the study of Janzen (1980b) in a lowland deciduous forest of Costa Rica. Of 975 species of dicotyledenous plants, at least 100 species usually had beetle seed predators. Of the 110 beetle species found, 75% were mono-phagous, and the remainder attacked only a few species. Similarly, 60% of plant species with seed predators were attacked by only one species of beetle and none by more than five species. The causes of specificity are not known in most individual cases, but secondary compounds can play a significant role. Many species of seeds have combinations of alkaloids and nonprotein amino acids lethal to unspecialized insects (Bell 1978).

6.3.3   Pollination Ecology

Most plants require pollen from another individual of the same species to complete sexual reproduction. The major means of pollen transfer are wind and flying animals. Most grasses, most gymnosperms, and many forbs and woody angiosperms are wind pollinated. However, in many lowland tropical forests, almost 100% of trees are animal pollinated. In contrast, a latitudinal gradient in temperate North America reveals a change from around 50% animal pollinated at 20°N to 20% or less at 60°N (Regal 1982).

The relationship between a plant and its animal pollinator is mutualistic, with the plant benefiting from outcrossing while the animal obtains food as nectar and, in some cases, pollen. Some tropical flower-pollinator systems, such as figs and their wasps or orchids and their euglossine bees, are closely coevolved, but caution is needed in attributing all apparently intimate relationships to coevolution. For example, Baker (1973) points out that papaya (or pawpaw, *Carica papaya*), native to the neotropics and with seemingly intimate pollinators, grows wild in alien Africa but is successfully pollinated there, despite having no coevolved pollinator. Nevertheless, a species-specific pollinator is of great potential benefit to plants. It is easy to see that the reproductive success of a plant is enhanced by pollen transfer between members of the same species. Pollen deposited in a different species of flower is wasted except in the rare cases of successful hybridization. This contrasts markedly with seed dispersal by animals. A suitable target for a seed to germinate and grow cannot be directly influenced by the parent plant. Consequently, there is less advantage in acquiring species-specific frugivores and seed dispersers.

Diverse adaptations are found in pollination systems, some of which are outlined in Table 6-6. Some plants have many pollinators that visit many species of flowers, while others tend to one type of pollinator (for example, bee, bat, or bird) or even one species, although the latter extreme is rare.

A comprehensive survey of pollination syndromes is being made in lowland seasonal forests of Costa Rica by Baker and colleagues (for example, Baker and Baker 1983). Comparable data from other areas do not exist, but regional differences do occur. For example, bat pollination is common in tropical America, Asia, and Australia, less common in West Africa, and relatively rare in East Africa. Hummingbirds are absent from the Old World but are replaced to some extent by the less diverse sunbirds of Africa and Asia and some of the honeyeaters of Oceania and Australia. In Costa Rica, 587 nectar-secreting plant species (from more than 90 families) of all growth forms have been surveyed along with more than 1200 animal species that regularly visit flowers. Some of the findings are summarized as follows:

1. Larger flowers produce more nectar and are usually pollinated by larger animals (the quantitative relationships are shown in Figure 6-10). Butterflies obtain less nectar than other animals of comparable size, possibly due to their lower energy demand for flight (see Heinrich 1975).

2. The lengths of mouthparts in hummingbirds, butterflies, and small bees are closely related to flower depth. Larger bees enter their flowers entirely and bats visit large open flowers and therefore show no such correlations.

3. Bees, bats, and hawkmoths obtain greater volumes of nectar relative to their body size than other pollinators, probably because they

Table 6-6 Some typical morphological and behavioral characteristics of flowers associated with common types of animal pollinators. Many variations from these generalizations occur.

FLOWER CHARACTERS	TYPE OF PRIMARY POLLINATOR					
	UNSPECIALIZED	BEE	BUTTERFLY	MOTH	BIRD	BAT
Size and shape	Small, radial symmetry	Large enough to admit bee, bilateral symmetry	Small, tubular	Small, deeply lobed or fringed	Long, tubular	Large, bowl or beaker shaped
Color	Various	Yellow, white, blue	Vivid red, white, blue	Drab or white	Red or strong contrasts	Drab or white
Perfume	Various	Fresh, weak	Fresh, weak	Sweet, strong	Absent	Fetid, strong
Nectar hidden when flower open	No	Yes	Yes	Yes	Yes	No
Time of nectar production) and/or flower opening	Most of time	Day	Day	Night	Day	Night
Examples	Dodson (1975)		Cruden and Hermann-Parker (1979)	Grant and Grant (1965)	Schemske (1980a)	Lack (1978)

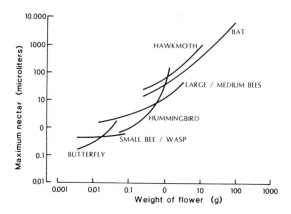

Figure 6-10   Relationship between nectar volume and flower weight for species of plants pollinated by different types of animal in Costa Rican forests. (From Opler, P. A. 1983. In: Bentley, B.; Elias, T. editors. *The Biology of Nectaries*. New York: Columbia University Press. Reprinted by permission.)

have a high-energy demand for thermoregulation and long-distance foraging.

4. Sugar concentrations in nectars vary in relation to the type of pollinator. Nectar viscosity increases with sugar concentration and some pollinators, notably bees, are better able to extract more viscous liquids. Bat flowers have the most dilute nectars.

5. Nectars are not merely sugar solutions but also contain amino acids and lipids. Higher amino acid concentrations are found in moth and butterfly flowers, probably because some of these adult insects obtain all their food from nectar. Bees, bats, and birds obtain extra protein from pollen, fruit, and insects, respectively.

6. Thirty-six percent of species have nectars containing amino acids that are not constituents of proteins and that may be a chemical defense against feeding by nonpollinators. Alkaloids and phenolics have also been detected in 12% and 50% of nectars, respectively. They seem to be more common in tropical than temperate plants (Baker 1978). Because nectar is such a rich source of energy, nectar "robbery" by nonpollinators is common, but secondary compounds may reduce it.

The generalizations presented in Table 6-6 and the more detailed information about nectars leave little doubt about the fact that plants have evolved to attract animal pollinators. Very often a particular group (such as bees) or size of animals (large bees, small bees) is attracted. Only rarely is there convincing evidence that the mutualism is obligatory and coevolved. One of these rarities is described next.

Pollination in Figs.   One plant species serviced by one pollinator species and vice versa is rare in most cases but is the norm in figs. The genus *Ficus* (belonging to the family Moraceae) contains more than 900 species found throughout the tropics and is represented by all types of woody growth forms. Their intimate relationship with parasitic agaonid wasp pollinators has been reviewed by Janzen (1979) and Wiebes (1979). Closely related *Ficus* species are pollinated by closely related wasp species, suggesting a coevolutionary history of speciation. The following is a general account of the relationship (based on Janzen's review), and to avoid specialized terminology, the plants will be called *Ficus* and their fruits, throughout development, figs.

One or a few adult female wasps, loaded with pollen from another *Ficus* of the same species, enter the unripe fig through a pore. These wasps pollinate the hundreds of single-ovule female flowers lining the fig and lay one of their own eggs in 50% or more of the fig ovaries. After a few weeks, wingless male wasps emerge from the ovaries that their larval stages have parasitized and mate with the females that remain in the fig ovaries. The winged females then emerge, collect pollen from unparasitized male fig flowers inside the fruit, and leave the fig through a hole cut by the male wasps, to find and pollinate another tree. The figs ripen and are dispersed by frugivores including birds, bats, and monkeys. Fig production on individual plants is highly synchronous so that self-pollination is usually impossible, but different individuals of the same species are not synchronized and are widely dispersed in the community. It is clear that the short-lived female wasps are very efficient at finding the correct species with receptive figs, although the way in which this is achieved is unknown.

6.3.4   Ant-Plant Mutualisms

Some plants produce nectars that are not associated with pollination. These "extrafloral" nectaries attract ants, which, because of their pugnacious attacks on other animals, afford some protection from herbivores to the host plant. Sixty-eight families of flowering plants are known to bear extrafloral nectaries on various organs (see Elias 1983 for a comprehensive review). Although ant-plant mutualisms occur in temperate communities, they are much more common in the tropics (Bentley 1977, Keeler 1981). The benefit of food for the ants is obvious, and the protective advantage for the plants can be dramatic, even when the mutualism is loosely organized (Table 6-7). The example in Table 6-7 shows a largely opportunistic relationship between a species of wild ginger and two species of ants that is probably not coevolved. Some other associations are species specific, with ants residing permanently on the plant. These intimate mutualisms, which are often coevolved, are known from forests throughout the tropics, but perhaps the best studied are those from the savannas of Central America and East Africa

Table 6-7   Reproductive success of the Central American coastal herb *Costus woodsonii* when ants (*Campanotus planatus* and *Wasmannia auropunctata*) are present and when they are experimentally removed. Reproductive success is greater when ants are present. The ants protect flower buds from the parasitic fly *Euxesta,* which leads to more flowers and seeds being produced. Pollination rates (by hummingbirds) are also greater in the presence of ants, and predispersal seed predation is reduced. Data are for the wet season from Schemske (1980b).

	ANTS PRESENT	ANTS EXCLUDED
Number of ovipositions by fly parasite	4	69
Number of flowers per inflorescence	55	44
Number of seeds per inflorescence	612	183
Flowers pollinated (%)	53	35
Seeds destroyed (%)	0.1	14

where *Acacia* trees and bushes have coevolved with *Pseudomyrmex* and *Crematogaster* ants, respectively.

The ants protect the acacias from herbivores and from competition with other plants. In return the acacias provide liquid food from extrafloral nectaries, solid food (from special Beltian bodies at the leaf tips in the neotropics), and swollen thorns to house ant nests, and remain in leaf throughout the year to sustain the ants. Janzen (1966) has documented the enhanced survival of Mexican acacias due to ants as a result of reduced herbivory by invertebrates. Less clearly established is the protection against browsing ungulates in Africa, since giraffes feed readily on *Acacia drepanolobium* despite the ants. More "thin-skinned" mammals may be deterred as illustrated by the avoidance of *Barteria* trees with ants by black colobus monkeys in Cameroon (McKey 1974b). However, this defensive strategy can backfire as some omnivorous monkeys such as the savanna-dwelling baboons and patas monkeys are attracted to and eat *Acacia* ants, often damaging the plants in the process. Nothing is known of the relative cost of using ant protection squads compared with chemical defenses. Presumably the continual production of nectar and solid foods consumes much energy, but the effectiveness of protection by ants seems to make the expense worthwhile. To avoid duplication of defense costs, *Acacia* defended by ants do not produce the secondary chemicals found in many nonant acacias (Rehr et al. 1973).

Another locally widespread mutualistic relationship found in the neotropics and Southeast Asia is that of ants "feeding" epiphytes with inorganic nutrients. The ants nest in the plants in which they dump the remains of their arthropod food and from which the plants extract a concentrated supply of nutrients. The ants benefit by the provision of nest sites in the plants but do not feed on the plants, nor do they attack herbivores (Huxley 1980).

6.3.5   Mimicry

Some individuals increase their chance of survival by resembling inedible objects such as stones or other organisms that are protected from predators. Such mimicry is especially common in the tropics as recorded by several 19th century naturalists, most notably Henry Bates and Fritz Müller. Descriptions of mimicry in some tropical butterflies follow, but many more examples can be found in Wickler's (1968) review.

Batesian Mimicry.   This is the classical form of mimicry originally described by Bates from his studies of Brazilian butterflies. A **model** species is avoided by predators either because of its aggressive behavior or because it contains distasteful chemicals (often sequestered from its food plant). One or more **mimic** species, which are palatable to predators, resemble the model in appearance and so are also avoided. Models are often distinctively colored, making their presence obvious to predators so that they can easily be avoided. Mimics obtain protection because their resemblance to the unpleasant model increases their chance of survival. Such relationships do not necessarily evoke an evolutionary response in the model and need not, therefore, be coevolutionary.

A mimic gains maximum advantage when it is rare relative to the model so that the predator, when sampling potential food, will learn rapidly to avoid organisms that are similar to the model. However, small differences in survival are all that is required for natural selection. Brower's (1960) experiments with predatory birds and their butterfly prey demonstrated protection for 17% of mimics, even with a ratio of nine mimics to one model.

Some mimics are polymorphic, with each morph resembling a different species of model. This situation evolves because a rare mimic is unlikely to be eaten, as just explained. Therefore, a new type of mimic will have a better chance of survival than one that coexists with many similar individuals. An ecological consequence of polymorphic mimicry is that total population density of all morphs is greater than a monomorphic mimic could sustain with the same degree of protection in the same environment. A remarkable example of polymorphic mimicry is the African swallowtail butterfly *Papilio dardanus*. Within most races the females are polymorphic, with each morph resembling a different species of model. Males are monomorphic and are not mimetic. The genetics are complex but well understood. More recent work on the Malaysian *P. memnon* has added generality to the earlier studies (Figure 6-11). The restriction of mimicry to females in many

MODEL   MIMIC

Figure 6-11   Polymorphic Batesian mimicry in the Malaysian butterfly *Papilio memnon*. Models are to the left, and mimics are to the right. Seven genes control most of the variations in pattern. (From Turner, J. R. G. [1983]. In: Vane-Wright, R. I., Ackery, P. R. editors. *The Biology of Butterflies.* Orlando, FL: Academic Press.)

species of butterfly is, at first, a paradox. Why should males forgo this form of protection? Intraspecific signaling using distinctive coloration in both territorial competition between males and attraction of females seems to override predator selection pressure and prevents the establishment or maintenance of male mimicry in these species (Turner 1984).

Contrary to some speculation, even a slight resemblance to a distasteful model confers significant protection from predators, and continued selection is likely to gradually improve this resemblance (Figure 6-12).

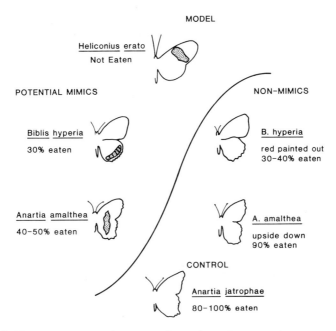

Figure 6-12   Experimental tests to show the selective advantage of incipient Batesian mimicry in Trinidadian butterflies. The model *Heliconius erato* is a black butterfly with a red band (stippling). It was always rejected by two species of predatory birds in the experiments, while the disimilar edible brown and white control butterfly *Anartia jatrophae* was eaten. Two edible butterflies, which resemble *H. erato* (that is, potential mimics) in being dark with red bands, were less acceptable to the birds. When the incipient mimicry was reduced (*B. hyperica* with red patches darkened), it still had a high rejection rate with its black appearance. When mimicry was totally removed (*A. amalthea* upside down, light color, no red), the birds readily ate it. Details of color patterns besides red and black are not shown. (Modified from Brower et al. [1971].)

Müllerian Mimicry. A second type of protective mimicry was described by Müller. All the species are unpleasant to predators, and they resemble one another in a mutualistic association. On average, each individual is better protected if many apparently similar but unpleasant prey coexist. A predator only has to eat one (or a few) individuals to reject all similar individuals of all species in the future. It is probable that coevolution is involved, with populations of each species changing genetically so that their appearances converge. Commonly Batesian mimics also become involved in such "mimicry rings." The Müllerian mimicry exhibited by butterflies of the genus *Heliconius* in Central

America and South America is extremely complex genetically, involving racial and polymorphic variation leading to a wide variety of color patterns. These mimicry rings may involve species from the same genus but often include some butterflies from different families (see Turner 1971, Gilbert 1983).

Although superficially similar, the Müllerian and Batesian types of mimicry are fundamentally different ecologically. Whereas Müllerian mimicry is mutualistic for all participants, the Batesian mimic is effectively "parasitic" on the model, since it enhances its own survival prospects at the expense of the model.

6.4   Does Evolution in the Tropics Differ from Evolution Elsewhere?

The processes of genetic variation, natural selection, and speciation are the same throughout the biosphere. Nevertheless, there has been speculation that tropical environments may lead to different ecological outcomes of these processes. Two papers entitled "Evolution in the Tropics" are separated by twenty years (Dobzansky 1950, Baker 1970). The first established many of the conventional wisdoms of tropical ecology, and the second set the scene for the blossoming of studies, particularly in Central America. Both addressed the problem of high species diversity in the tropics, which is dealt with at length in the next chapter. Dobzhansky's primary thesis was that selective forces in the tropics arise from biotic environmental factors to a greater extent than at higher latitudes where abiotic factors tend to predominate. There is no question that in many tropical ecosystems the biotic environment is more complex structurally and in the number of species. That other organisms have been important selective forces is shown by the greater frequency of interspecific mutualisms, some of which are obligatory, a point stressed by Baker. Mutualisms between individuals of the same species are also more prominent in most communities. Examples include the complex social systems of such diverse groups as ants, termites, birds, mole rats, carnivorous mammals, and primates.

Whether antagonistic interspecific interactions are more common in the tropics is less clearly established. A greater frequency of some chemical defenses and higher rates of predation have been observed in some groups (Figure 6-8; section 5.4.1). The proposal that interspecific competition is more intense is more contentious. It is supported by much circumstantial evidence but few direct demonstrations (see sections 5.1.1 and 7.4.1).

The evolution of life history patterns cannot be seen as a simple divergence between tropical and temperate populations. A superficial examination of Table 6-1 suggests that K-selection might predominate in the tropics. This impression is reinforced by information on clutch size and survivorship in Chapter 4. However, all communities contain a spectrum of life history adaptations. Even highly unpredictable environments such as deserts contain many K-tendency populations, a point elaborated in section 8.3.3. The greater diversity of species in the tropics is reflected by a greater diversity of life history adaptations rather than by a predominance of any one type.

High species diversity in the tropics indicates that speciation may have occurred there more frequently. Some suggest that the number of species is so high that it is difficult to imagine that geographical isolation is necessary for new species to form. However, the refugium theory described in section 7.2.3 is a convincing scenario for speciation as a result of geographical isolation caused by climatic fluctuations and fragmentation of communities. In addition, the high incidence of specialized pollination systems involving animals is likely to enhance reproductive isolation and speciation to a greater degree than the more frequent wind pollination of the temperate zone.

Summary

A knowledge of the processes of evolution is essential to an understanding of ecological interactions. Evolutionary changes affecting ecological processes are brought about primarily by natural selection (including sexual and kin selection), with chance and group selection playing lesser and possibly temporary roles. The effects of variation between individuals on population dynamics deserves more attention than previously given. Theories of life history evolution provide a useful way of comparing organisms in different taxa and different environments if used judiciously. Evolutionary interactions between prey and predator populations lead to adaptations in defense or offense, respectively, or in some cases to the development of mutualisms. The extent to which coevolution is involved in these interactions is difficult to determine.

Study Questions

Review

1. Why do ecologists need to understand evolutionary processes?
2. Write an essay on coevolution.
3. Why are some plant-animal mutualisms more intimate than others?
4. Why do we find a mixture of r- and K-tendency populations in most ecosystems?

Related Topics

1. Discuss the effects of mimicry on the population dynamics of the species involved.

2. Review the proposition that evolution in the tropics is primarily in response to the biotic environment, whereas temperate organisms evolve mainly in response to abiotic factors.

3. Investigate the possible roles of secondary compounds in plants other than that of defense against herbivores.

4. Why is body size an important factor in ecology?

Further Reading

To supplement the brief sketch of evolutionary processes given, two general texts are recommended. The first is an excellent (and inexpensive) introductory account; the second is much more detailed and technical.

Dobzhansky, T.; Ayala, F. J.; Stebbins, G. L., Valentine, J. W. (1977). *Evolution.* San Francisco: Freeman.
Maynard Smith, J. (1975). *The Theory of Evolution.* 3d ed. Harmondsworth, Middlesex, U.K.: Penguin.

The following ecology text does an excellent job of integrating ecology with evolution:

Ricklefs, R. E. (1979). *Ecology.* 2d ed. New York: Chiron Press.

Sections of the following two books deal with different aspects of the integration of population ecology and evolution:

Shorrocks, B. editor (1984). *Evolutionary Ecology.* Oxford: Blackwell. An important symposium volume from the British Ecological Society. Includes chapters on genetics and population dynamics, life history evolution, mutualism, and mimicry.
Southwood, T. R. E. (1981). Bionomic strategies and population parameters. In: May, R. M. editor. *Theoretical Ecology.* 2d ed. Oxford: Blackwell, pp. 30–52. Deals with life history evolution.

Following are two very different books on coevolution. The first is a brief (nonmathematical) analysis of the conditions likely to produce coevolution. The second is a more voluminous multiauthored account of all aspects of coevolution including genetic factors, mathematical models, pollination, mimicry, parasitism, and many more.

Futuyma, D. J.; Slatkin, M. editors. (1983). *Coevolution.* Sunderland, MA: Sinauer.
Thompson, J. N. (1982). *Interaction and Coevolution.* New York: Wiley.

Descriptive reviews of two types of biotic evolutionary interactions are contained in the following:

Forgri, K.; van der Pijl, L. (1979). *The Principles of Pollination Ecology.* 3d ed. Oxford: Pergamon.

Wickler, W. (1968). *Mimicry in Plants and Animals.* London: Weidenfeld and Nicholson, World University Library.

High diversity in tropical rain forests can be explained in part by the varied growth form of trees and the spatial heterogeneity that they create. This buttressed tree trunk and sinuous surface root system are from Panama. (Photograph by Eldridge S. Adams.)

CHAPTER 7

The Ecology
of Communities

7.1   Delimitation and Description of Communities

A community was defined in Chapter 1 as a group of populations of different species in a given area. Community ecology has grown rapidly in the last 20 years, yet it remains the most problematic level of organization recognized by ecologists. These problems are accentuated in the tropics because communities close to the equator generally have more species than those at higher latitudes. Some gradients toward higher diversity at low latitudes are documented in Table 7-1. Exceptions to this trend do occur such as some aphids in Australia, ichneumonid wasps, and some yeasts on fruit flies, but no biologist doubts its generality. In a book of this size, it is impossible to tackle all aspects of community ecology in detail, so this chapter hinges on reasons for high diversity in the tropics. However, it is important to remember that other features of communities besides the number of species also differ in different climatic zones as illustrated in Figure 7-1.

At the outset it is important to recognize at least two levels of species diversity. Table 7-1 shows **regional diversity** of whole nations or parts of continents within which many different communities exist. However, even within comparable communities (**local diversity**) the latitudinal trend is equally clear. For example, in Australia, tropical rain forests typically contain 140 species of plants in a small plot (0.1 ha), while a diverse Australian temperate forest has less than 50 species in

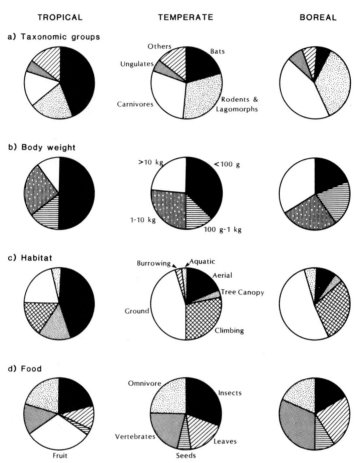

Figure 7-1   Besides the number of species being greater in the tropics, the types of species also differ as illustrated by this comparison of mammals in a Panamanian rain forest (9°N, 70 species), a deciduous temperate forest in Michigan, (42°N, 35 species), and a boreal conifer forest in Alaska (65°N, 15 species). The proportion of bats and consequently small aerial species decreases as latitude increases, an extra category of fruit-eating specialists is present, and ground-dwelling species are relatively scarce in the tropical forest. (Data from Fleming [1973].)

the same area (Rice and Westoby 1983). Similarly, a shrubland in Panama contained 42 species of resident birds compared with only 18 in a similar community in Illinois (Karr and Roth 1971). This is not to pretend that there are no species-poor communities in the tropics. Some grass savannas have comparable numbers of plant species to their temperate counterparts. Parts of equatorial Africa subjected to flooding have as few as 11 tree species with one of those strongly dominant.

The following account places emphasis on tropical rain forests because they are the most diverse of terrestrial communities. Reference

Table 7-1 Species richness at various latitudes. In many taxonomic groups, species richness increases as latitude decreases.

DEGREES (°) NORTH OF THE EQUATOR	FLOWERING PLANTS (AMERICAS)	TERMITES (SOUTHEAST ASIA)	ANTS (AMERICAS)	PSOCOPTERA* (BRITAIN AND PANAMA)	SNAKES (AMERICAS)	NESTING BIRDS (AMERICAS)	FOREST BATS (AMERICAS)
0–10	—	77	—	295	—	1359	70
10–20	—	90	222	—	293	469	—
20–30	2500	15	165	—	—	143	—
30–40	—	6	103	—	126	—	—
40–50	1650	—	59	—	—	195	35
50–60	31	—	2	51	—	118	—
60–70	11	—	—	—	22	56	1
Source	Clarke (1954)	Abe (1978)	Fischer (1960)	Broadhead (1983)	Fischer (1960)	Fischer (1960)	Cox et al. (1976)

*Psocoptera are foliage living insects that feed on epiphytic lichens, fungi, and algae.

to other biomes is made wherever adequate information is available. Before looking at patterns of diversity it is essential to clarify two questions: (1) do communities exist as identifiable entities (if so, how do we delimit them)?, and (2) what is species diversity and how is it measured? Two special aspects of species diversity close the chapter, that of oceanic islands and the concept of community stability.

7.1.1   Do Communities Exist?

Subjectively everyone is aware that a forest community is different from a grassland community in species composition and vegetation structure (**physiognomy**). For communities to be objective realities as discrete collections of populations, the expectation is that certain groups of populations tend to recur in similar environments and that these groups differ from adjacent communities. This expectation can be tested by **gradient analysis,** in which the distribution of species along an environmental gradient is described to see whether the ranges of different species coincide to produce distinct communities (Figure 7-2).

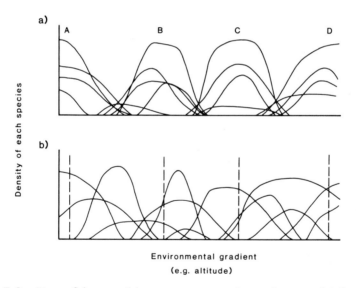

Figure 7-2   Two of the possible arrangements of populations of different species along environmental gradients. (Modified from Whittaker [1975a].) *a,* Communities are clearly defined (A, B, C, D) with groups of populations having more or less coincident distributions. The areas of overlap are called ecotones. *b,* Each population is distributed along the gradient independently of the other populations, and no clearly defined communities are formed. Note that if samples were taken at sites indicated by the broken lines, it would appear that discrete communites were present, but this would be an artifact of sampling.

Pioneering work by Robert Whittaker in North America indicated that each population was distributed according to its own environmental requirements, which did not usually relate closely to those of other populations. However, the generality of this conclusion is not well established.

Few strictly comparable studies have been made in the tropics, but the available evidence suggests that at least some populations are closely associated in distribution and therefore form recognizable communities. Both of the patterns stylized in Figure 7-2 occur (Figure 7-3). In Figure 7-3*b*, populations are distributed independently along an equable gradient but form rather distinct communities along a harsher gradient, even though many species are common to both sites. Even along subtle catenary gradients, herbaceous plants form distinct communities in the Serengeti savannas of Tanzania (McNaughton 1983; see also Figure 3-8).

Very few gradient analyses of animal populations have been made, but here again both coincident and independent distributions occur. In an analysis of tree canopy insects along an altitudinal transect in Hawaii, some taxa of unrelated predators formed distinct communities, while fungus beetles and twig-boring beetles had independent distributions (Gagné 1979). Along the eastern slope of the Peruvian Andes, Terborgh (1971) found that community formation accounted for the distribution of fewer than 20% of bird species. Of the remainder, about two-thirds of species were distributed independently of one another, with the rest limited by interspecific competitive exclusion. With so few examples and so much variation, it is impossible to decide which of the two patterns in Figure 7-2 is more common.

7.1.2   Classification of Communities

Even ecologists who doubt the existence of closely coinciding groups of species do not doubt that communities can often be recognized by difference in **dominant species** that by their abundance, size, or other attribute have a major influence upon other species occurring at the same locality. For example, in Figure 7-3*a*, the grass *Andropogon intermedius* is clearly dominant at 40 m along the transect as it covers 80% of the ground surface.

Classification of communities is hierarchical, the highest level being the subdivision of world vegetation into recognizable physiognomical categories, or **biomes,** the distribution of which is governed primarily by global climatic patterns, Biomes cannot be recognized by species composition because different species are usually dominant in different parts of the world. A low-resolution classification of terrestrial biomes based on temperature and precipitation is given in Figure 7-4. Holdridge and colleagues (1971) have devised a more detailed scheme, which they developed particularly for the classification of tropical forests (Figure 7-5). Their method uses more complex climatic variables and includes both

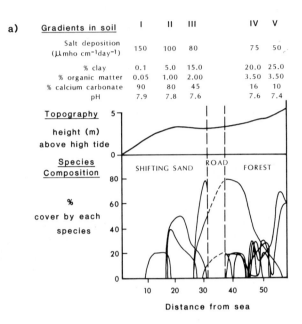

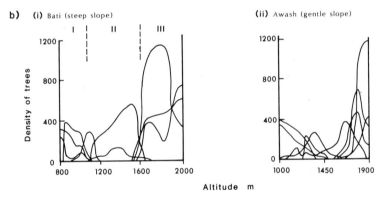

Figure 7-3   Distributions of populations of plant species along environmental gradients in the tropics. *a,* Beach transect in Maycock's Bay, West Barbados in the Caribbean. The main features of the gradient are the soil factors given; only a few species are tolerant of sea salt and mobile sand in particular. Some groups of species have very similar distributions (centered around 25 m, 42 m and 50 m), but others are independent. (Data from Randall [1972].) *b,* Two altitudinal gradients in Ethiopia. Most species are common to both transects. (Data from Beals [1969].) (i) Bati has a steep slope (800 m rise over 20 km) and rugged topography forming a Rift Valley escarpment. Groups of species tend to have similar distributions as indicated by the subdivisions I, II, III. (ii) Awash has a more rolling, gentle topography (900 m rise over 300 km) in the floor of the Rift Valley. Species are distributed largely independently of one another.

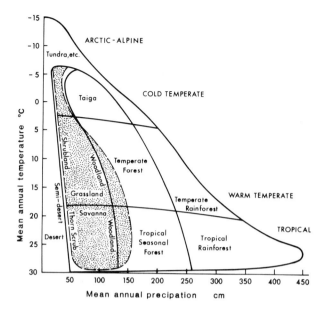

Figure 7-4 Whittaker's classification of terrestrial biomes in relation to mean annual temperature and precipitation. Boundaries are approximate, and within the shaded area, soil, grazing, or fire determine the relative abundance of woody and herbaceous plants. (Modified with permission of Macmillan Publishing Company from *Communities and Ecosystems* by R. H. Whittaker. Copyright 1975 by Robert H. Whittaker.)

latitudinal and altitudinal gradients. Holdridge's classification shows a greater number of different biomes at low latitudes, and this alone is likely to lead to greater regional diversity in the tropics (see section 7-3).

A complementary method of classification is that of **life forms,** in which plants are categorized in relation to their growth and reproductive patterns. The Raunkiaer system is illustrated in Figure 7-6 to show how the position of perennating organs or tissues vary with latitude. In colder climates a higher proportion of buds or other perennating organs are at or below the soil surface, and aerial parts die back in the winter. Similar trends occur along aridity gradients as the proportion of trees declines and the proportion of shrubs and herbs increases until annuals predominate in very dry environments.

Within these biomes (or **formations**), many different plant communities (or **associations**) are sometimes recognized, which can be defined by structure and species composition. Various techniques exist for defining such communities. Ecologists in continental Europe favor the Braun-Blanquet method for large-scale survey, whereas many English-speaking ecologists from Britain and North America prefer more objective techniques. In the Braun-Blanquet method, a sample is chosen subjectively within a community that includes all major species and that

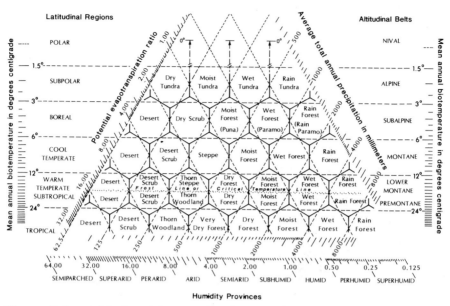

Figure 7-5   Holdridge's life zone classification of biomes in relation to climate. Mean annual biotemperature is the mean annual temperature in which all values below 0° and above 30° C are regarded as zero in calculating the mean (these being assumed to bound the temperature range of physiological activity in plants). Potential evapotranspiration ratio is an index of biological humidity. It is the mean annual potential evapotranspiration divided by the mean annual precipitation; thus a ratio of 1 means that potential evapotranspiration is exactly balanced by precipitation on an annual basis. (From Holdridge, L. R.; et al. 1971. *Forest Environments in Tropical Life Zones.* Oxford: Pergamon Press Ltd.)

is as uniform and homogeneous in plant cover as possible (see Mueller-Dombois and Ellenberg 1974). The vegetation is described structurally and floristically by visual observation, and details of soil and topography recorded. The principal criticism of this method is the subjective selection of samples, which, because of the selection criteria, give no indication of heterogeneity in the vegetation and leads to the impression of distinct communities whether or not they exist. Nevertheless, in the hands of a skilled botanist, the method is unequaled for rapid ground survey as a prelude to more objective studies. Objective methods involve the random selection of samples and the use of various statistical procedures to test similarities between these samples and thereby to identify and classify communities. These techniques allow for heterogeneity and commonly use more precise counting methods than does the Braun-Blanquet technique. Many such methods are available, as ably reviewed in Mueller-Dombois and Ellenberg (1974) or Chapman (1976).

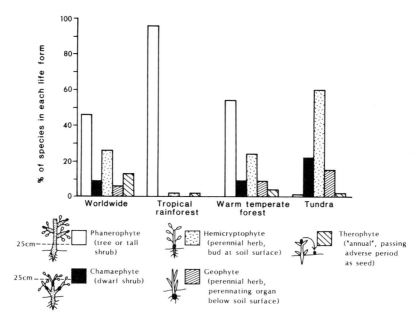

Figure 7-6   Raunkiaer system of life-form classification of plants used to illustrate differences in composition of communities in moist environments in relation to latitude (data from Whittaker [1975a]). The life-forms are illustrated in the key with perennating organs (buds, seeds, underground organs) in heavy shading; other perennial tissues, in light shading; and tissues present seasonally, unshaded. This is not a complete system of possible life forms (see Mueller-Dombois and Ellenberg 1974).

The classification of animal communities can in principle use similar methods to those used by vegetation ecologists. However, in practice few animal ecologists are concerned about precisely delimiting their communities spatially because of the mobility of individuals. Instead, field studies of animal communities are usually delimited taxonomically, with, for example, a study of the termite community or the ungulate community of a savanna but rarely both. A qualitative approach to delimiting animal communities is the habitat classification of Elton (1966). This subdivides ecosystems into vertical strata of vegetation, such as canopy and field layers, and includes a set of general habitats such as dung, carrion, and macrofungi. Together with trophic and biome information, this enables verbal description of an animal community as, for example, "the upper canopy leaf-eating beetle community of a tropical rain forest," which allows comparison with other animal communities. This system designed for temperate communities, can be adapted for the tropics by adding more habitats such as extra canopy strata and more general categories such as lianes, termite mounds, and tree buttresses.

7.1.3   What is Species Diversity and How is it Measured?

The number of species in a community (**species richness**) is not adequate to descibe diversity as illustrated by Table 7-2. Equally important is the **relative abundance** of each population. Diversity is greater where **equitability** is higher; that is, where populations are similar to one another in abundance rather than some very common and others very rare. When individuals vary greatly in size, some measure of **relative importance** is better, such as proportional cover in herbaceous plants, basal area in trees, or biomass in any organisms.

In natural communities there are usually a few abundant species and many rare ones, although the exact pattern differs when communities are compared. Some of these patterns are shown in Figure 7-7. Various theories have been advanced to explain these curves (May 1975a), but they are best regarded as descriptive until their biological bases are better understood. Of interest is the contrast between tropical forests and temperate forests, with trees of the former showing greater equitability and species richness (Figure 7-7a). The log-normal distribution of relative abundance has been found to take a canonical form in many temperate communities, with a predictable standard deviation for a given mean of individuals per species (Preston 1948, May 1975a). However, some tropical communities do not fit that specific model even though they show a log-normal pattern. For example, the bird communities of Figure 7-7c have "too many species" and moth communities of montane New Guinea have "too high a degree of equitability" for the canonical model (Hebert 1980).

Another hypothesis is that communities are random assemblages of populations. Any collection of nonidentical objects (such as species)

Table 7-2   Comparison of two hypothetical communities each with five species and one thousand individuals. In community I, species A is abundant and all other species rare (low equitability), while in community II, all species are equally abundant (maximum equitability). Community II has higher diversity.

SPECIES	COMMUNITY I	II
A	996*	200
B	1	200
C	1	200
D	1	200
E	1	200
Total	1000	1000

*Number of individuals

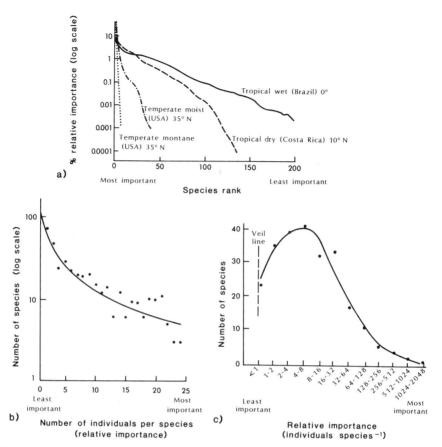

Figure 7-7   Different ways of representing relative importance patterns of species in communities. *a,* Comparison of tropical and temperate forests shows that both species richness and equitability are higher in the tropics (the most important species has a value of 5% at the equator, 11% at 10°N, 36% in the moist temperate forest, and 65% in the montane forest). (From Hubbell, S. P. 1979. *Science* 203:1306. Copyright 1979 by the AAAS.) *b,* The "log series" model applied to butterfly collections from Malaysia. There are many rare species and few common ones. (From Hutchinson, G. E. 1978. *An Introduction to Population Ecology.* New Haven, CT: Yale University Press.) *c,* The log-normal distribution of Preston applied to bird censuses in Panama (Preston 1980). Here the importance scale is in natural logarithms or "octaves" such that half the species with one individual are centered in the 0–1 octave; the second octave (1–2) has half the species with one individual and half of those with two individuals; the third octave (2–4) has half of those with two, all with three and half with four individuals, and so on. In this model, most species are of intermediate importance, and there are few very rare species. Theoretically the distribution should be symmetrically bell-shaped, and the veil line merely indicates that some of the very rare species were not found in the sample. (From "Noncanonical distribution of commoness and rareness" by F. W. Preston, *Ecology,* 1980, 61:88–97. Copyright ©1980 by Ecological Society of America, Reprinted by permission.)

appear structured statistically, even if they differ in a random manner and do not interact. While evolution tells us that species differences are not random, it is possible to postulate that a given community is a random selection of the species available to colonize the area. This is clearly not the case for birds on tropical islands (Figure 7-8), but there is a significant random component in the recolonization of defaunated mangrove islands by arthropods (Figure 7-9).

Mathematical models produce equally divergent views of community structure. Caswell (1976) suggests that communities are less diverse than would be expected if there were no interactions at all between species (that is, interactions reduce diversity). In contrast, Sugihara (1980) finds that a sequential subdivision of total niche space available to the whole community produces the distributions predicted by Preston's canonical model. Clearly, ideas about community structure are in a state of flux and will remain so until stronger links are established between the ecology of single populations and the interactions of the

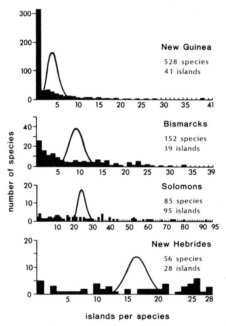

Figure 7-8   Comparison of actual bird species and randomly expected distribution of islands per species in four tropical archipelagos. The curve shows the expected distribution if each island had the same number of species as actually found, but the species were drawn randomly from the total species pool. The histograms show observed species distributions that are strongly divergent from the random hypothesis. (From Diamond, J. M. 1979. In: Anderson, R. M.; Turner, B. D.; Taylor, L. R. editors. *Population Dynamics*. Oxford: Blackwell.)

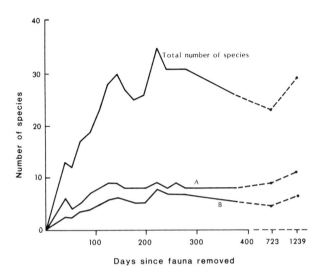

Figure 7-9   Recolonization of a mangrove island by arthropods in southern Florida after faunal removal shows a large random component in the particular species that become established. Line A is the observed number of the species common with the fauna before removal, and line B is same expected using a random-draw hypothesis. Dashed lines indicate periods with infrequent censuses. (From Simberloff 1978. In: Mound, L. D.; Waloff, N. editors. *Diversity of Insect Faunas.* Oxford: Blackwell.)

many that make up a community. The fact that certain community patterns are repeated in time and space indicates that the search for consistent organizing principles will not be fruitless.

Measurement of species diversity presents many practical problems. Even determination of species richness is difficult, since the number of species in diverse communities increases as the sample size increases (Figure 7-10). Various methods have been devised to combine species richness and relative importance in a single figure. The most commonly used is the Shannon-Wiener index (H). Others include Simpson's index (D) and alpha (α), a statistic of the log-series model of relative abundance. Some of the problems of their use and methods of computation are briefly described in Appendix C.

7.1.4   Factors That Promote High Species Diversity

Conceptually, an autotrophic ecosystem can reach equilibrium with one plant population and one decomposer population, and yet most communities contain many species linked by a complex food web. Sections 7.2 to 7.7 review various mechanisms that may increase species diversity and therefore account for the high diversity of tropical

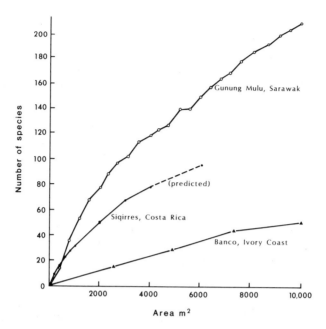

Figure 7-10   Tree species richness in three tropical regions. As bigger plots are sampled, the number of species found increases in moist tropical forests. Although different forests in each region have different numbers of species, it is generally agreed that the most species rich are in southeast Asia, followed by the neotropics and then Africa. (Data from Proctor et al. [1983], Holdridge et al. [1971], and Unesco [1978], respectively.)

communities. At the outset, it is important to realize that many of these mechanisms are not mutually exclusive and that they interact to produce the diversity that we observe.

Some ecologists believe that the species composition of communities represent equilibria, largely mediated by interspecific competition. Others suggest that such equilibriums are rarely attained because of disturbances from within or outside the community, some of which may be predictable but others not so. These views are not as irreconcilable as they seem if due regard is paid to problems of scale in time and space as summarized in Table 7-3. Because of climatic, geological, and evolutionary changes no community is in equilibrium on geological time scales. In the short term (months, years, perhaps hundreds of years), large communities may be in equilibrium with a dynamic balance of interactions between populations. Similar disparities arise from spatial scaling, since dynamic equilibria are only likely over relatively small areas given the spatial heterogeneity of most environments. At a particular time a large area of forest may appear to be in equilibrium as a whole, and many local populations within it may be at equilibrium, but at an intermediate scale, the relationship between patches in that forest may be continually changing.

Table 7-3   Temporal and spatial scales and the realms of study relating to organism-environment interactions. Ecological equilibria are only likely on short time scales, and we only know about processes likely to produce them on a small scale in space. On longer time scales and on larger spatial scales, something is known about pattern, but little is known about the processes producing those patterns.

TIME SCALE

SHORT ──→ LONG

	ONE YEAR	5–50 YEARS	HUNDREDS OF YEARS	THOUSANDS OF YEARS PLUS
Climate	Seasonal	Fluctuations (of varying predictability)		Paleoclimates (glaciations, arid phases)
Geology	Land-slides	Volcanic eruption Changes in fluvial patterns		Tectonic changes (continental motion uplifting, etc.)
Evolution		Microevolution	Intense coevolution	Speciation Origin of major taxa
Ecology	Population and interpopulation dynamics (pattern and process)		Successions (pattern only in most cases)	Inferences from paleoecology (pattern only)

SPATIAL SCALE

SMALL ──→ LARGE

	INDIVIDUALS AND LOCAL POPULATIONS	COMMUNITIES	BIOMES	CONTINENTS
Climate	Microclimate	Mesoclimate	Macroclimate	
Geology	Local substrate	Geological formations and topography		
Evolution	Microevolution, coevolution, sociobiology	Potential for isolation and speciation		
Ecology	Population and interpopulation dynamics (pattern and process)		Dispersion and abundance patterns (pattern only in most cases)	Inferences from biogeography (pattern only in most cases)

7.2   Temporal Changes and Their Effects on Diversity

7.2.1   Seasonal Time Scales

All terrestrial environments are seasonal to some extent, but the direct effects of variation in temperature and rainfall on species diversity are obscure. As mean annual rainfall decreases, there is a tendency to increased seasonality and decreased predictability of rainfall (section 1.2.1). A parallel observation is that forest tree species richness declines from around 150 species at 3000 mm to 50 species at 1000 mm mean annual rainfall in the neotropics (Gentry 1982). Temperature fluctuations, measured as the annual range, are correlated with species richness in a similar way along a transect from Alaska (65°N) to Panama (8°N) in western North and Central America. In gastropod mollusks and several groups of birds, the number of species decreases as temperature range increases (MacArthur 1975). This trend is also found on tropical mountains, where diversity decreases as altitude increases and temperature fluctuations become more pronounced. However, all aspects of diversity do not show the same climatic trend. The diversity of life forms of plants decreases in areas with little rainfall seasonality, primarily because of the increasing predominance of trees in wet areas (Givnish, in May 1975b). Also, ungulate diversity in African savannas is not related to rainfall reliability (Figure 7-11).

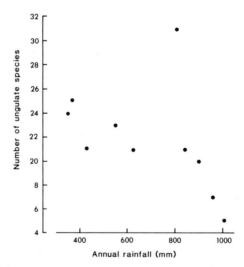

Figure 7-11   High primary production and predictability in East African savannas does not lead to high species richness of ungulates. Annual rainfall is used as an index of predictability and quantity of primary production as established in Chapter 8. (Data on ungulate species from Western and Ssemakula [1981]; data on rainfall from Coe et al. [1976].)

7.2.2   Successional Time Scales

The term succession, as it applies to vegetation, is used to define the changes that occur in plant communities following an environmental upheaval that produces bare ground. The classical scheme envisages a predictable series of communitites replacing one another through time, ending with an equilibrium **climax** community with no further changes. It is now clear that the sequences of species replacements are not as deterministic as was once thought and that climaxes are anything but constant in time or space. Nevertheless, the concept of **succession** is useful if defined broadly to include temporal changes of community structure brought about by interactions between populations and their local environment.

Primary Successions. When both vegetation and soil are absent, usually as a result of geological disturbance, a **primary succession** ensues. One example on a coastal sandy beach has already been described in Figure 7-3a. Other tropical primary successions that have been studied are those following volcanic eruptions. The island of Krakatau, near Sumatra in Indonesia, erupted violently in August 1883 and became covered with pumice and ash to a depth of 30 m. All life was probably destroyed, although a few plant propagules may possibly have survived. The subsequent revegetation of the island is outlined in Figure 7-12. In less than 50 years, species typical of secondary successions on Sumatra were well established and more than 270 species of flowering plants had reached the island.

Some lava flows in Hawaii show similar patterns of succession but with fewer (and different) species and without a distinct grass savanna stage. It has been estimated that these successions take around 400 years to reach a climax (Mueller-Dombois and Ellenberg 1974), which is quicker than most high-latitude primary successions on a rock substrate. Environmental changes during succession make it progressively easier for higher plants to colonize. A soil begins to develop due to erosion and the addition of dead organic matter from the early colonizers. Nutrients are added in rainfall and by the nitrogen-fixing abilities of cyanobacteria and other bacteria. As higher plants become established, many more microhabitats are created, allowing more specialized species to colonize. On Krakatau the reproductive characteristics of the dominant plants changed with wind-dispersed pioneers (including cyanobacteria, ferns, and grasses), followed by those floating over from the mainland and then steadily increasing proportion of animal-dispersed species. As a proportion of the species of flowering plants, woody species dominated throughout the succession, but grasses dominated in abundance between 1900 and 1920.

Primary succession as described above seems to be highly predictable. However, a detailed study of 18 different lava flows in Hawaii paints a different picture (Eggler 1971). The rate of colonization and the species

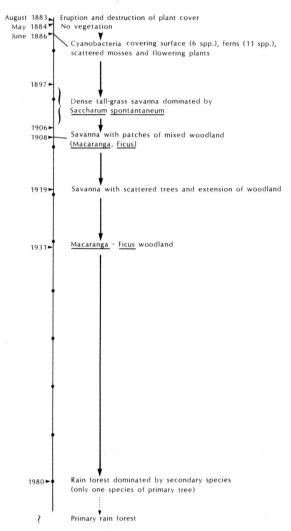

Figure 7-12   Primary plant succession on Krakatau following volcanic eruption. The sequence described is that occurring above the shore zone and below 400 m. (Modified from Richards [1952] and Thornton [1984].)

of pioneers differed markedly on different substrates and with different rainfall regimes. Four 60-year-old flows had 10.9, 6.9, 2.3 and 0 t ha^{-1} of plant biomass with 19, 13, 5, and 2 plant species, respectively. These flows had the same type of substrate, but annual rainfall ranged from 2500 mm to 500 mm in the same sequence. Different pioneer plants colonized different substrates with lichen predominating on the surface of lava flows and only a few cyanobacteria in contrast with Krakatau. Open crevices contained the tree *Metrosideros,* while crevices filled with lava fragments had a variety of ferns, herbs, or trees.

The potential climax vegetation in an area is determined primarily by climate (**climatic climax**) as illustrated by the biome classifications of Figures 7-4 and 7-5. Locally, different subclimax vegetation may persist due to soil or biotic factors. Where soil characteristics are the determining factor, such as the waterlogged peaty soils on which moor forest grow, an **edaphic climax** results. A **biotic climax** is where the community is maintained at an earlier stage in the climatic (or edaphic) succession by the action of such factors as herbivores, humans, and fires. Many savannas are biotic climaxes (see "Arrested Succession in Savannas" and section 8.2.1).

Secondary Successions.   Only small areas of the earth are undergoing primary successions, but most vegetation communities are at some stage (or combination of stages) of **secondary succession,** whether through natural processes of gap formation and refilling in forests or as a result of grazing or fire in savannas. Secondary successions follow similar courses to primary successions, but without the initial stages of colonization on an inorganic substrate and subsequent soil formation.

Gap Regeneration in Forests.   When trees fall, whether through death or external causes (hurricanes, landslips, shifting cultivators), a gap in the canopy is created that will eventually be filled by other trees. Treefall gaps in an otherwise undisturbed Panamanian forest vary in size from 20 m² to 360 m² (Brockaw 1982), but larger gaps may be produced by the processes listed previously. The size of a gap has an important effect on subsequent events, with smaller ones often being filled by surrounding canopy trees. Larger gaps lead to the establishment of pioneer trees such as *Cecropia* and *Ochroma* in the neotropics, *Macaranga* and *Musanga* in Africa, and *Macaranga* in Southeast Asia. These pioneers regenerate from dormant seeds, although such seed banks are smaller than in temperate forests. Most tropical canopy trees produce seeds that rapidly die and decay. Their regeneration is usually dependent on the arrival of fresh seeds from the surrounding forest. Regeneration from coppice shoots and root suckers may also be significant, as Stocker (1981) has shown in an Australian rain forest.

Subsequent events leading to the recreation of a closed canopy are extremely variable, with progressively larger (but slower growing and shade-tolerant) species eventually growing taller than pioneers and contributing to their death by shading. An example of a five-year period at the beginning of a forest secondary succession is shown in Figure 7-13.

Arrested Succession in Savannas.   Most savannas are not climatic climaxes. They would often develop different vegetation cover if fire and grazing were excluded. In many cases the cover of woody plants would increase significantly, in some to produce a closed canopy forest, in

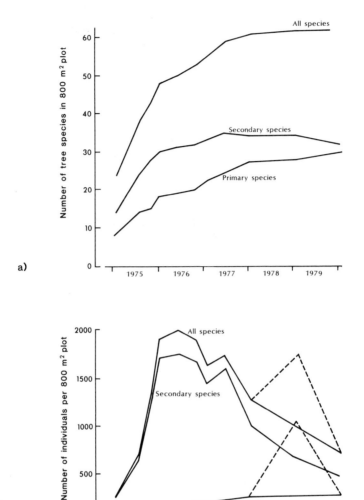

Figure 7-13   Early secondary succession on cleared evergreen forest land in Ghana. The events described are for an 800 m² plot cleared in 1974. (Data from Swaine and Hall [1983].) *a,* Change in the number of tree species present; primary species are those able to germinate in shade, while secondary species cannot. Even in these early stages many primary species are present. *b,* Number of individual trees present. Secondary species predominate initially, with primary species gradually increasing in density. The total numbers decrease as the trees get bigger and secondary individuals thin out. The dotted line represents a flush of germinating *Albizia zygia* seedlings, most of which later died. This flush emphasizes the occurring of unpredictable events in succession.

others to produce a woodland or a thorn scrub. Regeneration of these plants is prevented because growing shoots are consumed by animals or fire, both of which are often connected with human activities. Examples of the effects of removing herbivores or preventing fires are shown in Figure 7-14 and Table 7-4, respectively. The two factors are closely inter-related in determining the type of savanna community in a given area, and altering the intensity of either produces changes that are not always easy to predict (see section 8.2).

Cyclical Changes in Vegetation.   Secondary successions are repeated to some extent as new disturbances occur, but some vegetation types are cyclic in a much more predictable manner because of inherent characteristics of the species involved. Several examples are known from the temperate zone. Recently, a comparable pattern has been described in a saline coastal community in Sri Lanka. *Anthrocnemum indicum* is a perennial shrub with both prostrate and upright stems that trap windblown sand to form hummocks. These hummocks eventually erode due to vegetation changes, allowing the building phase to begin again (Figure 7-15).

Heterotrophs and Succession.   Little is known about microfloral changes during tropical plant successions. In the early stages of primary successions, dead organic matter often increases as soil development proceeds. This organic accumulation indicates that active decomposer

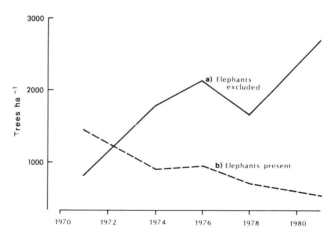

Figure 7-14   The change in tree density in semiarid savanna at Ndara, Tsavo National Park, Kenya. — = Plot protected from elephants; - - - = adjacent control plot. (Data from Oweyegha-Afundaduula [1982].)

Table 7-4   Experiments on fire-climax savannas in Olokemeji Forest Reserve (Southwestern Nigeria). Plots were cleared in 1929 and subsequently subjected to different fire treatments. The results are from a survey in 1957 and show that complete protection causes reversion to climatic climax forest. Early burns (less intense, at the beginning of the dry season) allow development of a wooded savanna with fire-resistant clumps of trees, while late burn (more intense, at the end of the dry season) allows only a few fire-resistant trees to grow. Data from Hopkins (1974).

	NO BURN	EARLY BURN	LATE BURN
Vegetation structure	Trees + shrubs, few grasses, forest succession	Savanna with clumps of closed canopy woodland	Grass savanna with scattered trees
Number of trees in 0.17 ha	433	163	98
Number of fire-sensitive trees in 0.17 ha	279	11	0

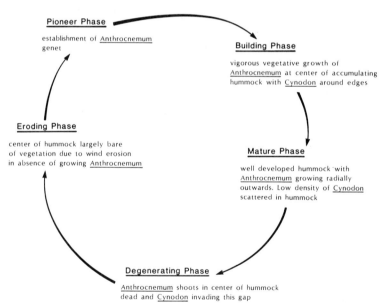

Figure 7-15   Endogenous cyclical changes in coastal vegetation in Sri Lanka. The shrub *Anthrocnemum indicum* traps sand, which causes hummocks of approximately 100 cm diameter and 20 cm height to form. These are invaded by the grass *Cynodon dactylon,* but both species decline as the hummocks are eroded from the center. (Modified from Pemadasa [1981b].)

populations are smaller than in the climatic climax when all dead organic matter is usually decomposed. Nevertheless, 23 years after the eruption the density of soil bacteria on Krakatau was comparable with many Javanese soils. All major groups were present, including those of importance to the nitrogen cycle (Ernst 1980). Mycorrhizal populations also increase during many successions. They are often absent from pioneer plants, but obligatorily associated with most lowland forest canopy trees (Janos 1980).

Most animals are dependent on plants, both as the base of food webs and for the creation of habitats. Therefore, they are often tied to successional changes in plant communities. Table 7-5 shows the invertebrate communities associated with different stages of a savanna succession in Central Africa. Overall density is similar in all stages, but taxonomic and trophic composition changes increasing regional diversity of the area as a whole.

Heterotrophs also undergo successions independently of plant species. For example, different decomposer populations predominate at various stages in the decay of organic remains. Similarly, the arthropod

Table 7-5   Density of herb-layer arthropods in different stages of an arrested savanna–to–woodland succession in Upper Shaba, Zaire. The data (individuals per 100 m^2) are from the wet season following a dry season fire in the savanna and one of the woodland sites. Data from Freson et al. (1974).

		SAVANNA	MIOMBO	WOODLAND
			BURNED	UNBURNED
HERBIVORES				
Chewers:	Short-horn grasshoppers*	342	124	118
	Long-horn grasshoppers*	80	29	29
	Crickets	66	190	164
	Caterpillars (larvae of moths and butterflies)	52	34	50
Suckers	Hompoteran bugs	69	71	58
CARNIVORES				
	Spiders	176	257	308
	Mantids	57	29	37
TAXA WITH MIXED FEEDING HABITS				
	Adult beetles	283	74	87
	Cockroaches	109	123	203
	Heteropteran bugs	60	86	208
Totals		1294	1017	1262

*The short- and long-term grasshoppers were the predominant groups within the Acridoidea and Tettigonoidea, respectively, which were each counted as a whole.

fauna of developing papyrus flower heads *(Cyperus papyrus)* changes with time. The buds have only a few insects, mainly sap-sucking bugs, but the mature heads have six times as many arthropods, most of which are ants and spiders (Sutton and Hudson 1981). These heterotrophic successions differ fundamentally from plant (autotrophic) successions because the end result is the disappearance of the community rather than any parallel with a climax. The persistence of such communities is entirely dependent on the creation of new colonization sites.

Succession, Disturbance, and Diversity.   The preceding review of succession shows that the process leads to high diversity in two ways. First, climax communities are more diverse than pioneer communities. Second, disturbance is a normal feature of communities, with many stages of secondary successions present simultaneously. Each stage has a characteristic mixture of species (or set of possible mixtures), leading to an increase in diversity of the community as a whole. Indeed, natural communities are a mosaic of successional stages progressing by various routes, rather than a definable climax, as shown conceptually in Figure 7-16.

Connell's (1979) "intermediate disturbance hypothesis" suggests that the high diversity of tropical communities with long-lived sessile

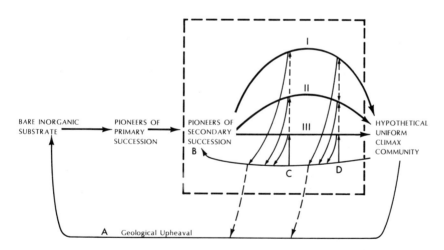

Figure 7-16   Relationship between succession, disturbance, and diversity of communities. Most communities are a mosaic of successional stages as outlined by the box of heavy broken lines. Various levels of disturbance (A, B, C, D) cause reversion to earlier successional stages. The heavy lines (I, II, III) represent various routes, progressing at different rates toward the climax. These may be deflected to an earlier stage at any point by further disturbances. High diversity results from a mosaic of successional stages proceeding by various routes and subjected to disturbances of various intensities.

dominants (rain forests and coral reefs, especially) results from the scale of disturbances in time and space. Many virgin rain forests have a low density of mature large trees as a result. The turnover time of gaps has been estimated as 80–140 years in Costa Rica, faster than is common in temperate forests (Hartshorn 1978). Rapid turnover is also evident in the natural leveling of pits and mounds caused by fallen trees. In a Panamanian rain forest, this takes only 10 years on average, compared with several hundred years in many temperate forests (Putz 1983). Attributes of vegetation structure in tropical forests may lead to intermediate rates of disturbance and gap formation, which enhance diversity. A possible mechanism results from the abundance of epiphytes (notably lianas) in tropical rain forests. A heavy burden of lianas seems to increase the mortality rate of trees that bear them. When such trees fall, they often bring down neighboring trees because of interconnections produced by the lianas (Strong 1977, Putz 1984).

Exogenous disturbances such as hurricanes and landslides are also more frequent in some parts of the tropics. The importance of hurricanes in maintaining high diversity in a montane rain forest in Puerto Rico has been demonstrated by Doyle (1981). Using a simulation model, he found that when hurricanes were not included, the predicted tree diversity was much lower than that observed in the forest. When the effects of hurricanes, which occur once every nine years on average, was included, the predicted diversity closely matched that of the forest. A high proportion of rain forests are in regions of high seismic activity, ranging from 38% in the Indo-Malayan region, through 14% in tropical America, to less than 1% in Africa (Garwood et al. 1979), but it is also true that many diverse rain forests appear not to be greatly influenced by such factors. Recently research in Amazonia shows that hydrological changes and fires may be important agents of disturbance, creating new successional mosaics every few hundred years. Newly discovered lakes in the lowlands of Peru have patterns of pollen deposition and sedimentation consistent with occasional flooding and changes in drainage patterns over the last few thousand years (Colinvaux et al. 1985). In southern Venezuela, layers of charcoal in soil cores indicate that fires have spread through the rain forest from time to time. Human causes are unlikely in some cases because the earliest charcoal deposits are dated before the period in which shifting cultivators are thought to have been present. Timing of the fires may coincide with somewhat drier climes than present, but postfire successions cannot be discounted even in the wettest of tropical forests (Sandford et al. 1985). In summary, there are many established causes of local disturbance in tropical forests. At present their relative and absolute importance as promoters of diversity is not known.

The local diversity of the herb layer of tropical savannas is probably no more diverse than in some temperate grasslands (for example, see Ricklefs 1977, McNaughton 1983). Both are subjected to similar disturbances—fire, grazing and unpredictable rainfall. However, the diversity of trees and shrubs is sometimes very high in Africa, probably because no species is able to achieve dominance in the face of repeated disturbance. African savannas are renowned for their diversity of large mammals, which is related to high habitat diversity, which in turn results from repeated disturbances to the vegetation.

7.2.3   Geological and Evolutionary Time Scales

Time scales involving significant geological and evolutionary changes stretch from a few years to many thousands (Table 7-3). Continental movements have profoundly affected the species, genera, and families that have evolved in regions isolated from one another at different times. Two hundred million years ago there was a single land mass, Pangaea. It later split into the northern Laurasia (North America, Europe, and northern and eastern Asia) and the southern Gondwana (South America, Africa, India, Australasia, and Antarctica). Twenty to fifty million years ago, Africa and India rejoined with Eurasia, and Australia split from Antarctica. The link between North America and Central plus South America did not form until about three million years ago. One of the many effects of these splits and connections is the predominance of trees from one family the Dipterocarpaceae in rain forests of Southeast Asia, where more than 500 species have arisen. In contrast many families contribute to the canopies of rain forests elsewhere.

Until recently, it was often proposed that high tropical diversity resulted from a long period of constant and benign climate that allowed time for niche diversification and specialization (see section 7.4.1). Meanwhile, higher latitudes were experiencing repeated glaciations, which meant that there was insufficient time for communities to become saturated with specialized species. An illustration of the importance of time for evolutionary diversification is the higher diversity of specialized woody plants, birds, and large mammals in African savannas compared with the much younger savannas elesewhere. However, it is now apparent that tropical regions have also experienced large climatic variations with humid and dry periods corresponding to warm and cold periods at higher latitudes. These variations have led to expansions and contractions of tropical forest vegetation. The present period is rather moist, with widespread forest. However, 20,000 years ago it seems that there were only small, isolated forest **refugia** surrounded by the dry country vegetation of savannas and deserts. Convincing evidence comes from paleoecological analyses of hydrology, soils, and pollen from the last 2.5 million years as well as from the present distribution of species. Pollen records from montane regions, fluctuations in lake levels and flooding patterns of rivers, and the apparent spread of desert sands combine to provide a coherent time scale for these climate and vegetation changes (Hamilton 1982, Prance 1982). There are dissenting voices, but this **refugium theory** is now widely accepted by paleoecologists. Figure 7-17 shows the effect of proposed forest contraction and expansion in equatorial Africa on bird diversity during and since the most recent glaciation in Europe.

Past forest refugia represent areas of present high diversity. Spread of populations from these areas is slow and may not be complete. It is possible, therefore, that the areas that were refugia may be saturated with species (equilibrium diversity), while surrounding forest areas and postglacial temperate areas have not yet become saturated. In theory, diversity could still increase in these latter areas (but is unlikely in practice because of human encroachment). The refugium theory also has important implications for conservation (see Chapter 11). However, the theory poses problems for ecologists who assume that most

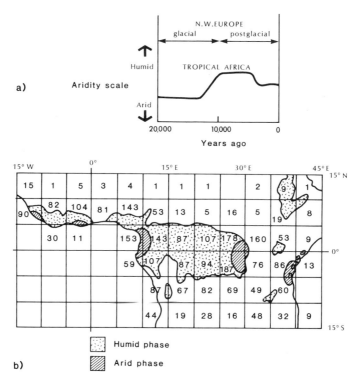

Figure 7-17   Spread of forest and forest passerine birds in relation to climatic changes during the Quaternary in equatorial Africa. (Modified from Hamilton [1982] and Diamond and Hamilton [1980].) *a,* Aridity variation in tropical Africa related to glaciation in northwest Europe. *b,* Proposed forest refugia 18,000 years ago (hatched area) compared with present extent of forest (stippled area). Numbers in grid squares are present numbers of forest passerine species recorded. Note that greatest species richness is found in areas of or adjacent to refugia, with strong gradients away from these areas.

communities exhibit an equilibrium diversity mediated by the balance between speciation and extinction (see, for example, Rosenzweig 1975). It also makes inessential the proposals relating high speciation or low extinction rates to high diversity (see below), although it does not preclude these processes as contributory factors.

Shorter-term evolutionary interactions are undoubtedly a major feature of tropical diversity. The prevalence of plant defenses has led to specialization in some herbivores, and this may have precipitated increased speciation through coevolution, although direct evidence is lacking (section 6.3.1). The common development of mutualistic relationships between populations and the diversity of ecological strategies found are certainly important contributory factors to high diversity as discussed in detail in Chapter 6. Mutualism, particularly when coevolution is implicated, are only likely to develop if the populations involved are in close contact over many generations. If tropical climate and vegetation changes were less drastic than those of the temperate glaciations, as seems probable, groups of species are more likely to have remained together. Such a proposal helps to explain the greater incidence of mutualism (and higher diversity) in the tropics (see section 5.3).

Faster rates of genetic change could produce more rapid divergence between populations and more frequent speciation if population turnover is faster in the tropics. However, the seasonality of breeding and tendency to lower reproductive rates make this unlikely despite the year-round potential productivity of the environment (see section 4.1.4). Species are, by definition, reproductively isolated from one another. For speciation to occur, it is generally agreed that a period of geographical isolation is necessary for most kinds of organism. It is possible that geographical isolation is more easily achieved in tropical populations because they are more sedentary. For example, the dipterocarp trees of Southeast Asia are able to disperse seeds over very short distances and cannot cross even minor geographical discontinuities (Ashton 1969). Even mobile organisms such as birds often have ranges strictly curtailed by major rivers, which would present no barrier to their temperate cousins (Mayr 1969). Janzen (1967) has suggested that mountain passes of comparable altitude form more effective dispersal barriers than in temperate regions. Put very simply, he argues that tropical lowland species are not adapted to cold weather, whereas temperate species are. Finally, the refugium theory provides an ideal scenario for speciation with repeated geographical isolation of small forest patches for long periods.

Fewer extinction due to climatically benign environments has been proposed as a contributory explanation for high diversity in the humid tropics. However, the biotic environment in such areas is anything but benign. Predation rates may well be higher, and in predictable environments, K-selection (with its implication of intense competition) tends to predominate (sections 5.4.1 and 6.2.1). Also, higher diversity means that many species are at low density, making their extinction more likely than high-density populations on a purely probabilistic basis.

7.3   Spatial Heterogeneity and Diversity

At the level of regional diversity, the presence of more biomes in the tropics will produce greater species richness even if each individual biome contains no more species than those at higher latitudes. (In fact most tropical biomes have more species as amply demonstrated in this chapter). The biome classification of Holdridge et al. (1971) shown in Figure 7-5 indicates that there are indeed more biomes in the tropics because of the greater range of climatic conditions suitable for life.

At a smaller scale, within a community, spatial heterogeneity undoubtedly increases spatial diversity. Karr and Roth (1971) show clearly that the diversity of birds increases as the vertical complexity of vegetation increases from grassland to forest. They also compared Panamanian bird communities with those in the temperate United States. In terms of its use by birds, the heterogeneity of vegetation was not greater in the tropics, but nonetheless, diversity of birds was greater in the tropical forests. In this case it seems that spatial heterogeneity had, at most, a small role in producing greater diversity in the tropics. In contrast, Ricklefs (1977) proposes that the abiotic environment in tropical forest gaps is more heterogeneous than in temperate areas, with steeper gradients of radiation, temperature, and humidity and patches of fertile soil resulting from rapid decay of the fallen trees. Small-scale variations in soil characteristics account for some of the observed pattern of tree dispersion in a Brunei rain forest, although much more evidence is needed to establish this as a primary cause of forest diversity (Austin et al. 1972; see also section 7.4.1).

As more plant species are added to a community, spatial hetero-geneity inevitably increases, both for other plants and for hetero-trophs, because new habitats are created. This introduces an impor-tant subdivision of local species diversity on different spatial scales. Within habitat or α diversity is the diversity within a particular habitat, while between habitat or beta (β) diversity, is the number of habitats within a particular community. The difference is illustrated in Table 7-6 by comparing equally diverse communities with different degrees of α and β diversity. Both undoubtedly contribute to high diversity in topical communities as illustrated in Table 7-7, but it is probable that high β diversity predominates as in Table 7-7a and the following examples. Tropical trees exhibit 21 "architectural models" (growth forms) above-ground and 23 in their root systems. Only a few of these are found in temperate trees (Jenik 1978, Tomlinson 1978). This diversity of archi-tecture greatly increases spatial heterogeneity and niche space for the

Table 7-6   Within and between-habitat diversity. High diversity in the tropics could result from (a) more species occurring in each habitat; (b) more habitats, each containing the same number of species; or (c) a combination of both.

	TROPICAL	TEMPERATE
(a) High within-habitat diversity		
Average number of species per habitat	50	10
Number of different habitats	10	10
Total number of species	500	100
(b) High between-habitat diversity		
Average number of species per habitat	10	10
Number of different habitats	50	10
Total number of species	500	100
(c) Combination of both types of diversity		
Average number of species per habitat	20	10
Number of different habitats	25	10
Total number of species	500	100

epiphytic flora and the fauna. Large tropical river systems that cause seasonal flooding also create more distinct habitats for birds than do comparable temperate rivers. Of terrestrial bird habitats, 15% are river based in the forested lowlands of Amazonia and 6% in the Congo Basin compared with 2% in the Mississippi Basin of the United States (Remsen and Parker 1983). β diversity is also of great significance in African savannas. The herbaceous vegetation of the Serengeti, Tanzania, shows greater diversity at this level than temperate grasslands even though α diversity may be similar (McNaughton 1983). East African savanna ungulates show their greatest diversity in varied landscapes. Where there is a mixture of vegetation structures, rivers, and rocky outcrops, both widely ranging species and those more restricted to specific habitats are represented (see Figure 8-11).

Spatial heterogeneity is difficult to separate from many other promoters of diversity such as succession, evolutionary change, and species packing, but it should not be underrated as a contributory factor.

7.4   Species Interactions and Diversity _____

7.4.1   Competition and Species Packing

Gause's hypothesis implies that populations of different species with the same resource requirements cannot coexist indefinitely and that this has led to ecological segregation of resource utilization (section 5.1.2). An important question following from this is how dissimilar do populations have to be to coexist? Hutchinson (1959) attempted an

Table 7-7 Comparisons of α and β diversities in tropical and temperate communities. In (a) α diversity is similar in comparable habitats, but β diversity is greater in the tropics. Data from Stanton (1979), where methods of calculation are given. In (b) β diversity is similar (the extra habitat in Costa Rica and Brazil is high canopy), but the α diversity is greater in the tropical communities. Data from Jeanne (1979).

(a) LITTER-INHABITING MITES		SUCCESSIONAL FIELDS	BROAD-LEAF FOREST	PINE PLANTATION
TROPICAL (COSTA RICA)				
α diversity:	Species richness	6	13	27
	Shannon-Wiener index	1.23	2.13	1.80
β diversity		8.8	6.3	4.0
TEMPERATE (WYOMING, UNITED STATES)				
α diversity	Species richness	4	15	14
	Shannon-Wiener index	0.96	2.00	2.02
β diversity		7.3	3.7	2.4

(b) ANTS ATTRACTED TO BAITS	UNITED STATES			COSTA RICA	BRAZIL
	NEW HAMPSHIRE	FLORIDA	MEXICO		
β diversity (number of habitats)	5	5	5	6	6
α diversity (average number of species per habitat)	8.2	11.6	17.2	16.5	17.1

answer by comparing the size of food-collecting apparatus (such as bill length in birds) in pairs of closely related sympatric vertebrates and found that the limiting ratio of similarity (larger : smaller) averaged 1.3. He proposed that this figure sets the approximate limit to which competing species can be similar without one species displacing the other. This ratio in turn sets a limit to the number of species that can, through ecological segregation, be **packed** along a resource base of finite dimensions. Such groups of populations utilizing the same type of resources in similar ways are called **guilds**. Two examples of ecological segregation and guild formation in animal communities are presented in Figures 7-18 and 7-19.

There are many examples of vertebrate guilds, in particular, that are apparently structured by competition, but for this to produce greater diversity in the tropics requires that species be packed more tightly than at higher latitudes. Comparative studies on similar guilds at different latitudes suggest that this may not be the case. For example, tropical

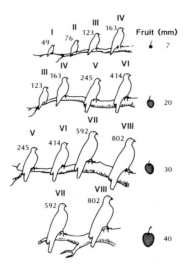

Figure 7-18   Niche relationships in a guild of eight fruit-eating pigeons in lowland rain forests of Papua New Guinea. Ecological segregation occurs in fruit size on different trees and in branch size, with smaller species being able to feed and perch on smaller branches. The species are indicated by roman numerals and size is given by body weight (grams). The average size ratio between successive pairs is 1:5. (From Diamond, J. M. 1973. *Science* 179:767. Copyright 1975 by AAAS.)

moths are more diverse in appearance than temperate moths in the Americas because of greater total niche space rather than finer subdivision of that niche space (Ricklefs and O'Rourke 1975). Similarly, a comparison of arthropod guilds on trees in South Africa and Britain shows a comparable range of species packing in both places (Southwood et al. 1982). However, a study of reptile and amphibian communities in Thailand found a greater degree of species packing in more predictable climates. Three areas were compared that decreased in climatic variation in the order evergreen forest, deciduous forest, agricultural land. Niche dimensions studied were habitat features of the type suggested in section 7.1.2. The evergreen forest had 137 habitats, the deciduous forest had 83, and the agricultural land had 59. Twenty-six of these were common to all three sites and contained 64, 57, and 46 species respectively, indicating the greatest packing in the least variable environment. A detailed statistical analysis showed that the niche breadth was similar in the two forests but that niche overlap and guild formation were most pronounced in the evergreen forest (Inger and Colwell 1977).

    Although they accept that guild structure may often be controlled by competition in vertebrate populations, Lawton and Strong (1981) have argued that this is rarely the case in herbivorous insects. They suggest that climate, plant phenology (including seasonal changes in chemical and physical availability, patchiness, and sometimes predators

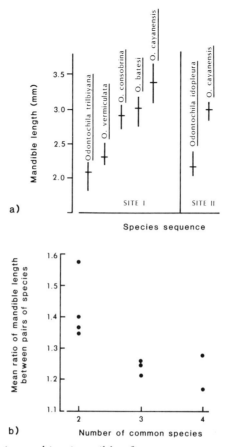

Figure 7-19   Species packing in guilds of congeneric tiger beetles. *a,* Mandible length (which is correlated with size of arthropod prey) is compared at two upland forest sites in Ecuador: (I) Rio Palenque and (II) Limoncocha. *O. batesi* was rare, which may account for its extensive overlap with two other species. Horizontal bars are means and vertical bars are ranges of sizes for 20 beetles of each species. (Modified from Pearson [1980].) *b,* As more species are packed into a guild, the differences between them is expected to decrease. Tiger beetles from eight moist upland forest guilds in central and south America and one in Papua New Guinea are charted. As the number of common species increases, the mean ratio mandible length between pairs of species decreases. (Data from Pearson [1980].)

combine to prevent these insects from competing in most instances. Among the supporting evidence is a study of the insect larvae that inhabit the water-filled bracts of *Heliconia* plants in Costa Rica. In 14 interactions between pairs of species from the 24 studied, no competition could be detected. Indeed, some of the interactions were mutualistic, a factor known to be of importance in tropical communities (Siefert and Siefert 1976).

Much less is known about niche specialization in rain forest trees, since so little is known about the precise ecological requirements of the myriad species. Given the frequent occurrence and variation of size of gaps in the canopy, Denslow (1980) suggested that ecological segregation occurs in terms of requirements for seedling establishment and the probability of propagules reaching a gap of appropriate size. Certainly, her pantropical survey shows that many species specialize in large or small gaps and that propagule size and density will produce differing dispersal patterns. Nevertheless, as Whitmore (1982) points out, it seems almost inconceivable that in very diverse forests each species has a well-defined and separate niche. Even in the relatively species-poor montane forests of Jamaica, Tanner (1982) was unable to find sufficient niche differentiation to explain the 56 species of trees present. Perhaps many species have more-or-less overlapping niches and there is an element of chance about which species actually establishes in any particular gap. A problem with this proposal is that rare species would have a high probability of extinction, thereby gradually reducing diversity.

Coexistence of many competing species in heterogeneous environments is predicted by Tilman's (1982) theory of resource use by plants (see section 5.3). According to this theory such coexistence is possible where several limiting resources are required in different ratios by the different species. A specific prediction, that diversity is greatest in moderately resource-poor environments, is circumstantially supported by the high diversity of tropical rain forest trees on nutrient-poor soils in Australia, Malaysia, and Costa Rica.

7.4.2   Predation and Diversity

Experiments, particularly with seashore invertebrates, have shown that predators can increase prey diversity by preventing a few species from becoming dominant and excluding the others (reviewed by Connell 1975). Similar effects are found in the interactions between vertebrate herbivores and their herbaceous plant prey. For example, after almost 15 years of protection from large grazers, grasslands of the Serengeti savannas in Tanzania were less diverse than comparable grazed areas (McNaughton 1979).

A widely quoted proposition is that tree diversity in tropical forests results from seed predation. It is hypothesized that specialized seed-eaters concentrate around their host tree and eat all seeds falling in the vicinity. As a result, seeds falling further from the tree have a higher probability of survival and mature trees of a particular apecies are widely spaced (Figure 7-20). Supporting evidence comes from the attack rates of bruchid beetles on rain forest palm seeds and the high density of surviving seedlings at intermediate distances from the parent tree (Wright 1983). The effects of fungal pathogens on seedling survival, rather than predation on seeds, may produce similar results. For example, mortality

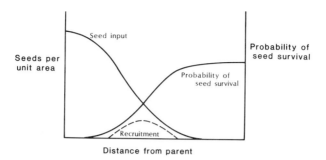

Figure 7-20   Seed predation and the relative spacing of tropical forest trees. Janzen's (1970) hypothesis is that trees are only recruited at an intermediate distance from the parent tree because of the balance between distance to which seeds are dispersed (seed input) and the probability of seed survival in the face of predation.

of seedlings of the canopy tree *Platypodium elegans* in Panama decreases with distance from the parent tree (Augspurger 1983). However, Hubbell (1980) contends that the statistics of seed predation and seedling survival do not produce a pattern of widely spaced conspecifics and do not necessarily enhance diversity. Indeed, it is known that many rain forest tree species exhibit clumped dispersions to some extent. Most work to date does indicate that seed or seedling survival is decreased by close proximity to a mature tree of the same species or by being in a dense aggregate of conspecific seeds or seedlings. Unfortunately, very few of these studies provide conclusive evidence for the pattern illustrated in Figure 7-20 (see Clark and Clark 1984).

A similar effect has been proposed as an enhancer of ant diversity. On Barro Colorado Island, Panama, swarms of nomadic army ants periodically clear areas of forest floor of their preferred species of ant prey. This clearance allows the establishment in the gaps of nonprey ants, which would otherwise be displaced (Franks and Bossert 1983).

7.5   Does High Primary Production Lead to High Diversity?

A clear difference between comparable tropical and temperate communities is that the tropical communities usually have greater primary production (Table 2-1). Can this greater supply of energy, ultimately from sunlight, lead to enhanced plant and animal diversity? Tree canopies in humid lowland tropical forests are deeper than their temperate counterparts, with several layers often being recognized. The perpendicular angle and greater intensity of sunlight allow trees with a greater range of heights to coexist in competition for light. However, greater canopy depth is unlikely to explain more than a little of high tropical diversity and is not applicable to nonforest communities. Diversity increased as production of the herb layer increased in a Zimbabwe savanna, the reverse of a trend found in a temperate grassland (Kelly and Walker 1976). At a higher trophic level, the spectacular diversity of herbivorous ungulates in African savannas is not related to primary production (Figure 7-11; note that rainfall is closely correlated with primary production).

The continuity of production throughout the year in the humid tropics is more significant than total production. This factor explains much of the high diversity of tropical birds in forests, since they use resources unavailable in temperate forests. Specialized fruit-eaters are common, making up 10%–20% of the tropical forest avifauna. Although most trees produce fruit seasonally, they do so at different times of the year, which means that some fruit is available at all times unlike in temperate forests (Karr 1971; see Figure 7-1 for a similar situation in mammals). Another niche not available in temperate areas but used by tropical birds is insectivores that follow and steal prey from army ant swarms. There are also birds that capture large insects from leaves and branches while hovering and that spend much of their time searching for prey from convenient perches. This group comprises 12%–33% of species in Costa Rican forests, but only 0%–15% in temperate American forests (Orians 1969). Such specialization is possible because of the year-round presence of these insects and because foliage insects are usually larger at low latitudes and are, therefore, more profitable food items (see Schoener and Janzen 1968, Southwood et al. 1982).

7.6   Diversity on Islands

As living laboratories, tropical oceanic islands fascinate many biologists. Most notably, Charles Darwin's observations in the Galápagos archipelago provided key insights into speciation. The *Theory of Island Biogeography* by MacArthur and Wilson (1967) has stimulated a later generation of ecologists to study islands with a view to understanding more about diversity and community organization. The emphasis is on oceanic islands, but the same principles may be applicable to isolated communities on continents such as the tropical alpine flora (see section 8.2.2). Some conservation areas can also be likened to islands in a sea of human activity (section 11.5).

7.6.1   The Equilibrium Theory of Island Biogeography

The equilibrium theory, developed largely by MacArthur and Wilson (1967), stems from the observation that islands have predictable species richness depending on (1) the area of the island (Figure 7-21), and (2)

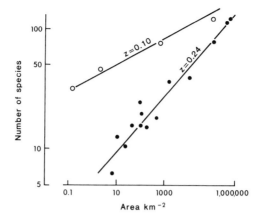

Figure 7-21 Relationship between island area and the number of terrestrial mammal species in southeast Asia. The Sunda Islands (Borneo, Sumatra, Java and adjacent smaller islands), represented as • , show a steeper decline than ∘, a series of forest reserves on the mainland of Malaysia, indicating that the latter may not be truly isolated or that the number of species has not declined to a new equilibrium after their relatively recent isolation. The fewer species for a given area of island results from extinctions during the last 10,000 years. Before that most of the islands were connected to the mainland. (Data from Heaney [1984].)

distance from the mainland (Figure 7-22). Number of species (S) is generally related to island area (A) by the formula $S = CA^z$, or $S = \log C + z \log A$, to give the straight line on logarithmic coordinates (Figure 7-21). C is a constant for a particular taxon in a particular location, and z is the slope of the line relating species richness to island area. A wide range of studies indicate that z is usually in the range 0.2 to 0.4. The relationship is of the same type as that between species number and sample size (see Figure 7-10 drawn on arithmetic coordinates).

Equilibrium theory predicts species number on an island from its size and distance from a colonizing source as depicted in Figure 7-23. The equilibrium species richness on any island is determined by the balance between the rate of immigration to and the rate of extinction of species on the island. Immigration rate is determined by distance from a colonizing source and extinction rate, by the area of the island. While larger islands have more spatial heterogeneity, the relationship seems to hold even for ecologically homogeneous islands. Experiments with mangrove islands in Florida illustrate this. The vegetation is fairly uniform, but islands that are experimentally subdivided lose species. For example, an island of approximately 1000 m^2 had around 75 species of foliage arthropods, but when reduced to 300m^2, it lost 20% of those species. (Simberloff 1978). A confirmation of the shape of the curves

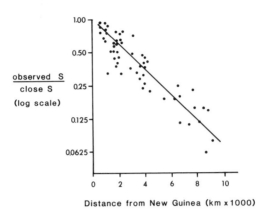

Figure 7-22   Number of species of resident land and freshwater birds on islands of the southwest Pacific. Islands close to a source of colonization have more species than those further away. The ordinate is the number of species (observed S) on each island divided by the number on an island of equivalent size close to New Guinea (close S). (Modified from Diamond and May [1981].)

in Figure 7-23 comes from studies of the avifauna of the Melanesian islands (Gilpin and Diamond 1976).

In terms of moving toward an equilibrium species richness after their creation, two different types of island occur. The principles of the subsequent maintenance of diversity are the same in both. True oceanic islands have never been connected to a mainland. They arise by geological events such as the eruption of Krakatau and are then colonized (section 7.2.2). In contrast, land-bridge islands, such as the Sunda Islands of Figure 7-21, result from fragmentation of a former mainland area caused by a rise in sea level. Initially land-bridge islands are "supersaturated" with species. Through time, some species become locally extinct until a new island equilibrium species richness is attained.

The changes in species composition on islands can be large. New species arrive continually, but few become established, even with repeated immigration. On a sandy cay 650 m² in area and 1 km east of Puerto Rico, 51 species of plants have been recorded, but only seven of these are established. Local extinctions result from lack of suitable habitats, hurricanes, and erosion (Heatwole and Levins 1973).

Whether a species reaches islands depends on its ability to disperse across oceans. Flying animals are able to cross all oceanic gaps, but large terrestrial vertebrates have not crossed more than 80 km and small terrestrial mammals not more than 800 km of ocean (Gorman 1979). The flora of remote islands reflects this pattern to some extent, with most of the plants dispersed by birds and very few by other animals or wind. Whether a species establishes or becomes extinct depends on suitable abiotic conditions, species already present (prey, mutualists,

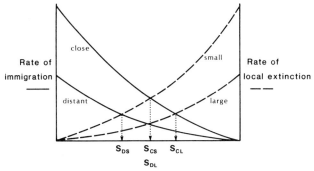

Equilibrium number of species on island

Figure 7-23   The equilibrium theory of island biogeography as proposed by MacArthur and Wilson (1967). The equilibrium number of species, S, represents the balance between immigration and local extinction rates, which themselves vary with island size and distance from colonizing sources. Small distant islands have few species (S_{DS}), large close islands have many species (S_{CL}), but close small (S_{CS}) and large distant (S_{DL}) islands have intermediate species richness.

competitors, predators), and chance extinction while the population is small. The debate between ecologists as to whether interspecific competition is the primary determinant of community structure is as relevant to islands as other communities. Many vertebrate ecologists contend that species composition on islands is largely predictable from the species characteristics of dispersal, life history pattern, and competitive abilities. Substantial evidence for this point of view has been collected for birds of Oceania and the lizards and birds of the Bahamas in the Caribbean (Diamond 1975 and Schoener and Schoener 1983, respectively). In contrast, some invertebrate ecologists suggest that while species richness and abundance may follow predictable patterns, the particular species mixture on a given island is largely determined by chance immigrations and extinctions (see Simberloff 1978 and Figure 7-9).

7.7   Conclusions About Species Diversity

It should be obvious from the preceding discussion that many of the mechanisms proposed to explain high diversity overlap and interact. Certainly no single factor is wholly responsible, and few factors are mutually exclusive, particularly when a range of communities is considered. Figure 7-24 shows how many of the factors discussed in sections 7.2 to 7.5 are interrelated and can combine to increase diversity.

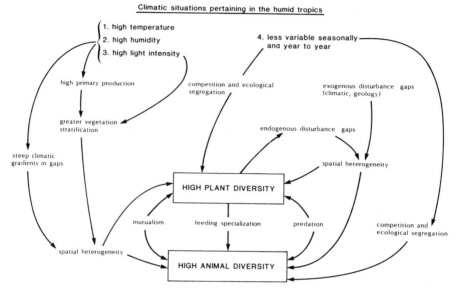

Figure 7-24   Many factors may interact to produce high species diversity in a moist tropical forest. Arrows represent "opportunities" for increased diversity rather than proven relationships (see Table 7-8). Many of these interactions also apply to other types of community but possibly to lesser degrees. Long-term interactions are omitted in this figure. (Based in part on Price [1975].)

However, all factors are not of equal importance and Table 7-8 attempts to rank them with respect to available evidence. The first column ranks in terms of likelihood of generating high diversity per se and the second is in terms of suggestive evidence for high tropical diversity in particular. While good evidence exists for seven of the factors leading to increased diversity, only four of these are supported by firm evidence as promoters of high diversity in the tropics. (It must be stressed that these rankings are based on the strength of evidence available now, and this does not preclude the possibility of new evidence substantiating the importance of other factors.) These four factors in combination may well be sufficient to explain high tropical diversity, particularly in animal communities. While year-round production and secondary succession probably have only additive effects on total diversity, spatial heterogeneity at both the biome and habitat levels is likely to be multiplicative. Furthermore, if coevolution of plants and herbivores as outlined in section 6.3.1 occurs, such a process could lead to a geometric increase in species richness through time.

Whether the ratings of Table 7-8 are adequate to account for the enormous diversity of tropical trees is more problematic. Definitive studies of tree populations in tropical rain forests are rare and extremely

Table 7-8   Assessment of the present standing of empirical evidence in support of various mechanisms proposed to explain high diversity, first, in general and, second, the greater diversity of tropical regions. While many factors are known to increase general diversity, many fewer are known to act disproportionately in the tropics. The taxonomic groups listed indicate where the strongest evidence lies and do not necessarily include all organisms within the groups indicated, nor do they exclude the possibility of other groups being added as more evidence becomes available.

	EVIDENCE FOR GENERAL DIVERSITY	EVIDENCE FOR HIGH TROPICAL DIVERSITY	TAXONOMIC GROUPS
TIME SCALES			
Seasonal to a few years	Slight	Slight	All groups
Primary succession	Good	Poor	All groups
Secondary succession (including disturbance)	Good	Good	All groups
Geological and evolutionary			
Specialization	Good	Fair	Mainly animals
Coevolution	Fair	Fair	All groups
Refugia	Fair	Fair	All groups
Higher speciation rate	—	Slight	All groups
Lower extinction rates	—	Poor	?
SPATIAL HETEROGENEITY			
More biomes	Good	Good	All groups
More habitats	Good	Good	Mainly animals
COMPETITION AND SPECIES PACKING	Good	Fair	Mainly vertebrates
PREDATION	Good	Fair	Plants and invertebrates
PRODUCTION			
High production	Poor	Poor	—
Year-round production	—	Good	Vertebrates

difficult because of potential longevity and the enormous life time reproductive output of individuals. Much more work is needed before we can generalize about the relative importance of the various factors assessed in the table.

7.8   Stability of Communities

Communities that do not change for a long time are thought of as being stable. At one time many ecologists believed that the most complex and diverse communities were the most stable. Although no such simple relationship exists, it is a notion that continues to be advocated. Much of the dispute about the relationship between diversity and stability stems from the multitude of uses of the term "stability," some of which are contradictory (see Orians 1975 for a review of this semantic muddle).

Elton (1958) used the term to mean **constancy** of population densities over several generations and supported his idea that complex tropical rain forests are the most constant (that is, stable) of communities with circumstantial evidence, much of which is now outdated. Clearly, constancy can only occur in some sort of equilibrium or climax community. Such constancy is rare in local patches of tropical forest as individuals and populations frequently replace one another within the life span of older trees. If one takes a large tract of forest, however, it may remain fairly constant overall in terms of species composition for long periods, although we have only circumstantial evidence to support this.

Resilience, the ability of a community to return to its former state after an exogenous disturbance, has also been labeled stability, even though its meaning is often contradictory to that of constancy. In the same abiotic environment, simple, early successional communities are clearly more resilient than climax communities, since the former are regenerated more quickly than the latter following vegetation removal.

Mathematical models of population interactions have been used to investigate aspects of stability in simple communities. Hypothetical food webs of increasing complexity suggest that resilience often decreases as diversity increases (May 1981c, Pimm 1982).

7.8.1   Are Tropical Communities More Stable Than Temperate Communities?

Elton's contention that populations are relatively constant through time in tropical forests, compared with larger fluctuations elsewhere, has been discussed already (section 4.3.6). Many insect populations in tropical forests fluctuate widely in short periods. Population studies of trees and vertebrates over several generations that show stable populations are virtualy nonexistent. Connell and Sousa (1983) stress the necessity of long-term observations if conclusions about population or community stability are to be drawn. Their analysis does not support the idea that populations in dynamic equilibrium are common, but no sufficiently long studies of free-living tropical populations could be found. An analysis of multi-species bird communities in North America failed to find convincing evidence of higher stability along the latitudinal gradient from northern Canada to the subtropical regions of the southern United States (Noon et al. 1985). Changes over several years of various stability indicators produced very weak latitudinal trends at best.

Similarly, little direct evidence is available on the comparative resilience of tropical and temperate communities. Shifting cultivators have used the successional resilience of tropical forests for generations, but until recently their impact was no greater than nonhuman disturbances such as hurricanes and landslides. It is also clear that savannas are characterized by resilience to climatic and biotic disturbances that are themselves essential determinants of these communities (section 8.2).

Finally, it must be stressed that all aspects of community stability are only reflections of the stability of constituent populations. Communities as a whole cannot evolve toward stability, although individual populations may be K-selected in predictable environments. Perhaps the only way that a form of stability can be easily recognized is through persistence of the same type of community in a given area, while recognizing that large temporal and spatial changes may be occurring within that community.

Summary

Because the different populations are genetically isolated from one another, complex communities are much looser assemblages than are populations. Nevertheless, there are properties of communities that can be measured and compared. One interesting attribute is species diversity, which increases markedly as one moves toward the equator. Many arguments have been put forward to explain this gradient. A critical look at the evidence suggests that the types and time scales of disturbances, the length of association between species, the spatial heterogeneity of environments, and a larger total niche space enhance diversity in the tropics. However, the generality of these factors (and the possible importance of others) is difficult to assess definitively. Especially difficult to explain quantitatively is the diversity of trees in tropical rain forests. Although they shed limited light on the diversity of mainland communities, islands are important living laboratories for the study of the dynamics of community structure. Concepts of community stability are subject to semantic and technical problems. There is little evidence that tropical communities are more stable than temperate communities.

Study Questions

Review

1. Discuss the role of disturbances in determining community structure.
2. How would you delimit and describe a community?
3. Describe how the choice of time scales affects different concepts of community stability.
4. Compare the relative importance of interspecific interactions on community diversity with that of abiotic variables.

Related Topics

1. Are communities random assemblages of populations?
2. Write an account of a named heterotrophic succession.
3. Review the validity of the equilibrium theory of island biogeography.
4. To what extent do trees in the same tropical forest occupy different ecological niches?

Further Reading

Sections of the following general textbooks deal particularly well with community ecology:

Krebs, C. J. (1985). *Ecology*. 3d ed. New York: Harper & Row, Chapters 19–24.

Ricklefs, R. E. (1979). *Ecology*. 2nd ed. New York: Chiron Press, Chapters 36–39.

Whittaker, R. H. (1975). *Communities and Ecosystems*. 2d ed. New York: Macmillan, Chapters 3 and 4.

Following are two small monographs describing theoretical approaches to community ecology not covered in this book:

May R. M. (1974). *Stability and Complexity in Model Ecosystems*. Princeton, NJ: Princeton University Press (Monographs in Population Biology 6).

Pimm, S. L. (1982). *Food Webs*. London: Chapman & Hall.

The following collections of papers deal with specific issues in more detail:

American Naturalist. November 1983. 122:583–705. Presents the debate between equilibrium/competition–based community theory and nonequilibrium/chance advocates.

Cody, M. L.; Diamond, J. M. (1975). *Ecology and Evolution of Communities*. Cambridge, MA: Belknap Press. An important book in tribute to Robert MacArthur. Stresses equilibrium and competition based ideas about communities and has a strong tropical flavor.

Mound, L. A.; Waloff, N. (1978). *Diversity of Insect Faunas*. Symposium of the Royal Entomological Society of London No. 9. Oxford: Blackwell. A mixed bag of papers reviewing diversity in the most diverse group of organisms.

The following books give insights into how community attributes are measured in different groups of organisms:

Mueller-Dombois, D; Ellenberg, H. (1974). *Aims and Methods of Vegetation Ecology*. New York: Wiley.

Southwood, T. R. E. (1978). *Ecological Methods with Particular Reference to the Study of Insects*. 2d ed. London: Chapman & Hall.

Two accounts of island ecology supplement the brief treatment given in this text:

Gorman, M. L. (1979). *Island Ecology.* London: Chapman & Hall. A short introduction.

Williamson, M. (1981). *Island Populations.* Oxford: Oxford University Press. Provocative, comprehensive account of biogeographical, ecological, and evolutionary aspects of island biology.

*Alpine savannas are a tropical biome that has some striking plant growth forms. This tussock grassland on Mount Elgon on the Kenya-Uganda border has giant rosettes in the foreground (*Senecio*) and giant heaths in the background. (Photograph by Ian Deshmukh.)*

CHAPTER 8

Tropical Terrestrial Biomes: Patterns in Time and Space

In a book that emphasizes ecological concepts, it is inappropriate to give detailed descriptions of all the tropical biomes. Instead, the major biomes are classified and a selected topic from each biome is then discussed. These topics are seasonal variation in forests, herbivory in savannas, and adaptations to aridity in deserts. Each topic is explained in terms of the interactions between populations and their environments, following the approach advocated in the previous chapter.

The biome concept is a useful one for comparisons across geographical regions. Rain forests (or any other type of biome) differ substantially in species composition when tropical America, Africa, southern Asia, and northern Australia are compared. Nevertheless, these forests are more similar physiognomically and in community attributes such as diversity or energy flow than are rain forests and nearby savannas on the same continent. These similarities are the result of evolutionary responses to similar environmental conditions such as temperature, rainfall, and the presence of similar major taxa on most land masses. Such convergence, however, should not obscure that local adaptations that appear similar superficially may involve very different species in different places.

Figure 8-1 shows the geographical distribution of tropical biomes based on the likely vegetation structure present before intensive modifications by human activities. Also shown is the approximate total area of each biome as estimated in the late 1970s. The boundaries are not precise, and the naming system is only one of several in use. In

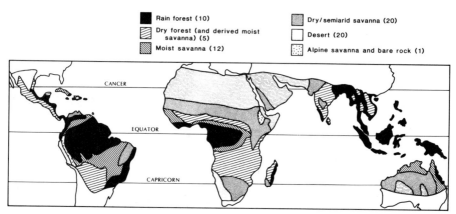

Figure 8-1 Distribution of major tropical biomes based mainly on climatic determinants and before extensive human modifications. Figures in brackets are estimates of the current area ($\times 10^6 \text{km}^2$) of each (Bolin et al. 1979). For comparison, total land area is about $150 \times 10^6 \text{km}^2$ and cultivated + urban area in the tropics $12 \times 10^6 \text{km}^2$. (Modified from Walter [1975] and Bolin et al. [1979].)

fact, there is a continuum of vegetation structure from rain forest to desert. This continuum is strongly influenced by rainfall (Figure 8-2), but local soil types or biotic influences may override climatic effects in some areas. As a rough guide, rain forests occur in climate types I, IIa, and IId of Figure 1-3 (annual rainfall is more than 1000 mm with, at most, a short dry season), and deserts in type III (less than 250 mm often spread unpredictably through the year). Between these lies a variety of drier forests and savannas, with the actual vegetation present

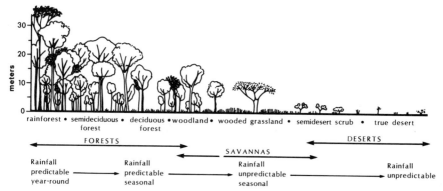

Figure 8-2 Tropical lowland ecocline in vegetation structure from rain forest to desert and its rainfall correlates.

often depending on the balance between climate, soil, grazing, fire, and human activity.

A **forest** is a community with a closed canopy in the tree stratum and a sparse herbaceous layer with few grasses. **Savannas** have a herb layer dominated by grasses and a broken tree or shrub canopy (in some cases a low shrub layer may predominate). **Deserts** have a scattered herb and shrub canopy interspersed with bare soil. Additional categories such as **woodlands,** with a dominant but incomplete tree canopy, and **semi-deserts,** with thorn scrub and scattered grasses, can be included to give transitional forms between forest and savanna and between savanna and desert, respectively (Figure 8-2).

8.1   Tropical Forests

Forests vary in species composition from continent to continent as a result of long geographical isolation, but forests in equivalent climates are remarkably similar in structure throughout the tropics. These similarities are the basis of Holdridge's "life-zone" classification of biomes (see Figure 7-5), which has been widely used in the neotropics. It is probably applicable in other regions but has not been used extensively and is not used here for that reason. Instead, a broadly based and somewhat imprecise classification is used that emphasizes similarities between forests of different regions.

8.1.1   The Altitudinal Sequence

Most ecologists agree that forests can be split into low and high altitude types. A generalized scheme of altitudinal zonation is given in Figure 8-3, but precise boundaries cannot be set. The lower limit for lower montane forest may be as low as 700 m, with upper montane forest occurring as high as 400 m. The boundary between upper and lower montane types is usually within the range 1500–2500 m. Above the forests, various scrub and grassy communities are found with an upper limit for vegetation of 4300–4900 m. Despite these variations in altitudinal limits, lowland, **lower montane** and **upper montane** forests have many recognizable characteristics, which are summarized in Table 8-1. However, lower montane forests differ in species composition in different regions. For example, oaks and laurels are common in southeast Asia while conifers such as podocarp and juniper are common in East Africa.

Throughout the tropics, the upper montane forests (also called "mossy forest" or "elfin forest") often contain trees of the heather family (the Ericaceae), which attain only dwarf shrub stature in many temperate regions. Some conifers such as podocarps are also common. All the trees

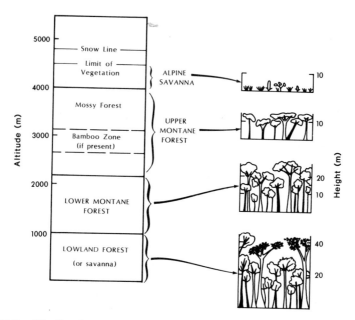

Figure 8-3 Idealized vegetation zonation on tropical mountains (left) and the vegetation structure associated with each zone (right).

tend to be gnarled and twisted, with abundant epiphytic lichens, mosses, and ferns.

The most obvious climatic change at high altitudes is the decline in temperature Mean annual temperature decreases fairly uniformly from 20°C or more at 1000 m to around 10°C at 3000 m and then to 2°C at 4000 m. Equally significant is the marked diurnal fluctuation in temperature, with very little seasonal variation. For example, Walter (1979) reports the annual maximum of 14.5°C at 1 PM on February 10, 1967 at 3600 m in a Venezuelan alpine savanna. The minimum for that year, 7.5°C, occurred only 41 hours later. Many upper montane forests are clothed in cloud much of the time, which ameliorates such extreme fluctuations to some extent.

The altitudinal changes in climate and vegetation lead to differences in soil also. Decomposition is slower because of the absence of termites and earthworms and because of reduced microorganismal activity at lower temperatures. This produces highly organic soils, which may become waterlogged and peaty where drainage is impeded. Nutrient cycling is slower at high altitudes because of reduced photosynthesis and slow decomposition. Many trees, particularly on ridges, are of small stature, probably because of the highly leached soils under the constantly humid conditions (Grubb 1977).

Table 8-1   Changes in the characteristics of vegetation associated with altitudinal zonation of rain forests on tropical mountains. Note that leaf area index is estimated as leaf area per unit weight and is not comparable with the usual method given in section 2.2.2. Modified from Grubb (1977) and Whitmore (1984).

CHARACTERISTIC	LOWLAND	FOREST TYPE LOWER MONTANE	UPPER MONTANE
Height of taller trees (m)	>30	15–35	<13
Leaf area index (area cm^2 g^{-1} of leaves)	90–130	about 80	about 70
Net production of wood (t ha^{-1} a^{-1})	3–6	about 1.4	<1.4
Buttresses on trees	Common	Rare	Absent
Predominant leaf size	Large	Large	Small
Abundance of conifers	Rare	Common	Common
Abundance of epiphytes Vascular Nonvascular	Common Rare	Very common Common	Common Common
Abundance of climbing plants	Woody common	Nonwoody common	Rare

Zonation of animal populations and communities is less pronounced. Species richness usually declines with altitude, and some species of rodents are known to be limited to one forest type. However, the altitudinal ranges of many species do not coincide with vegetation boundaries. For example, some birds of the Peruvian Andes and some rodents of the Rwenzori Mountains in Central Africa, are distributed independently of the vegetation zones (Terborgh 1971; Misonne 1963, respectively). Indeed, some species, ranging in size from rats to elephants, are known to occur in all zones from lowland forest to alpine savanna. Much less is known about invertebrates, although Janzen et al. (1976) found greater species richness in arthropod communities and larger average sizes of individuals in montane forests than at lower elevations in the Venezuelan Andes.

8.1.2   Lowland Forests

Lowland forests can be classified largely on the basis of rainfall and its seasonal distribution. **Wet forests** are evergreen with average monthly rainfall greater than potential evapotranspiration throughout the year. Such environments are widespread in Southeast Asia, equatorial South America, eastern Central America, and eastern Madagascar, but only small pockets exist in Africa, India, and Australia. **Humid forests** have several consecutive months in which potential evapotranspiration regularly exceeds average rainfall (roughly three months with as little as 50–100 mm rainfall). Some of the larger trees may be deciduous (but not necessarily leafless simultaneously), while most of the lower canopy is evergreen. Such forests are widespread in West and Central Africa as well as northeastern Australia, the neotropics, and Southeast Asia. Collectively wet forests, humid forests, and wet montane forests are usually called **rain forests** (note this is a different usage than that in Figure 7-5 and includes rain, wet, and moist forests of that classification). **Dry forests** in the tropics (often called monsoon forests in the Old World) have a prolonged dry season, often of six months or more. They are either deciduous, with most trees having a leafless season, or semidecid-uous, with a seasonally leafless high canopy while much of the lower tree canopy remains in leaf. This biome occurs in many parts of Central and South America, Africa, and India and grades into woodlands or more open savannas. Such boundaries can be quite distinct, particularly where human activity is prevalent.

Fresh water **swamp forests** are periodically inundated by water when rivers are in flood. They are locally abundant in some coastal regions near river outlets or alongside major river systems. By far the largest area of this sort is the 10% or more of the Amazon Basin in which the rain forest is subject to extensive flooding. Local soil peculiarities produce other forest types such as **heath forests** on soils derived from rocks of high-silica concentration and **moor forests** on peaty waterlogged soils. **Gallery forests** occur as an evergreen strip alongside rivers in dry forests and savannas where they often form important dry season refuges for the fauna. The peculiarities and distribution of these forest types are discussed by Richards (1952) and Whitmore (1984).

8.1.3   Phenology of Tropical Forests

Phenology is the study of periodic ecological phenomena on seasonal time scales or over a few years. Such patterns are usually thought to be correlated with annual climatic cycles because of studies carried out in temperate regions where reduction in day length and temperature reduce the activity of many organisms in winter. These particular changes are absent or less pronounced in the tropics and do not dictate the virtual cessation of biological activity. Seasonality in rainfall is a much more prominent feature of tropical climates. However, even in the wettest forest with plentiful rain throughout the year, many organisms exhibit phenological patterns. In the temperate zone, pronounced seasonal events include bud break, leaf fall, and flowering in plants; hibernation in rodents; migration in birds; and diapause in insects. The ultimate cause and the proximate cues for these changes are decreases in day length or temperature. In contrast, a primary cause and probable proximate cue of phenological patterns in the tropics is rainfall seasonality. Another distinct and influential factor is a biotic seasonality dictated by plant-herbivore interactions. Even in drier tropical forests, which have regular seasons of water stress, biotic factors such as the dispersal of seeds by animals may dictate fruiting seasons and lessen the effect of climate.

It is probable that even the wettest forests experience short periods of water stress that are not apparent from averaged rainfall data. For example, Singapore has an annual average of 2400 mm, with no month receiving less than 150 mm on average but in some years, some months may have as little as 50 mm, causing a brief soil-water deficit (Nieuwolt 1965). In such wet forests some rhythms are related to rainfall patterns, but the peaks in plant behavior are much less pronounced than in seasonal forests and show considerable variation from year to year.

Medway (1972) studied tree phenology in a wet dipterocarp forest of western Malaysia for several years (Figure 8-4). Rainfall patterns were not very consistent from year to year, but generally January to March was drier and October to December wetter. He examined 61 canopy trees representing 45 species of which 13 species were dipterocarps. Leaf flush and peak flowering occurred toward the end of the drier season, followed by the major fruiting period. There was a secondary peak of leafing and flowering at the height of the wetter season. However, only

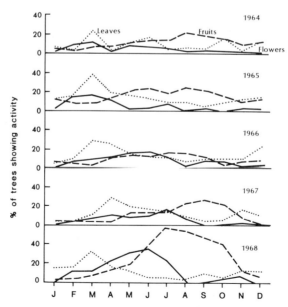

Figure 8-4   Phenological activity of canopy trees in Ulu Gombark forest, western Malaysia. Note the variation from year to year and the mass flowering and fruiting (of dipterocarps) in 1968. "Leaves" refers to flushing of new leaves. (Data from Medway [1972].)

10 species produced flowers and of these six set fruit in all years of the study. Of the 10 species, only five showed an annual and synchronized peak of flowering by different individuals of the same species. The year 1968 was particularly dry, and this triggered widespread synchronous flowering in the dipterocarps, increasing the percentage of species flowering and fruiting from 45% and 30% observed in other years, to 79% and 68%, respectively. These spectacular flowering periods in the dipterocarps occur every two to five years following a relatively dry spell, although not all mature individuals flower even then.

Phenological paterns in a wet Costa Rican forest do not correspond closely with those of Malaysia. Leaf flush is early in the drier season, and flowering shows two or three peaks in the upper and lower canopies, respectively (Frankie et al. 1974). The difference between these regional phenologies may indicate that climate is not the imperative in wet forests, but a general conclusion to this effect is premature. Patterns of water stress can be subtle, with shortages occurring during vigorous growth on wet soils, whereas even during long dry spells, sufficient rehydration of trees may follow leaf fall to allow bud break (Reich and Borchert 1984).

In drier forests, phenology is more closely tied to rainfall patterns. Barro Colorado Island in Panama supports a humid evergreen forest, which normally receives less than 50 mm of rain in February and March. Most tree species flush leaves, flower, and then fruit in the late dry season to early wet season. Such tendencies are even more pronounced in dry semideciduous forests (Figure 8-5). In the Costa Rican forest depicted, 83% of canopy trees are deciduous. Understory plants have less pronounced peaks and are not closely synchronized with canopy trees. Flowering in the understory occurs before canopy trees fully expand their leaves while shading is minimal. A study of wood growth in trees of a Nigerian semideciduous forest showed some seasonal variation but less distinct phenological patterns than in leaf or shoot growth (Hopkins 1970).

8.1.4   Phenology in Heterotrophic Populations

If not limited by other factors, populations of herbivores are constrained by periods of lean food supply. Nectar and fruit eaters obviously experience peaks of abundance of food from particular species. For leaf-eaters in evergreen forests, such limitations are not apparent until one realizes that most herbivores prefer young leaves, which are usually flushed in a seasonal pattern (Figure 8-4; Chapter 6). The influence of these seasonal effects has been studied in the humid forests of Barro Colorado Island, Panama. The island was cut off from adjacent forest in 1914 when the Chagras River was dammed. It seems that many animal populations are now limited by food availability during the dry season (see Leigh et al. 1982). For example, rodents experience a lean season

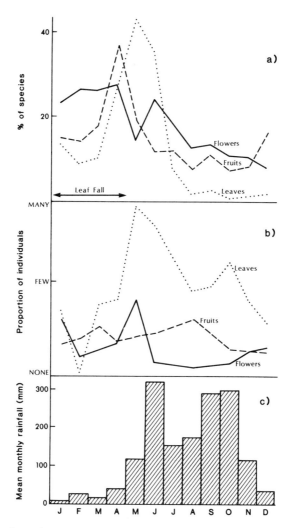

Figure 8-5 Phenological patterns in a dry forest in Guanacaste, Costa Rica. "Leaves" refers to leaf flush. Note the pronounced peaks compared with Figure 8-4. *a)* Deciduous canopy trees. (Modified from Frankie et al. [1974].) *b)* Understory treelets (small woody plants with single stem) and shrubs, most of which are evergreen. The ordinate is an index of the proportion of individuals of all species exhibiting a particular activity. (Modified from Opler et al.[1980].) *c)* Rainfall. (Modified from Frankie et al. [1974].)

when few fruits fall to the forest floor. During this period, agoutis *(Dasyprocta punctata)* subsist on seeds stored during periods of abundance, while pacas *(Agouti paca)* turn to low-quality leaves and as a consequence have to mobilize body fat reserves. Nutritional stress is apparent from reduced growth rate in juveniles, longer periods spent foraging, and resulting increased susceptibility to predation. The foraging pattern of howler monkeys *(Alouatta palliata)* maximizes the intake of high-quality food (young leaves, fruit). At the height of the wet season, older leaves are the abundant food. The monkeys minimize activity (and energy expenditure) during this period because of the dearth of high-quality food (Milton 1980). Young leaves are also preferred by sap-sucking insects such as the Homoptera, which have a seasonal peak early in the wet season. Different species reach peak abundance sequentially, suggesting temporal ecological segregation (Wolda 1982).

Phenological patterns have a marked influence on interspecific competition among three species of cercopithecine monkeys in a Gabonese humid forest. All species are primarily fruit-eating during the wet season when such food is abundant. When fruit is scarce, diets diverge. *Cercopithecus pogonias* continues to be a fruit specialist, and the other two species *(C. nictitans,* and *C. cephus)* take more leaves (Gautier-Hion 1980).

Even in wet forests, herbivore populations exhibit seasonal fluctuations, which affect higher trophic levels. For example, insectivorous bird populations in Sarawak have strongly seasonal patterns of molting and dispersal as well as of breeding (Fogden 1972; section 4.1.4). These rhythms result from the periods when trees have few young leaves and herbivorous insects are correspondingly scarce.

Much less is known about phenology in decomposer populations. In wet forests, microbial action occurs at a similar rate year-round but slows noticeably during dry seasons in drier types of forest. In a dry Nigerian forest, earthworms produce casts and microarthropods migrate into surface litter only during the wet season (Madge 1965). The distribution and abundance of litter arthropods in the humid forest on Barro Colorado Island are also influenced by moisture. Experimental watering of litter during the dry season reduced the total density of arthropods mainly due to fewer psocids but increased the number of Collembola. Further experiments suggest that woodlice, millipedes, ants, and Collembola may migrate into moist litter, while spiders and psocids move to drier areas (Levings and Windsor 1984). Of 13 arthropod groups, nine showed their major annual increase in density in the wetter part of the year. Fluctuations in other groups were not strongly correlated with obvious environmental changes except in the case of thrips. These insects were most abundant in litter during the peak of flower fall. There were also substantial differences in density of most groups for the same season in different years. In a three year study, lowest densities occurred in 1976 following the harshest dry period (Levings and Windsor 1985).

In Chapter 6 the prevalence of pollinating and seed-dispersing mutualism in the tropics was stressed. Indeed, it appears that flowering and fruiting phenologies are often adapted to maximize the benefits of such associations. Coevolved interactions both between plant species and between plants and their herbivores are likely in some instances. Tropical plants have greater diversity of flowering phenologies than their temperate counterparts. In the latter most exhibit the "cornucopia" syndrome with a large number of flowers produced over several weeks. Additional flowering patterns commonly found in the tropics are "continuous," or throughout the year; "steady state," with a few flowers each day for a month or two; and "mass flowering" for a few days only, with a high degree of synchrony between individuals. Some vines also have "multiple mass flowering" with several short periods of abundant flowering at different times of the year. All these syndromes have been identified by Gentry (1974) in a single family, the Bignoniaceae, in Central America. It seems that flowering phenology is closely related to the type of pollinator.

Different species of the wild plantains *(Heliconia)* of Costa Rica display flowering phenologies that are thought to result in part from ecological segregation among both the plants and the guilds of hummingbirds that pollinate them (Figure 8-6). Most species flower in the wetter part of the year, but some species prevent cross-pollination by attracting hummingbirds in other seasons. In drier forests, such flexibility in flowering time is not always physiologically possible because of the overriding effects of climate (Borchert 1983).

Trees and shrubs of the genus *Miconia,* which grow in neotropical lower montane forests, show remarkable fruiting patterns. Even in areas of Trinidad with strongly seasonal rainfall, 19 species fruit sequentially each for periods of one and a half to four months such that at least two species but not more than six are fruiting at any one time. Snow (1965) suggests that this pattern has evolved because of competition between the plant species for fruit-eating and seed-dispersing birds. Another benefit may be the reduced interspecific competition that results from seeds germinating at different times. A consequence of this competitive interaction between the plants is to provide a year-round food supply for the frugivorous birds.

In general, it seems that even the wettest forests have "lean seasons" that effectively limit heterotrophic populations, although these are less pronounced and shorter than those in more seasonal environments and are not always directly related to climatic events. All environments occasionally experience unusually wet or dry periods, which may temporarily alter phenological patterns. Dry years can clearly result in food shortages for herbivores, but very wet conditions can produce similar effects. For example, on Barro Colorado Island fruit-eaters experienced a famine in late 1970 and early 1971 because of low-fruit set following the absence of a dry season in 1970. Similar events occurred in 1931 and 1958. It is thought that the dry season is used as a proximate cue for flower and fruit production (Foster 1982).

Biotic upsets in phenological patterns also occur in animals. An omnivorous bulbul *(Andropadus latirostris)* of a humid Gabonese forest usually breeds seasonally each year. However, in 2 out of 10 years the birds did not breed at all, despite no discernible change in climate or

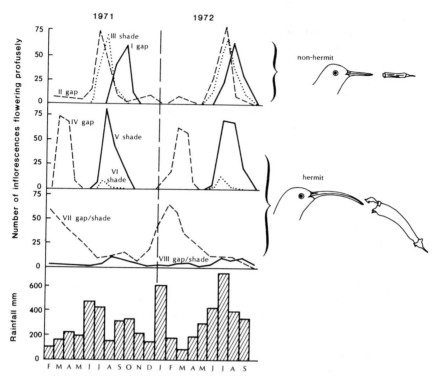

Figure 8-6   Phenological flowering patterns of *Heliconia* species (I–VIII) in a Costa Rican rain forest as an adaptation to prevent cross-pollination by hummingbirds. The species show ecological segregation in one or more of the following ways: (1) peak flowering in different seasons (e.g., I and II, IV and V); (2) use of different types of hummingbird (hermits are long-distance foragers, nonhermits territorial to a few plants); (3) occupation of either gaps or shady areas (e.g., II and III, IV and V); (4) Deposit pollen on different parts of the birds' heads (as in species pairs I and II, V and VI). (Modified from Stiles [1975]; Hummingbird/flower pictures from Feinsinger, P. 1983. In: Futuyma, D. J.; Slatkin, M. editors. *Coevolution*. Sinaur, p. 302.)

food supply. Following the two years of nonbreeding, many fewer nests were destroyed by predators, and fledging success was substantially higher. Brosset (1981) postulates that this occasional cessation of breeding has evolved as an adaptation to reduce predation.

8.2   Savannas

Used in its broadest sense, the term savanna describes a range of vegetation types from humid woodlands to dry grasslands. Such formations cover a large proportion of the southern continents (approximately 65% of Africa, 60% of Australia, and 45% of South America). Trees and shrubs are often present, but all have an extensive grass cover for most of each year. Because savannas are the result of a suite of interactions, it is preferable to use a simple structural classification in the first instance rather than one dependent on the supposed origins of particular savannas. Figure 8-7 gives such a classification based largely on the type and abundance of woody vegetation. These savanna types do not necessarily follow climatic gradients, although woodlands and tall grasslands are likely to occur in wetter conditions than bushlands and short grasslands. Although many vernacular names exist for savannas such as "veld" and "nyika" in Africa and "chaco" and "cerrado" in South America, such terms should not obscure their common characteristics.

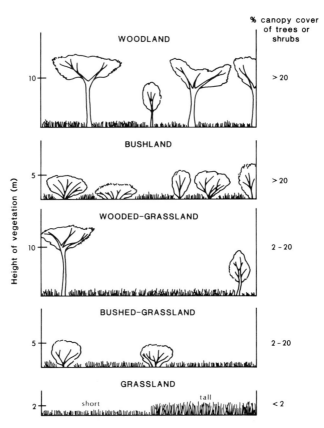

Figure 8-7   Classification of savanna vegetation based on the relative cover of woody and herbaceous vegetation. Additional composite categories such as wooded–bushed–grassland can be derived. (Modified from Pratt et al. [1966].)

8.2.1   Determinants of Savanna Vegetation

Much argument has surrounded the question of whether savannas occurred independently of human activity, particularly in tropical America. It now seems indisputable that savannas were present in many areas before human populations were present, that their extent has increased and receded as long-term climatic changes have occurred, but that they have become much more widespread as a result of human influences (see, for example, Sarmiento 1984). Grasses seem to have evolved about 50 million years ago, at roughly the same time as the modern mammals. Savannas are therefore much more recent than forests, which have been extant in one form or other for more than 300 million years.

The present distribution and maintenance of savannas results from the interactions between the following factors:

Climate. In the absence of people or soil conditions unfavorable to forests, savannas occur in regions with at least one long dry season. Woodland begins to replace closed forests where rainfall is less than 1000 mm per year, with thorn bushlands occurring in still drier conditions. Savanna environments can be classified in terms of annual rainfall, length of dry season, and reliability of rainfall (Table 8-2). Moist savannas would undoubtedly support forest vegetation if other environmental factors did not intrude. The categories of Table 8-2 sometimes form distinct zones along a rainfall gradient as in West Africa. Between the forest belt in the south and the Sahara Desert in the north is the sequence moist Guinean savanna, dry Sudan savanna, and semiarid Sahel savanna. Some neotropical savannas may be relicts of former drier climates in the Pleistocene, which forest vegetation has failed to recolonize. For example, some Amazonian savannas on white sand soils with peculiar hydrological conditions persist even though there are adjacent forests in apparently identical climatic conditions (Huber 1982).

Table 8-2 Classification of savanna environments based on rainfall patterns. The geographical distribution of the categories can be roughly gauged from Figure 1-3, although the categories are slightly different. Based on Harris (1980).

RAINFALL PATTERN	TYPE OF SAVANNA		
	MOIST	DRY	SEMIARID
Mean annual rainfall (mm)	1000–2000	500–1000	250–500
Length of dry season (months)	2.5–5.0	5.0–7.5	7.5–10.0
Normal deviations from mean annual (%)	15–20	20–25	25–40

Soils.   Edaphic savannas are those resulting from local soil conditions. Soil moisture and its ramifications are usually the key factors. Seasonal flooding with intervening intensely dry seasons restricts tree growth. Moisture also affects soil nutrients, with moist savannas more acidic and nutrient deficient because of leaching than most dry or semiarid savannas. In very acidic soils, high concentrations of soluble aluminum occur, which are toxic to many tree species. This element prevents proliferation of root cells and thereby limits root growth. It also causes the precipitation of phosphorus as insoluble aluminum phosphate, reducing soil fertility.

Defoliation: Herbivores and Fire.   Grasses are adapted to defoliation because their shoots grow from or below ground level. In contrast, most forbs and woody plants grow from the shoot tip and are vulnerable to defoliation. The relatively recent success of grasses in savannas may well be due to frequent defoliation by large mammals and the increasing frequency of fires caused by people. The effects of fire and large herbivores in the suppression of woody plants are illustrated in Table 7-4 and Figure 7-14, respectively.

The myriad interactions between herbivores and African savanna vegetation are becoming well known, but it is important to realize that there was once a diverse fauna of large mammals in American savannas also. These mammals were common in grasslands at subtropical latitudes in both North and South America. In the Pleistocene, northern species invaded the southern areas and vice versa. They must have occupied the intervening tropical savannas at least temporarily while the interchange was occurring. Most species are now extinct, although the reasons for this are not clear (Webb 1978).

Fire is a feature of all savannas. Natural fires, resulting from lightning or vulcanism are infrequent compared with those set by people. Many savannas throughout the tropics result from deliberate and regular burning. In moist areas, forest and woodland may be replaced by open savannas, while in drier climates, the predominance of grasses has increased at the expense of shrubs. Fires are most frequent and spread most easily in moister areas because a larger mass of fuel (plant litter) accumulates where primary production is high (see pp. 239). Plant species that are not adapted to frequent fires may be replaced by those that are. For example, the grass *Themeda triandra* does not spread very effectively vegetatively. It has seeds that bury themselves in the soil, thereby escaping from fires. The seeds germinate after the soil is exposed to high-light intensities and temperatures—conditions that prevail following vegetation removal by burning (Lock and Milburn 1971).

Because of the complex of factors that interact to produce any given type of savanna, these communities are variable in time and space as the balance of these factors changes. As a result, ever-changing landscapes are found in many savanna environments.

8.2.2   Alpine Savannas

Using the structural criteria in the introduction to this chapter, the vegetation communities above the tree line on tropical mountains can be regarded as savannas (Figure 8-3). Tussock grasses provide most ground cover, with scattered bushes and other tall plants. Where human populations occupy high altitudes, they may create savannas below the alpine zone by felling forests and subsequent grazing and burning.

There is a remarkable parallelism between some of the peculiar and spectacular growth forms on the "moorlands" of equatorial mountains in Africa and the "paramos" and "puna" of the South American Andes (Figure 8-8). Similar tussock grasslands also occur on the high mountains of Papua New Guinea and Irianjaya (Indonesia). The plant growth forms are important adaptations to the maintenance of

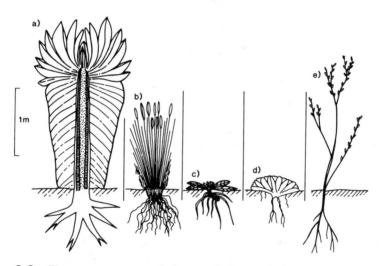

Figure 8-8   Five common growth forms of plants inhabiting tropical alpine savannas and representative genera of each type. Temperature (and probably moisture) conditions are ameliorated by these growth forms. Note that many of the genera are common in temperate regions. (Hedberg 1964; Hedberg and Hedberg 1979). *a,* Giant rosettes with stems surrounded by insulation of dead leaves. The rosettes close at night (e.g., *Espletia* in Andes and *Senecio* in East Africa). *b,* Tussock grasses. As in *a,* dead and decaying shoots remain attached (e.g., *Muehlenbachia* in Andes and *Festuca* in East Africa). *c,* Small rosettes close to the ground (e.g., *Hypochaeris* in Andes and *Ranunculus* in East Africa). *d,* Cushion plants. Usually comprise a dense aggregation of small rosettes (e.g., *Plantago* in Andes and *Agrostis* in East Africa). *e,* Shrubs with small leaves (e.g., *Hypericum* in both regions). (Reproduced by permission of O. Hedberg.)

water balance in these extreme climates of intense daytime radiation and rapid cooling at night. Experiments with the Andean giant rosette plant *Espletia schultzii* show that the closing of the rosette at night protects buds from freezing and that removal of the insulating dead leaves surrounding the stem increases the probability of death (Smith 1974, 1979).

Because these mountain tops are often isolated from one another by very different environments, they show high levels of endemism. For example, 80% of species from the alpine vascular flora in eastern Africa occur nowhere else, while (on average) 64% of the species above 3000 m occur on only one of the mountains in the region (Hedberg 1969).

Animals in the alpine zone are a mixture of those foraging opportunistically from the forests below and those specially adapted to montane environments. Insects tend to be flightless (or reluctant to fly), a common characteristic in isolated environments such as islands and mountain tops. In such situations, dispersal to new areas is unlikely to be successful and the advantage of flight consequently reduced. The alpine insects may also have less need for flight as they remain in sheltered spots among the vegetation to avoid the large fluctuations in temperature, radiation, and wind (Cloudsley-Thompson 1969). Many animals are also darkly pigmented as a protection against the high intensity of ultraviolet light. In the Afroalpine savannas some of the larger mammals of lowland savannas occur, but few are permanent residents. Elephants, duikers, bushbucks, klipspringers, lions, and leopards have all been recorded. In addition, the Ethiopian mountains have a specialized grass-eating primate, the gelada baboon.

8.2.3   African Savannas: Rain, Grass, Fire, Termites, and Large Herbivores

Many aspects of savanna ecology can be accounted for by rainfall and its variation, some features and effects of which are illustrated in Figure 8-9. In dry and semiarid savannas, production of herbaceous forage is closely correlated with rainfall (Figure 8-9*a*). This relationship controls food supply to large herbivores, allowing their community biomass to be predicted from mean annual rainfall (Figure 8-9*b*). However, rainfall (and forage production) varies enormously from year to year even at adjacent sites (Figure 8-9*c*), with drier areas experiencing greater fluctuations (Figure 1-4). Seasonal patterns of rainfall are also very important, particularly as they affect forage quality and herbivore distribution.

Despite the spectacular local diversity and abundance of large herbivores, it is clear that most primary production is decomposed by invertebrates, microorganisms and fire (Figure 8-10). Fire is more common in wetter savannas (see p. 237) where it is likely to occur every

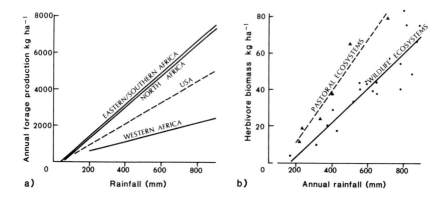

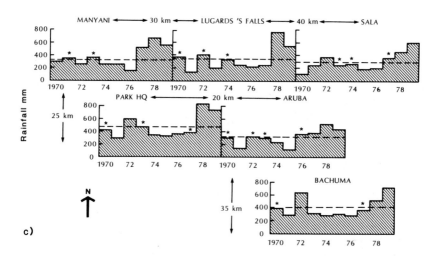

Figure 8-9   Factors relating to and variations in rainfall in African savannas. *a*, Relationship between rainfall and herbaceous forage production in different regions of Africa (solid lines). Broken line is a similar relationship for temperate grasslands in the United States. Forage production is a rough (but low) estimate of net primary production. (Modified from Deshmukh (1984). *b*, Large herbivore biomass in eastern and southern Africa increases with rainfall because of the increase in forage production. The line for pastoral ecosystems (▲ - - -) includes livestock and is steeper than that for wildlife ecosystems (· —). (See Chapter 9 and 11.) (Data from Coe et al. [1976].) *c*, Spatial and temporal variation in annual rainfall in Tsavo National Park, Kenya. Six rain gauge sites, on average 30 km apart, showed large differences from one another in the decade begining in 1970. Actual rainfall was within 20% of the mean (broken like) in only 17 (*) out of 60 possible occurrences and those years rarely coincided at the different sites.

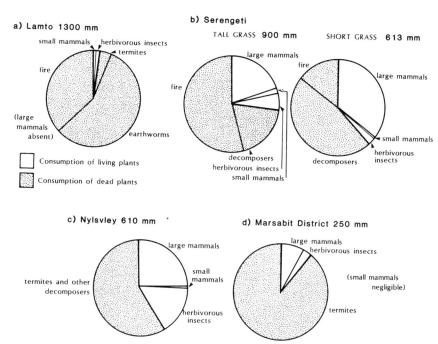

Figure 8-10 Comparison of forage consumption by herbivores and decomposers (including fire) in a variety of African savannas. *a,* Moist Lamto savanna, Ivory Coast. (Data from Lamotte [1982].) *b,* Dry Serengeti savannas, Tanzania. (Data from Sinclair [1975].) *c,* Dry Nylsvley savanna, South Africa. (Data from Gandar [1982].) *d,* Semiarid/arid savannas in Marsabit District, northern Kenya. (Data from Bagine [1982].)

year, often consuming 50% or more of the annual forage production. Animals play a bigger role in decomposition of plants in semiarid and arid environments than in most other terrestrial ecosystems. Termites in concert with their microorganismal mutualists are the most important decomposers in such environments. They often surpass the biomass of the more obvious mammalian herbivores. Estimates of termite biomass in African savannas range from 10–500 kg ha^{-1} (Wood and Sands 1978) compared with the 5–100 kg ha^{-1} of mammals recorded in Figure 8-9b. The proportion of net primary production consumed by termites ranges from 10% in a semiarid savanna in Senegal to more than 80% in a semiarid region of northern Kenya (Lepage 1974; Bagine 1982, respectively). Earthworms are the major decomposers in the moist Lamto savanna (Figure 8-10), but they are ill adapted to drier areas. Many large herbivores also act as decomposers in the dry seasons as they eat significant amounts of dead grass. This fact is often overlooked, but the distinction between eating live and dead plants is not trivial, as pointed out in Chapter 2.

It is possible to calculate total consumption of the large herbivore communities of Figure 8-9*b* from information about species composition, mean body weight of each species, and rate of food intake (Figure 2-8), and to compare the result with forage production in the same area, predicted from Figure 8-9*a*. Such calculations reveal that these wildlife communities consume between 2% and 10% of the forage produced in years of average rainfall over a wide range of savanna ecosystems. Herbivorous insects and small mammals usually consume less than 5% of the food available, even where larger mammals are absent (Figure 8-10). In general, average primary consumption falls into line with that of other terrestrial ecosystems (Table 2-5). It might be supposed that fire seriously limits the amount of forage available to herbivores, but this is not normally the case, because fire consumes the residue of vegetation and may even be absent where grazing pressures are high.

Despite low proportional consumption by herbivores in years of average rainfall, there is convincing evidence that most populations of large herbivores are periodically limited by their food resources. One line of evidence concerns seasonal variations in forage quality, particularly with respect to the protein content of grasses. During the wet season, protein concentration may reach 20% in young grass leaves and 15% in their stems and sheaths, but in dry seasons it falls to 3% or less. Wildebeest, for example, require 5%–6% protein in their diet for maintenance. Sinclair (1975) concluded that the Serengeti wildebeest population was limited by this shortage of protein in the dry season. During the five months when insufficient protein was acquired from forage, the animals mobilized body reserves accumulated during the prior wet season. In addition to this seasonal limitation in food quality, the quantity available in very dry years may be inadequate because of low forage production. Given the large variability in annual rainfall, the calculated requirement for large herbivores may exceed the sustainable forage offtake in dry years, especially when trampling and wasteful feeding methods are taken into account.

An important paradigm of ecology that recurs in this book is that interspecific competition leads to ecological segregation. Among the closely related ungulate species of African savannas, ecological segregation manifests itself in several ways (Figure 8-11). Segregation through difference in size is common. Body weight varies from 4 to 1700 kg if the nonungulate elephant is included. Tiny dikdiks are better able to penetrate into thickets and select small items of browse than are giraffes. Similarly, small muzzled gazelles are able to select specific plant parts from a grassy sward more effectively than broad-muzzled wildebeest. Other differences include habitat and food preference, but these do not eliminate all overlap, particularly between medium-size grazers. For example, topi, wildebeest, and zebras are commonly found grazing in

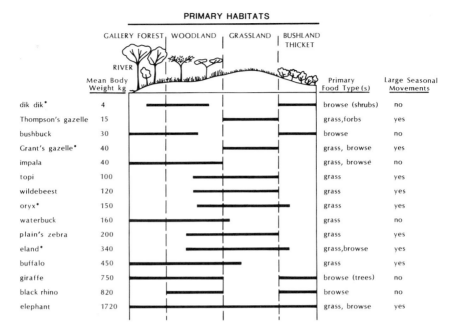

Figure 8-11   Ecological segregation between African savanna ungulates (plus elephants) in terms of habitat and food preference, size difference, seasonal movements, and adaptations to arid environments(*). Habitats largely from Jewell (1980); body weights from Coe et al. (1976).

the same area. They feed almost exclusively on grasses but choose differently between leaf, sheath, and stem, particularly during dry seasons (Bell 1970).

Some species, ranging in size from Grant's gazelles to elephants, reduce competition and increase food quality by eating a larger proportion of browse during dry seasons. Another adaptation is to move in search of better forage, a spectacular example being the wildebeest migration in the Serengeti-Mara ecosystem of Tanzania and Kenya. In fact, many other species are widely dispersed during wet seasons but concentrate in dry season refuges, which have permanent water supplies and often have higher primary production.

Some seasonal movements have been called grazing successions, in which larger species such as buffalo and zebra enter an area first, to be followed by medium-size wildebeest and topi, and finally the small Thompson's gazelle. The larger species, which need less protein in their diets, trample down and graze the coarse grass stems, followed by the leaf- and sheath-eating wildebeest and topi. In some instances, intense grazing by migratory wildebeest herds stimulates fresh grass growth that is then eaten by the gazelles which have a higher requirement for dietary protein (see Jarman and Sinclair 1979; McNaughton 1979). Such grazing successions have been described in parts of Serengeti and the Rukwa Valley of Tanzania, but it is not clear how common the phenomenon is elsewhere.

8.3   Deserts

Following the physiognomical classification on p. 225, deserts are communities with a broken vegetation cover of herbs, shrubs, and sometimes scattered trees, with extensive areas of bare soil between (usually 50% or less cover most of the time). Generally, deserts are in regions with a mean annual precipitation less than 250 mm and with a severe water deficit limiting the abundance of plants and other organisms. **Extreme deserts** occupy very arid areas with less than 100 mm of rainfall per year. The trends in rainfall patterns described for savannas are even more pronounced in arid lands, with rainfall being unpredictable, sometimes to the extent of being effectively aseasonal (Figure 8-12).

Hot deserts are commonly found at the boundaries between the tropical and subtropical zones because of the pattern of global air circulations. Near the equator, hot air rises from the land surface, gradually cooling and losing moisture as rain as it gains altitude. These cool air masses move north and south, descending at around 30° north and south. As it descends, dry air becomes warmer and therefore absorbs moisture, which is carried back toward the equator to replace the air ascending there. The largest hot desert is the Sahara and its extension into the Arabian Peninsula. Other large pockets are in central Australia, the Sonoran and Chihuahuan of the United States–Mexico border region, the Atacama in southern Peru and northern Chile, the Kalahari and the Namib of southern Africa, parts of the horn of Africa, and the Thar of northwest India.

Edaphic deserts occur where adverse soil conditions (particularly excessive salt) prevail even though rainfall is adequate for more dense vegetation. **Cold deserts** in the tropics are found above the altitudinal limit of vegetation on mountains. In addition, there are areas where human activity has reduced plant cover to produce **biotic deserts** irrespective of climate. Many urban areas fit into this category as do those rural areas that have been degraded by human pressures.

8.3.1   Desert Soils

Desert soils result mainly from geological, climatic and hydrological processes rather than biological ones. A notable exception is the rearrangement of finer soil particles by termites. For example, *Odonto-termes* in northern Kenya move 13 t ha^{-1} a^{-1} of soil in covering plant litter before eating it (Bagine 1982). Physical processes produce

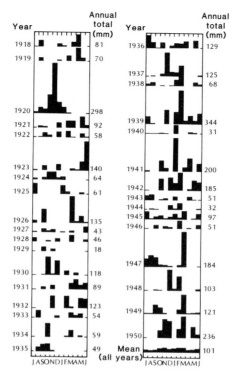

Figure 8-12   Unpredictable seasonal and annual variations in rainfall in the central Australian desert at Mulka between 1918 and 1950. Compare with the fairly equable misrepresentation given by the monthly means for the period at the foot of the second column. (From Walter, H. 1971. *Ecology of Tropical and Subtropical Vegetation.* Edinburgh: Oliver & Boyd.)

a variety of soil textures that differ considerably in their retention of sporadic, heavy rain. Walter (1971) indicates that a storm penetrating 10 cm in clay will soak down to 50 cm in sand and even further in stony soils. In clay, evaporation is rapid from the upper 5 cm, and 50% of the water dries out quickly. Sandy soils retain 90% beneath the upper layers in the short-term and stony soils, even more under the same conditions. This helps to explain the distribution of plants with sparse vegetation found on desert clays, ephemeral grasses on sandy soils, and shrubs and trees on stony soils. Soil texture also affects the species composition of burrowing animal communities (see section 8.3.3).

Many desert soils, particularly those of fine texture, contain high-salt concentrations, originating from parent rock or deposited in rainfall or inflowing surface water. Surface accumulation, sometimes to the extent of a surface crust, is pronounced because of rapid evaporation and lack of deep penetration. In contrast, in most humid areas or those with low evaporation, excess salts are washed downward and eventually out of the ecosystem in stream flow. The problem of salt accumulation is exacerbated when irrigation of hot deserts is attempted. As a consequence of salty soil, some desert plants contain more than twice the ionic concentrations as those from tropical forests. To prevent osmotic stress the plants have specialized physiological mechanisms to secrete salts or to deposit them in metabolically inactive and dead tissues.

Calcium and magnesium carbonates are also common in desert soils, leading in many instances to subsurface hardpans that limit the rooting depth and the downward leaching of surface salt accumulations.

8.3.2   Fog Deserts

The Namib and Atacama deserts are adjacent to cold ocean currents where local low pressure over the land draws in moist fog. Both deserts have areas with extremely low rainfall (less than 20 mm per year), and fog precipitation is, therefore, a potentially important source of water. However, while condensed fog may double annual precipitation, it is relatively evenly spread through many days, penetrates the soil only slightly, and is lost rapidly through evaporation when the fog clears.

Fog condenses most readily on contact with solid surfaces, such as rocks or plants, thereby providing locally enhanced moisture conditions. In addition, the reduced temperature and increased humidity reduce water stress in these otherwise extremely arid areas. There is no convincing evidence that aerial plant organs take up fog precipitation directly in the Namib, but by providing condensation surfaces they do supply water to the roots. Similarly, beetles of the Namib have no specialized structures for atmospheric water absorption but do have behavioral adaptations for obtaining condensed water from their body surfaces, vegetation, and the sand (Seely 1979). In the Atacama, the bromeliad *Tillandsia* does absorb condensed fog directly into the foliage.

8.3.3   Adaptations to Desert Environments

Community metabolism, as reflected by production, respiration, and nutrient cycling is low in deserts because of the shortage and unpredictability of water supply. Individual organisms are adapted to these and other stresses such as high radiation and temperature, but it is the intermittent nature of water stress that is the primary influence on community structure. Table 8-3 lists the ways in which organisms are adapted to arid environments. Anatomical, morphological, and physiological adaptations are not elaborated here, as detailed accounts can be found elsewhere (see Cloudsley-Thompson and Chadwick 1964; Walter 1971; Crawford 1981).

Some organisms have refuges from the extremes of dessication found in the atmosphere and at the soil surface. Plants may obtain some shelter in rock crevices or beneath the canopy of others, although

Table 8-3 Classification of adaptive responses in desert organisms to arid environments. See text for examples.

RESPONSE	MAJOR MANIFESTATIONS	PLANTS	ANIMALS
TOLERANCE OF DEHYDRATION			
Normal body tissues able to dehydrate	Physiological	Some algae, ferns, and lichens	Some nematodes
TEMPORALLY REDUCED ACTIVITY			
Ephemerals with special dormant phase (which may be tolerant of dehydration)	Ecological, physiological	Annual* angiosperms as seeds, bulbs, or rhizomes	Encysting Protozoa, eggs or pupae of annual* insects, and few reptiles
Large seasonal reduction in active metabolism without dormancy	Physiological, anatomical	Perennial angiosperms shedding some leaves, stems, and rootlets	Rodents, snails, some reptiles that estivate
CONTINUALLY ACTIVE			
Maintain high level of metabolism through dry periods			
With no refuge	All types	Sclerophyllous shrubs, succulents	Large vertebrates
Local refuge	Behavioral	Window algae	Most surface-active arthropods and vertebrates
Migrate to moister area	Behavioral, ecological	None	Locusts, birds, larger mammals (including humans)
ENVIRONMENTAL MODIFICATION	Ecological, behavioral (including social and cultural)	None	Social insects, humans

*Annuals in this context may not be active every year if rainfall is insufficient.

competition for water limits the effectiveness of the latter behavior. The curious window algae are found beneath translucent quartz pebbles where moisture conditions are improved by condensation and reduced evaporation (Walter 1971).

Small, surface-active desert animals, including many arthropods, reptiles, and rodents, often forage in the evening or at night when humidity and temperature are more equable. They spend the day under stones or organic debris or in burrows.

Because of the harsh abiotic environment, deserts are important ecosystems for testing various ecological hypotheses about population dynamics and community structure. Unfortunately, insufficient evidence is available for unequivocal answers to most of these questions as illustrated by the following examples.

How Important is Rainfall in Controlling Ecological Characteristics?   It is sometimes argued that the brief, intermittent nature of rainfall and plant growth means that populations do not have time to build up to levels at which competition is important. While this conclusion is thought erroneous by many (see below), there is no doubt that the amount and unpredictability of rainfall have profound effects. For example, a relationship between rainfall and plant production similar to Figure 8-9a predicts zero production at 11 mm in annual grasslands of the Namib desert (Seely 1978).

Many aspects of animal population dynamics are also correlated with the amount of rainfall, often that received during the prior wet period. Such effects are mediated by the availability of plant food and its effect on reproductive success (Figure 8-13; see also Wagner 1981). Because of the enormous variation in rainfall, many populations undergo large fluctuations. In central Australia, eruptions of rodent populations occur infrequently and irregularly. They seem to follow when two periods of high rainfall happen within three to five months of one another (Newsome and Corbett 1975).

Does Competition Determine the Structure of Desert Communities?   An alternative to the hypothesis that competition is unimportant in deserts is that, given the shortage of resources, competition is likely to be at its most intense. The wide spacing between plants has been attributed to competition for water, and rooting depth has been suggested as a means of niche differentiation between different types of plant (Figure 8-14). Although there is much inferential evidence in support, direct demonstration of competition between desert plants is lacking (see Barbour 1981).

Evidence for interspecific competition between animal populations is also mostly indirect, coming largely from observations of ecological segregation. The best examples include segregation with respect to food size in seed-eating ant and rodent communities in the North American

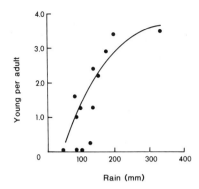

Figure 8-13   Number of gambel quail *(Lophotyx gambeli)* in the autumn population of the Sonoran Desert is strongly correlated with the amount of rainfall in the preceding December to April. The population fluctuates in response to rainfall fluctuations but after a time lag of approximately nine months (From Wagner, F. H. 1981. In: Goodall, D. W.; Perry, R. A. editors. *Arid-land Ecosystems: Structure, Functioning and Management.* Cambridge: Cambridge University Press.)

deserts (see, for example, Davidson 1977) and the burrowing abilities of different species of Namibian scorpions in different soils (Lamoral 1978). In lizard communities of Australia, North America, and southern Africa, niche differentiation results from different combinations of food choice, microenvironment, and activity patterns in the three continents (Pianka 1975).

What Types of Life History Patterns Predominate?   From the predictions of section 6.2.1, for spatially and temporally unpredictable environments, r-tendency life histories might be expected to predominate in deserts. Is this the case? Do most populations consist of small, short-lived individuals with the potential to track patterns in rainfall or food availability? At this point, it is important to reemphasize that such predictions at best are only one set of a complex group of possible responses in life history evolution. Indeed, deserts serve to confirm the difficulty of comparing simple life history traits in different species.

As rainfall decreases, annual grasses play an increasing role in herbaceous production. Also, large population fluctuations occur with zero production in some years and a short-lived but lush carpet of grass in others. However, perennial trees, shrubs, and herbaceous succulents are long-lived and relatively constant in density.

Arthropods exhibit many life history patterns from ephemeral to a longevity of several years. In fact, many well-studied species such as scorpions and colonies of ants and termites have K-tendency characteristics. Perhaps the best known r-tendency insects that periodically take advantage of temporarily improved conditions are the locusts. The

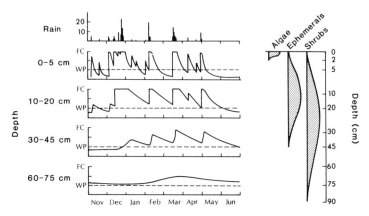

Figure 8-14   Relationship between rainfall received (millimeters), soil water at different depths (line graphs), and the rooting depth of different growth forms of desert plants (vertical hatched graph). Deep rooting plants have an adequate water supply throughout, but shallow rooting plants experience periods of water shortage. (WP = wilting point; water supply inadequate. FC = field capacity; maximum amount of water in soil after drainage). (From Noy-Meir, I. 1973). Reproduced, with permission, from the *Annual Review of Ecology and Systematics,* Volume 4, ©1973 by Annual Reviews, Inc.)

reasons for outbreaks are by no means clear. A plausible explanation is that most of the time high juvenile mortality is caused by poor-quality diet that is low in nitrogen. Following a prolonged wet period, forage quality improves, mortality is reduced, and a plague is triggered (White 1976). A similar effect of diet quality (in this case, shortage of vitamin A) is thought to be the cause of population fluctuations in the gambel quail implied in Figure 8-13.

Among desert mammals, rodents seem to constitute a greater proportion of total biomass than in the less variable savannas where ungulates often predominate. However, smaller mammals predominate in tropical forests also (Figure 7-1). It cannot be argued, therefore, that there is a simple trend to small mammals because of aridity.

At the more detailed level of comparisons between populations, it is necessary to compare like with like. There is little point in comparing population growth rate (or other attribute) of a desert annual grass with that of a forest tree, but comparisons between a desert and a forest antelope could yield fruitful insights. Unfortunately, rigorous and comparable data of this sort are lacking, although Wagner (1981) reaches two tentative conclusions. First, latitudinal gradients (such as lower fecundity in the tropics) exert a stronger influence on life history characteristics than does the unpredictability of the desert environment. Second, desert populations often seem to be less productive (and by inference more K-tendency) than those of similar species at the same latitude.

Another relevant prediction of r-K theory is that organisms may escape in time or space when harsh conditions prevail. Examples such as dormant stages, nocturnal foraging, and local refuges are common (see Table 8-3 and "Further Reading"). Large mobile animals are able to migrate to more equable areas in dry periods, a pattern found in many lizards, birds, kangaroos, ungulates, and domestic livestock. Locust and rodent plagues also migrate when their local food supply is used up. It is the subsequent invasion of agricultural areas that give them their pest status.

The most elaborate way of persisting in deserts is also the least common—to exert control over and moderate the environment. Only social insects and humans have mastered this approach. Individual termites are not well adapted to withstand dessication, but the environment in their nests and galleries is kept relatively constant—warm and moist rather than hot and dry—by the behavior of the colony. Some Australian termites build mounds along a north-south axis to minimize heating and evaporative effects of the sun. Foraging worker termites cover their food with soil to avoid exposure to the ambient environment, and some workers pentrate deeply to moist soil or even to the water table to keep the colony humid. Indeed, the complex system of galleries, chambers, and ventilation holes indicates that some mounds are effectively air-conditioned, a level of sophistication equaled only recently in a few human dwellings.

Summary

Division of major vegetation community types into biomes is useful for making ecological generalizations. In the tropics, total annual rainfall and its seasonality permit a clear distinction to be made between rain forest and desert biomes. More than half of the tropical zone comprises intermediate vegetation types ranging from dry forests and woodlands to semiarid savannas. The dominant vegetation in such areas depends on the interactions between climate, fire, soil, herbivores, and people. All tropical forests exhibit some degree of seasonality. In drier climates the pattern of wet and dry seasons is important. In wet areas with no distinct dry season, interactions between species may determine phenological rhythms. Africa has older and more widespread savannas than other tropical land masses. The structure and resilience of these communities is determined by the predictability of rainfall, the varying effects of large herbivores, and the frequency of fires. In hot deserts the unpredictability of rainfall determines the life history characteristics of many populations. Some populations are ephemeral. Others have adaptations enabling them to persist in active or semiactive states through the long periods without rain.

Study Questions

Review

1. How does rainfall seasonality determine the patterns of animal activities in different types of forest?
2. Discuss the various interactions between fire and vegetation in savannas.
3. Are deserts simple communities?
4. Why do we sometimes find different types of vegetation in areas with the same climate?

Related Topics

1. Review the ecological adaptations found in the organisms of alpine savannas.
2. Investigate competitive interactions between seed-eating ants and seed-eating rodents in hot deserts.
3. Report on the effects of seasonal flooding on rain forest plant and animal communities.
4. Discuss the interactions between herbaceous and woody vegetation in savannas.

Further Reading

Walter, H. (1971). *Ecology of Tropical and Subtropical Vegetation.* Edinburgh: Oliver & Boyd (English edition). A detailed descriptive review of the vegetation of tropical biomes.

Forests

Leigh, E. G.: Rand, A. S.; Windsor, D. M. editors. (1982). *The Ecology of a Tropical Forest.* Washington, DC: Smithsonian. A collection of papers about a Panamanian rain forest, many of which are about phenology.

Richards, P. W. (1952). *The Tropical Rain Forest.* Cambridge: Cambridge University Press. Remains the outstanding account after more than 20 years.

Unesco (1978). *Tropical Forest Ecosystems.* Paris: Unesco/UNEP/FAO. A useful source of information but not very readable.

Whitmore, T. C. (1984). *Tropical Rain Forests of the Far East.* 2d ed. Oxford: Clarendon Press. An excellent book with much information applicable to other regions.

Savannas

Huntley, B. J.; Walker, B. H. editors. (1982). *Ecology of Tropical Savannas.* New York: Springer-Verlag. A collection of papers, which brings together information on savannas from all the southern continents, including accounts of energy flow in some African savannas.

Sarmiento, G. (1984). *The Ecology of Neotropical Savannas.* Cambridge, MA: Harvard University Press. The first comprehensive account of American savannas, mainly from a plant ecophysiology viewpoint.

Sinclair, A. R. E.; Norton-Griffiths, M. editors. (1979). *Serengeti: Dynamics of an Ecosystem.* Chicago: University of Chicago Press. A multiauthored account of a Tanzanian savanna, which stresses the role of large mammals.

Unesco (1979). *Tropical Grazing Land Ecosystems.* Paris: Unesco/UNEP/FAO. Same comments as for Unesco (1978).

Deserts

Goodall, D. W.; Perry, R. A. editors. (1981). *Arid Land Ecosystems: Structure Functioning and Management.* Cambridge: Cambridge University Press. A comprehensive review of desert ecology (two volumes).

Louw, G. N.; Seely, M. K. (1982). *The Ecology of Desert Organisms.* London: Longman. Deals with adaptations and community structure, with much information on the Namib desert.

Noy-Meir, I. (1973). Desert ecosystems: environment and producers. *Ann. Rev. Ecol. Syst.* 4:25–51.

Noy-Meir, I. (1974). Desert ecosystems: higher trophic levels. *Ann. Rev. Ecol Syst.* 5:195–214. Two papers with an ecosystem approach to the structure and processes of desert communities as determined by unpredictable moisture availability.

Walter, H. (1971). See previous listing. Strong on ecophysiology of desert plants.

HUMAN ECOLOGY

Nomadic pastoralism is a common feature of African savannas. Maasai and cattle in Tanzania. (Photograph by Dana Slaymaker.)

CHAPTER 9

Human Food: From Foraging to Fossil Fuel

The hominids, of which *Homo sapiens* is the sole surviving species, split from other primate lineages more than four million years ago. Present evidence indicates that this divergence occurred in eastern Africa, with the first hominids being omnivorous savanna dwellers who hunted and scavenged animal prey and gathered plant food. This existence, as one species among many mammalian foragers, persisted for most of human history with agriculture arising not much more than 10,000 years ago. The change to agriculture was not an abrupt transition, the first steps probably being the tending of particular plants to ensure their continuing productivity and the taming of animals. **Domestication** carries this process much further because it involves gaining control of reproduction of the organisms involved. This in turn allows selective breeding, which has produced the domestic species that we have today. Agriculture not only involves the exercise of evolutionary control over the species selected, but also the creation of special environments in which to keep them. The resulting combination of people, domestic plants, domestic animals, and their environments are called **agroecosystems** to distinguish them from natural ecosystems in which people play no special part.

Figure 9-1 outlines the food procurement systems used by humans and the evolution of different types of agroecosystems. All seven systems at the foot of the figure persist in the tropics. Foraging, shifting cultivation, some forms of continuous cultivation, and nomadic pastoralism are practiced primarily for subsistence rather than to produce excess food for sale. Agroecosystems differ from natural ecosystems because

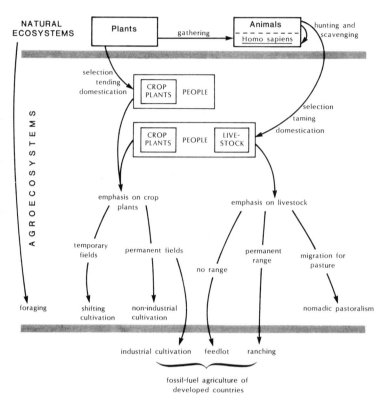

Figure 9-1   Comparison of the means by which human populations obtain food. The vertical axis qualitatively represents time in that the sequence of relationships between people and other organisms follows the supposed historical sequence. Human foragers in unmanaged ecosystems consume a small proportion of the total biota and domestication has selected even fewer species. All the feeding systems at the foot of the figure exist in the tropics, although there may be overlaps between them. The horizontal lines represent major qualitative changes, the upper from natural ecosystems to agroecosystems and the lower from largely subsistence economies to those dependent on fossil fuel.

their existence is supported by **energy subsidies** whereby work initiated by humans is performed to create and maintain the special environments required. In subsistence agriculture, energy for this work comes from human labor and draft animals, but in **industrial agroecosystems**, in which the primary objective is to produce food for sale, most of the energy subsidy comes from fossil fuels. The subdivisions illustrated in Figure 9-1 are not necessarily distinct in any given population, with many people combining various portions of them or being in a state of transition from one to another (see sections 9.2.5 and 9.3).

Sections 9.2 and 9.3 outline some features of ecological interest in the various systems without attempting to account for the social, economic, and political nature of the human cultures. A basic premise is that the subsistence cultures are at least minimally adapted to their environments, a supposition supported by their lengthy persistence. Section 9.4 looks at the development of agroecosystems in the tropics in response to the pressing demands of human population growth.

9.1   Human Nutrition

The gross qualitative requirements of human diet are well known. The basic needs are supplies of energy, amino acids, and a variety of vitamins and minerals. Energy and protein are the usual measures of nutritional adequacy because relatively little is known about vitamin and mineral content of subsistence diets unless specific deficiency diseases are present. A 60 kg person can live on 8 g of protein per day, provided this has the exactly correct balance of essential amino acids (Blaxter 1973). An adult eating ordinary foods in a well-balanced combination normally needs 30 g of protein each day. Any protein above the basic necessity may be used as a source of energy, but carbohydrates and fats are much richer sources of dietary energy (Chapter 2).

To meet the basic metabolic needs, an inactive 60 kg mammal needs about 6500 kJ of assimilable energy from food each day. Any type of activity increases this basic requirement, but there is no good consensus as to how much extra energy is needed for a "normal" activity schedule. Table 9-1 shows the daily energy intake in relation to body weight recommended by the United Nations and compares it with that observed in a New Guinean population. The observed intakes of energy and protein are well below those recommended, yet no obvious signs of malnutrition were evident among that group of people. The protein/energy balance of staple foods can be quite different in neighboring populations. Many subsistence crop-growers eat low-protein, high-carbohydrate food, while adjacent subsistence pastoralists have a diet rich in animal protein and fat but low in carbohydrate. Nevertheless, malnutrition is rarely attributable to a simple portein deficiency even when starchy grains are the staple. In most cases, when protein intake is inadequate, so is energy (Wheeler 1980).

Figure 9-2 shows some variations in energy intakes of two populations. Children, because of their lower body weight, have a smaller daily intake (but a greater intake per gram body weight). Old people also have a lower energy requirement because they are less active (Figure 9-2a). In many subsistence economies, food supply varies seasonally, and people may lose weight during the crop-growing season when farming activity is high. After the harvest, weight is regained to

Table 9-1   Daily energy and protein requirement per person and per kilogram of body weight recommended by the United Nations compared with the observed intake of the Kaul shifting cultivators of Papua New Guinea. Kaul data from Norgan et al. (1974).

	UNITED NATIONS		KAUL PEOPLE	
	MEN	WOMEN	MEN	WOMEN
TOTAL INTAKE				
Body weight (kg)	65	55	56	48
Daily energy intake (kJ)	12,750	9218	8145	5966
Daily protein intake (g)	53	41	37	24
INTAKE PER KILOGRAM OF BODY WEIGHT				
Daily energy intake (kJ)	193	168	145	124
Daily protein intake (g)	0.82	0.75	0.66	0.55

maintain an annual energy balance (Figure 9-2*b*). Daily energy intakes as low as 5030 kJ for men and 3350 kJ for women have been recorded during the wet season in Ethiopia without any clinical signs of starvation or reduced ability to perform essential work (Miller and Rivers 1972).

Leslie et al. (1984) have devised a model of human energy requirements that pays due regard to thermal environment, stature and body composition, age structure, reproductive patterns, and activity patterns of different populations. Predictions fit more closely the field studies of food intake in the Nuñoan indians of Peru, the !Kung hunter-gatherers of Botswana, and the New Guinea populations of Table 9-1 than do previous models including those of the United Nations. It seems that many subsistence cultures are extremely conservative in the use of metabolic energy, in contrast with the profligacy of modern European cultures upon which dietary standards are based. This is not to imply malnutrition is absent in such populations. Seasonal shortages of food may have drastic effects on young children and pregnant or lactating women, and this undoubtedly contributes to the high rates of child mortality in these societies. Mass starvation is also likely to occur periodically in drier climates because of the unpredictability of rainfall and primary production (see Chapter 8 and Figure 9-11).

This discussion illustrates some of the problems in establishing universal dietary standards. Populations are culturally, physiologically, and probably genetically adapted to very different diets. At present we have inadequate evidence to assess the amount of variation in human nutrition that is possible without detrimental effects. In such a situation, it is desirable to set requirements too high rather than too low.

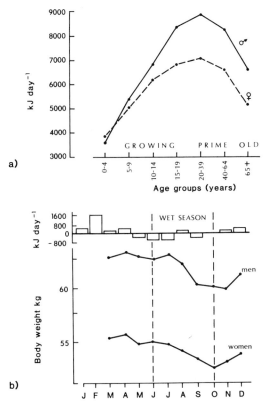

Figure 9-2 Temporal variation in daily energy intake in human populations. *a,* Variation with age and sex in the Nuñoan cultivator-pastoralists of the Peruvian Andes. (Data from Thomas [1977].) *b,* Seasonal energy balance (top) and changes in body weight (below) of adults in a Gambian agricultural village. (Modified from Weiner [1980].)

9.2   Subsistence Systems

Few food procurement systems are entirely devoted to subsistence, but many people in the tropics forage for or produce food primarily for their own needs and have only peripheral contact with cash economies. This does not preclude trade between different subsistence groups in the form of food, labor, or other goods (see section 9.2.5).

9.2.1   The Human Forager: Hunting and Gathering

A well-documented example of human foragers is the Gidjingali people of the coastal woodlands in northern Australia (Jones 1980). Foraging patterns are largely determined by the seasonal availability and profitability of different types of food from the ocean and from the land (Figure 9-3). In addition to foraging, these people also buy significant amounts of high-quality carbohydrate as flour and sugar. Nevertheless, most of their energy comes from foraging for more than 50 species of animals, including marine mollusks, fish, and terrestrial vertebrates. Roots, tubers, and fruits are also gathered from more than 20 plant species.

The seasonal patterns of rainfall and foraging are shown in Figure 9-3a. Early in the year during the monsoon rains, food is scarce and people are relatively inactive, doing little more than meeting basic energy needs. Although this is the period of most plant growth, much of it is as foliage and therefore unsuitable as human food. A bonus at this time is mollusks thrown up on to the seashore by storms. These are collected by the women. As the rains subside, a plentiful supply of energy is available from plants (particularly yams), which the women gather. This

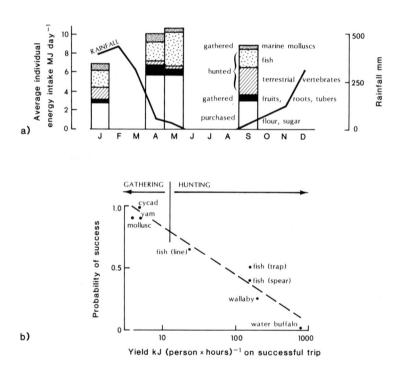

Figure 9-3   Subsistence foraging among the Anbarra group of the Gidjingali people of coastal northeastern Australia. (Data from Jones [1980].) *a,* The phenology of energy intake in relation to rainfall, food type, and feeding strategy. See text. *b,* The relationship between energy yield and the probability of foraging success for a variety of food types. Although hunting yields high quality items that feed many people, the effort in obtaining them is much more than from gathering or fishing.

is also the peak fishing season for men and mollusk-collecting season for women among the newly flooded swamps. It is no coincidence that this period of plenty is when many energy-consuming social events also occur. As the dry season proceeds, easily gathered foods diminish and hunting for terrestrial vertebrates by the men assumes greater importance. Hunting is made easier because animals tend to concentrate around permanent water sources and in the flush of green grass brought about by fires set by the Gidjingali in June and August. These fires are carefully managed to remove dead grass but to minimize damage to patches of woodland and forest.

Food choice seems to be determined by similar criteria to those employed by nonhuman foragers (see Chapter 5). Large items of high-quality food are rare and difficult to obtain (Figure 9-3b). This rarity is reflected in the frequency with which foraging for different foods is attempted with only one day of hunting for water buffalo, 30 days for wallaby, 60 days of spear fishing, 250 days of line fishing, and 210 days of mollusk collecting during the year of study (Jones 1980).

Are the Gidjingali typical of human foragers? Judging by the review of Hayden (1981), they seem to be, despite their partial dependence on purchased food. Among Hayden's many conclusions are the following:

1. Most human foragers have a high quality diet with much of the energy derived from animal protein.

2. In many environments, seasonal food shortages occur. One response is to reduce activity and energy demand as in the Gidjingali. Another is to increase foraging distances as in the G/wi of Botswana who increase their average individual foraging area from 0.001 km^2 after the rains to 0.4 km^2 during the dry season (Silberbauer 1981). A third approach is to engage in mutualistic trade with a neighboring population (section 9.2.5).

3. Food sources are chosen using criteria that maximize returns relative to effort (see Figure 9-3b).

4. There may be distinct divisions of labor, with women gathering and men hunting, for example. This pattern is not always pronounced as it is in the Gidjingali, with !Kung men in Botswana gathering plant food as well as hunting.

5. Sufficient food is often collected in only two to five hours each day and, in rich environments, less than half the population may be engaged in foraging.

6. To maximize foraging success, most populations undergo seasonal migrations. The Gidjingali move from inland dunes early in the dry season to the coast for the wet season.

7. Many foragers exercise some management and conservation of their food resources. One method employed by the Gidjingali is burning. Others include careful removal of the edible parts of perennial plants and the sparing of pregnant animals to maintain recruitment in prey populations.

9.2.2   Shifting Cultivation

Shifting cultivation (otherwise known as swidden agriculture or "slash and burn") was the predominant mode of crop production in the tropics before European colonization. Harris (1977) emphasizes that there is no sudden switch from foraging to crop growing. His studies in northern Australia, Papua New Guinea, and the islands between show a rich diversity of foraging, tending, and planting of some species of tubers, sometimes within the same human population. Shifting cultivation remains a widespread technique, particularly for subsistence farmers. It is well suited to nutrient-poor soils in areas of low human density. The natural plant community is destroyed by cutting and burning, crops are planted for a few seasons, and then the fields are abandoned and the process begun elsewhere. Eventually the original plot may be reused after a fallow period. Provided that the fallow is long enough relative to the cropping period, the cycle is sustainable indefinitely. The cultivation period is just another disturbance in the continuing pattern of gap creation and secondary succession (see section 7.2.2).

Because of their great age and highly leached condition, many soils of the wet and moist tropics are too poor to grow crops without fertilization. Cutting and burning the vegetation provides this fertilizer in the form of a rich inorganic ash. This fertility does not last long (Figure 9-4a) because of nutrient losses due to cropping and leaching and because of a rapid buildup of pests and weeds. After the fields are abandoned, soil nutrients are rapidly reestablished (Figure 9-4b). Numerous studies have shown that the system causes no permanent damage to soils, provided that the ratio of cropping period to fallow period is not too high. Nye and Greenland (1960) suggest that one to two years of cropping followed by 10–20 years of fallow is usually sustainable in wet and humid forest areas, and four years of cropping with 10 or fewer years of fallow is usually sufficient in drier areas.

In terms of energy subsidy, shifting cultivation is often very efficient. Far more energy is harvested as food than is invested as labor and other energy inputs. The ratio of energy output to energy input is 15:1 for the Tsembaga cultivators of the mountains of New Guinea (Figure 9-5). These people also hunt and raise pigs to supplement protein intake. Although pig production produces only a small return on energy invested, these animals have important ritual significance and also are a means of storing food when crops are plentiful for use in poorer seasons.

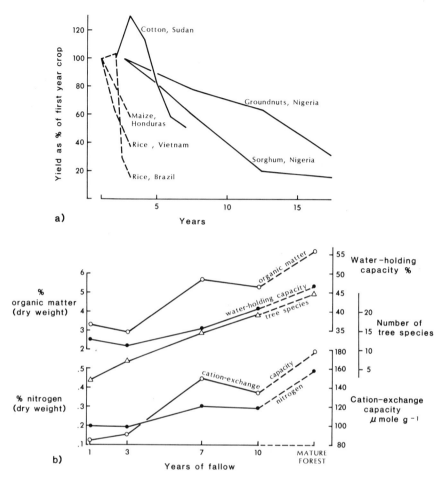

Figure 9-4 The decline and restoration in fertility associated with shifting cultivation. *a,* Decline in yield of crops as a result of nutrient depletion, weeds, and pests. Solid lines are drier areas (less than 1500 mm annual rainfall); broken lines are wetter areas. (Data from Nye and Greenland [1960], Kunstadter and Chapman [1978], Sanchez et al. [1982].) *b,* Restoration of tree vegetation and soil fertility in a 10-year fallow period in the humid forest zone of Nigeria. Substantial increases in tree species, soil nutrients, organic matter, and waterholding capacity occur during the time period; the extrapolations to the values of mature forest are on an unknown time scale. (Data from Aweto [1981a, b].)

As with many other shifting cultivators, the Tsembaga grow a wide range of crops (Figure 9-5), most of which are planted together in the same field. This practice is now known to increase yield and reduce pest problems (see sections 9.3.1 and 9.3.2).

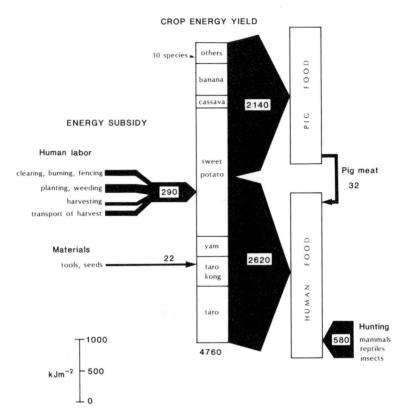

CROP ENERGY YIELD

Figure 9-5   Energy flow in the primarily shifting cultivator Tsembaga people who live in the mountains of Papua New Guinea. (Units = kJ m⁻²a⁻¹.) Depicted are energy inputs to cultivation, crop yields, and consumption by people and domestic pigs. Note the high diversity of crops (36 species). The people supplement their agricultural diet by hunting. (Data from Rappaport [1971] and Pimentel and Pimentel [1979].)

9.2.3   Nomadic Pastoralism

Pastoralists are people who obtain most of their sustenance from domestic animals, be it in the form of flesh, milk, blood, or the sale of animal products. Because of the need for pasture, they mostly inhabit the savanna and desert zones, which necessitates seasonal migrations in search of pasture and water. These movements are similar to those of wild ungulates, with the herds being widely dispersed during wet periods and concentrated around permanent water sources during drier periods (section 8.3.3). Given the distribution of climatic savannas and the major centers of livestock domesticaton, it is not surprising that pastoral economies have arisen mainly in Africa, Asia, and Arabia. The most important species—cattle, camels, sheep, and goats—were all domesticated from old world ungulates. A similar type of nomadic pastoralism seems to have occurred in the Andes of central Peru using the llama and alpaca up until around 500 AD, but these people also obtained much of their food from crops (Browman 1974). Many old world pastoralists cultivate no plants at all, although they may obtain plant food by trade or by incidental foraging while performing other tasks.

The interspecific interactions between people and their livestock cannot be characterized simply in ecological terms. The feeding relationship is predatory, more specifically carnivorous when meat is eaten or parasitic when milk (and blood) are consumed. In addition, the interaction is mutualistic, with both participants gaining from their association. The human benefit is obvious, but the livestock also gain by being guided to pasture and water and by protection from other predators.

Nomadic pastoralists generally occupy areas too dry to sustain rain-fed agriculture (Figure 9-6a), although they may move into wetter areas during dry periods. The density of people and livestock and the species composition of herds is also related to rainfall (Figures 9-6b and 8-9b). When camels, cattle, sheep, and goats are present, they represent a range of sizes and feeding habits not dissimilar from the wild ungulate communities with which they share the savannas.

Human densities are low because of low and unpredictable primary production and because of the dependence on warm-blooded herbivores and the resultant small amount of energy available to human secondary consumers. Nevertheless, pastoralism allows the conversion of low-quality plant food (grass), which is unsuitable for people, into high-quality food (meat, milk) in areas that can otherwise support few people.

Brown (1971) calculated that a typical East African pastoral family of eight (equivalent in consumption to 6.5 adults) needs 35–40 cattle to support sustainably its daily energy need, assuming a diet of 75% milk and 25% meat. This estimate is certainly too high because it does not allow for the richness of the milk (3700 kJ l^{-1} compared with 3000 kJ l^{-1} from cattle in industrial agriculture; see Little 1980), plant food obtained by trade or foraging, or the energy conservatism of many subsistence cultures (section 9.1). Using Brown's calculation, such a family herd consumes less than 52,000 kg of forage per year. This consumption is obtainable from less than 100 ha of savanna receiving 700 mm of rain, assuming that only 10% of forage production is eaten (using calculations of the type in section 8.2.3). Such a cattle density is equivalent to supporting one family per square kilometer, which is similar to the human density corresponding to 700 mm in Figure 9-6b. Given average rainfall evenly spread, such a system is easily sustainable. However, as stressed in Chapter 8, savannas are subjected to large spatial, seasonal, and annual fluctuations in rainfall, and this leads

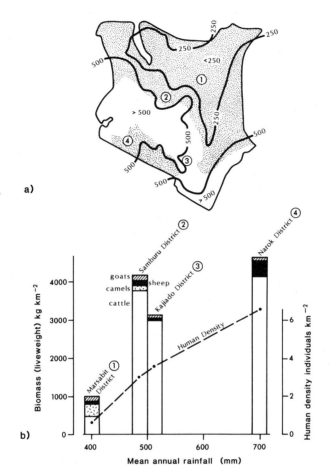

Figure 9-6   Nomadic pastoralism in Kenya. *a,* Most pastoral areas (stippled) are semiarid savannas (below 500 mm isohyet) and arid savannas (below 250 mm isohyet). (Modified from Dyson-Hudson [1980].) *b,* Livestock biomass and human population density in rural areas related to rainfall in four pastoral districts of Kenya. (Data from Kenya census [1969], Coe et al. [1976], Dyson-Hudson [1980].)

to periodic shortages in forage, high livestock mortality, and, without external support, high human mortality.

During wet periods, livestock populations may increase more quickly than wildlife because husbandry leads to a higher potential rate of recruitment. Each herd comprises 50%–75% adult females, 10%–40% calves and only 5%–30% adult males—an age and sex structure allowing for rapid population increase if sustained over several years. In favorable periods, families may build up livestock populations much greater than their subsistence needs, with 120 cattle and 150 sheep being common among the Maasai of Kenya and Tanzania (Jacobs 1975). The corollary is that during a series of dry years, livestock mortality is very high, reaching 70% among Maasai cattle during droughts of the early 1960s and mid-1970s. The emphasis on large herds during wet periods is a means of ensuring that at least some will survive through periods of drought.

This pattern of increase and crash is predicted by models of populations tracking changes in environmental carrying capacity (Figure 4-15b). It suggests that pastoral husbandry leads to a greater predominance of r-tendency characteristics than is normally found in large mammals, as an adaptation to unpredictable environments (see Chapter 6). While it may seem a harsh way of life, it allows for the maintenance of a higher livestock biomass, on average, than wild herbivores attain (Figure 8-9b) and therefore feeds more people than could otherwise be supported in these environments.

Nomadic pastoralists generally limit their manipulation of the environment to selection of grazing and watering sites, digging of wells, and the setting of dry season fires. When used judiciously, this latter practice improves pasture by stimulating grass growth and preventing bush encroachment.

9.2.4   Continuous Cultivation

Although shifting cultivation is the most widespread form of subsistence agriculture in the tropics, there are large areas, notably in India, where fields are used year after year or with only occasional and short fallow periods under grass. Similar systems were also used in temperate countries before the advent of industrial agriculture. To counteract the decline in crop production on tropical soils, increased labor can help to control pests and weeds, but nutrient losses remain a major problem. Careful management may return nutrients removed in crops, but losses due to soil erosion are much greater than under natural vegetation and cannot be eliminated entirely (Table 9-2).

Continuous subsistence cultivation in the tropics is, therefore, only possible on unusually fertile soils (such as those of recent volcanic origin), where there is significant import of nutrients (such as from periodic flooding), or where intensive management minimizes losses. In the last case, crop rotations, intercropping, and the use of all available manures help to maintain adequate yields, but as livestock dung is usually an integral part of such systems they are considered under mixed agriculture (section 9.2.5).

Import of nutrients is an important feature of seasonally flooded river valleys, and much of the rice paddy cultivation of India and Southeast Asia is partially dependent on nutrients in the alluvium that is deposited. Such irrigated rice cultivation is also enhanced by nitrogen-fixing cyanobacteria in the paddy fields. More than 30 million

Table 9-2   Effects of crop growing on soil erosion and nutrient losses in West Africa. The top portion of the table shows that soil erosion under crops or from bare ground is much greater than under natural vegetation. Higher rainfall and steeper slopes increase the rate of loss. The bottom portion of the table shows the nutrient losses in runoff water and eroded soil at Samaru, Nigeria, under the same cropping regimen as the top portion and the approximate quantity of nutrients removed in a typical maize crop. Data from Kowal and Kassam (1978).

			EROSION (t ha^{-1}a^{-1})		
LOCATION	MEAN ANNUAL RAINALL (mm)	SLOPE (%)	NATURAL VEGETATION	CROPS	BARE GROUND
Abidjoss, Ivory Coast	2100	7	0.03	0.1–99.0	108–170
Sefa, Senegal	1300	1–2	0.2	7.3	21
Samaru, Nigeria	1071	0.3	Negligible	4–21	4
Ouagadougou, Burkina Faso	250	0.5	0.1	0.6–8.0	10–20

	kg ha^{-1}a^{-1}				
NUTRIENT LOSSES	N	P	K	Ca	Mg
From soil under crops by runoff and erosion	14	8	9	8	3
In crops assuming a 9000 kg ha^{-1} harvest	225	27	130	28	27

hectares of India is now farmed in this way and the sustainability of production is illustrated in areas that have been farmed for more than 2000 years (Sopher 1980). Continuous cultivation also seems to have been a feature of the Maya in some parts of Central America. In southern Mexico and norther Belize, archeologists have found the remains of systems of canals and permanent fields dating back almost 3000 years (see Flannery 1982).

The net energy gain (crop yield : energy subsidy) is usually much less than in shifting cultivation because more energy is applied as labor and in the manufacture of implements, and the crop yield is usually less. In Vietnam, shifting cultivators commonly achieve twice the rice harvest of their neighbors using continuous cultivation (Condominas 1980). Similarly, a comparison of maize production systems across a wide geographical range of subsistence farmers shows that shifting cultivators achieve twice the yield and two to four times the net energy gain of continuous cultivators (Pimentel and Pimentel 1979).

9.2.5   Mixed Subsistence Economies

Few human populations obtain all their food from the exclusive pursuit of one of the strategies outlined in the preceding sections. Nearly all have some relationship with cash economies, as in the purchase of food by the Gidjingali foragers. Some barter goods or labor. Many employ a mixture of food procurement techniques, such as the shifting cultivation, pig rearing, and hunting of the Tsembaga.

Trade does not require access to cash economies or sophisticated cultures but arises where mutual benefit can obtain. Milton's (1980) study of two neighboring populations in the Amazonian rain forest illustrates this clearly. The Maku are nomadic forest hunters who gather little plant food but have recently begun to cultivate small amounts of cassava. In contrast, the Tukanoa live close to large rivers in which they fish; they hunt in the forest only occasionally but grow cassava as their dietary staple. It seems that the Maku obtain adequate protein throughout the year but experience a seasonal shortage of food energy, while the Tukanoa have an adequate energy supply from their cassava year-round but are short of protein when fish are scarce in June and July. A mutualistic relationship has developed in which the Tukanoa obtain animal protein from the Maku hunters in exchange for cassava. The Maku women also provide labor in the Tukanoan fields in return for cassava during the lean season.

Mixed agriculture, with crops and livestock integrated into a single agroecosystem is widely practiced. Besides food, the animals provide fertilizer and domestic fuel in the form of dung and labor for ploughing, operating irrigation systems, and transport. In the moist forest zone of East Bengal, India, Odend'hal (1972) described the role of cattle in a 15 km² agricultural community. Rice is the staple food of the 16,000 people, but in addition 3800 cattle, 32 water buffalo, 1800 goats, 4 pigs, and 3000 chickens and ducks are kept. Because of religious beliefs, cattle are not eaten but do supply milk, draft power, fuel, and fertilizer while consuming agricultural waste products for the most part. The energy budget for this cattle population is shown in Figure 9-7.

Dung as a fertilizer is particularly important on nutrient-poor tropical soils. The Quechua and Aymara are mixed farmers in the alpine savannas of the Peruvian Andes where they employ a crop and grazed fallow rotation (Winterhalder et al. 1974). During the grass fallow, llamas, sheep, and cattle deposit feces on the soil while grazing in the daytime. Prior to cultivation, sheep dung from overnight corrals is also spread on the fields (llama and cattle dung are used as domestic fuel). A typical application of 18,000 kg ha⁻¹ of sheep dung reduces soil acidity (pH 5.1 to 6.8) and produces threefold to fourfold increases in nitrogen, phosphorus, and exchangeable cation concentrations. Dung fertilizer is thought to have been used by farmers such as the Ufipa of Tanzania prior to the rinderpest epidemic of the 1890s in Africa. This disease wiped

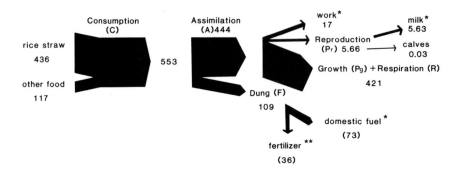

Figure 9-7   Energy budget of domestic cattle in East Bengal, India. (Units = kJ m^{-2}a^{-1}.) The cattle eat mainly waste products but provide useful energy (*) to their owners as milk, domestic fuel, and work as draft animals. They also provide fertilizer in the form of dung (**). "Other foods" are mustard oil cake, wheat bran, and rough pasture. (Data from Odend'hal [1972].)

out perhaps 80%–90% of livestock. Without this fertilizer, the people were forced to move their field sites and settlements much more frequently to maintain adequate food production (Kjekhus 1977).

9.3   Industrial Agriculture: Dependence on Fossil Fuels _____

Industrial agriculture is, by analogy with the industrial revolution of 19th century Europe, the substitution of much human and animal labor with energy from fossil fuels. In the systems reviewed previously, fossil fuels played little or no part beyond the manufacture of simple implements. Most of the energy subsidy in industrial ecosystems is supplied by fossil fuels, although human or animal labor still play an important role. The fossil fuels are used to manufacture agrochemicals (fertilizers, herbicides, pesticides), to manufacture farm machinery, and to run this machinery as it prepares soils, plants seeds, harvests crops, pumps water, applies agrochemicals, transports produce, and performs numerous other tasks. Major socioeconomic changes accompany this transition as local populations change from being largely self-supporting to being dependent on national and international economies.

The energy budget of the Egyptian national agroecosystem is shown in Figure 9-8. Rainfall is low throughout the country, and agriculture is therefore dependent on the irrigation waters of the Nile. The land alongside the river and in the delta is intensively farmed, with most fields yielding at least two harvests per year. Less than one fourth of the land is without a crop at any given time. Overall, the system produces more than 45×10^6 kJ ha^{-1}a^{-1} of energy and 330

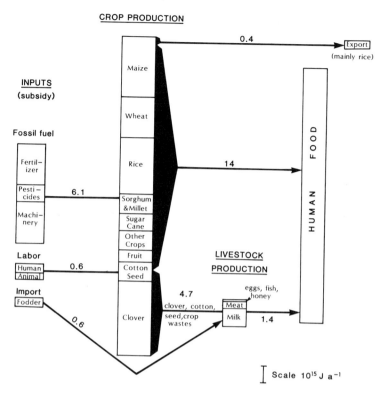

Figure 9-8 Annual energy flow in the Egyptian national agroecosystem of 3 million hectares. Fossil energy inputs include production and running costs of agrochemicals, farm machinery, and transportation. In addition, 14% of land is used to produce 0.8 kg ha^{-1} of cotton lint, most of which is exported. "Other crops" includes beans, groundnuts, onions, watermelons, and potatoes. (Data from Stanhill [1979].)

kg ha^{-1}a^{-1} of protein as human food. Ten percent of this energy and 20% of the protein is in animal products. Only a small proportion of the produce, mainly rice, is exported, but 15% of the cultivable land is devoted to cotton production, which makes only a small contribution to human food as cotton seed oil or indirectly as cotton seed fodder for livestock. If the total food production was evenly distributed to Egyptians, which it is not, every person would receive as adequate daily supply of 10,000 kJ of energy and 70 g of protein.

As the degree of industrialization (measured as the fossil fuel subsidy) increases, so does food production. However, the net energy gain (food production : energy subsidy) is negative in most developed countries (Figure 9-9). In the extreme case of Israel, 4 J of energy input is needed to produce 1 J of food. Even more remarkable is the 78 J subsidy needed to produce 1 J of beef protein in feedlot systems in the United

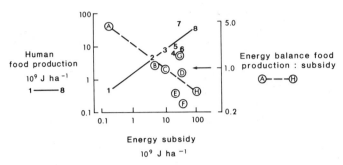

Figure 9-9   Comparison of the agroecosystems of eight states. A larger energy subsidy in industrial agriculture increases food production. However, the net energy gain decreases as the subsidy is increased. Numbers and solid line represent food production; letters and broken line represent net energy gain. 1,A, Australia; 2,B, United States; 3,C, California; 4,D, United Kingdom; 5,E, France; 6,F, Israel; 7,G, Egypt; 8,H, the Netherlands. (Data from Stanhill [1979]; logarithmic scales.)

States (Pimentel and Pimentel 1979). Such practices are dependent on cheap and plentiful supplies of fossil fuel. Since this resource is nonrenewable, these systems are not sustainable in their present form.

A simple transfer of agroecosystems from developed to developing countries will result in feeding fewer people at greater cost. For example, if Egypt was to copy the agroecosystem of the United States, it would need to spend much more on machinery and redirect much of the primary production to feeding livestock (Table 9-3). Such a change would reduce per capita intake of food, leading to the necessity of imports. It would also release many people from agricultural pursuits to look for other work. Already the large-scale irrigation control enabled by the Aswan Dam has necessited large energy subsidies for fertilizer manufacture (Table 9-3). Prior to damming the Nile, seasonal flooding provided a nutrient-rich alluvium, but most of this "free" fertilizer is now retained behind the dam.

9.3.1.   From Subsistence to Industrial Agriculture in the Tropics

There are fundamental differences between the major energy and matter flows in industrial agroecosystems and subsistence agroecosystems (Figure 9-10; Table 9-4). Subsistence systems are largely self-contained, not greatly different from natural ecosystems, and potentially sustainable indefinitely at low population densities. In contrast, industrial agroecosystems are totally dependent on energy and materials from other sectors of the industrial economy, which in turn are dependent on nonrenewable fossil energy (Figure 9-10*b*). This does not mean that industrial economies will of necessity collapse as fossil fuels become scarce and

Table 9-3   Comparison of some features and consequences of the industrial agroecosystems of three states in the early 1970s. Note the greater inputs of machinery, some agrochemicals, and the greater consumption of animal products and fewer people employed in California and Israel compared with Egypt. Data from Stanhill (1979).

	EGYPT	CALIFORNIA	ISRAEL
FOSSIL FUEL INPUTS (ha^{-1}a^{-1})			
Inorganic fertilizers (kg)	146	156	458
Pesticides (kg)	70	6	20
Tractors	0.007	0.04	0.04
OUTPUT OF HUMAN FOOD (ha^{-1}a^{-1})			
Animal origin (%)	8	38	33
Persons fed	11.3	6.8	7.8
LABOR INPUTS			
Persons employed (ha^{-1})	1.9	0.1	0.2
Persons fed per person employed	6	85	39

expensive, but they will have to substantially modify current farming methods to reduce this dependency. Developing countries have the opportunity to devise sustainable agroecosystems that address other priorities to those of the developed regions and that are better suited to tropical environments. New methods are essential as population growth outstrips the potential of subsistence systems. In terms of cultivable land, many countries of Central America, the Caribbean, Africa, and Asia have gone beyond the 50 or so people that can be supported on every square kilometer by shifting cultivation.

Some ecologists, such as Janzen (1973) argue cogently that temperate farming systems are inappropriate for tropical environments and economies. The remainder of this section enumerates some of the problems and suggests some solutions that take account of tropical ecological factors. Ecologists do not have all the answers; what is needed is a coalition of agriculturists, plant breeders, engineers, soil scientists, and others familiar with tropical environments to devise appropriate farming methods. These scientists and technicians should not, however, ignore the local knowledge accumulated over many generations of subsistence peoples.

Energy Subsidies.   Modern European styles of agriculture have very high yields at the cost of enormous fossil energy subsidies (Figure 9-9). To increase crop production in the tropics, greater subsidies are probably inevitable, but the fossil fuel component can be kept at low levels. Draft animals can be used to operate much agricultural machinery, particularly

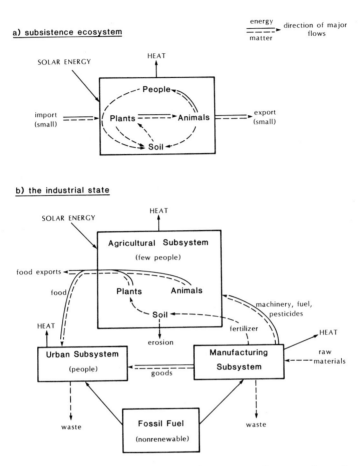

Figure 9-10   Contrast between idealized energy and matter flows in subsistence ecosystems exploited by people and industrial agroecosystems. The decomposer system has been omitted for clarity as it is little used as human food and is actively suppressed in industrial agroecosystems. *a,* Subsistence systems are largely self-contained and not greatly different from natural ecosystems. *b,* Only the agricultural subsystem of an industrial state is easily compared with the subsistence system. However, this portion is not self-supporting and is interdependent with other subsystems. Nutrients are not cycled, with crops removed to centers of population and soil lost by erosion. Replacement of nutrients is by manufactured inorganic compounds from the industrial subsystem. All subsystems are directly or indirectly dependent on energy from fossil fuels. Urban and industrial wastes are usually dumped as they are not in easily cyclable forms without further processing.

on small holdings. Indeed, Bhatia (1977) could find no evidence that their replacement by tractors led to increased crop yield or financial return in India. The use of animal and human labor always means a smaller energy subsidy than modern machinery, although it usually takes much longer to perform a given task. For example, hand spraying of

Table 9-4   Comparison of some characteristics of different types of food procurement systems. Numerical estimates are only rough approximations.

CHARACTERISTICS	FORAGING	SHIFTING CULTIVATION	NOMADIC PASTORALISM	INDUSTRIAL AGRICULTURE
HUMAN POPULATION				
Density (km^{-2})	Less than 10	25-50*	1-5*	2-500
Mobility				
Settlements	Temporary	Temporary or permanent	Temporary	Permanent
Food-collecting area	Temporary	Temporary	Temporary	Permanent
More probable food limitation	Energy	Protein	Energy	Protein, energy, or none
PRODUCTION SYSTEM				
Food production (kJ m^{-2}a^{-1})	Inapplicable	250-500*	5-75	100-5000
Energy balance (output : input)	+ve, less than 5:1	10:1-30:1	(Marginally +ve?)	+ve 5:1 to -ve 1:5
Fossil energy subsidy	Negligible	Negligible	Negligible	Massive
Nutrient cycles	Undisturbed	Medium-term balance	Medium-term balance	Highly disrupted
EFFECTS ON ECOSYSTEMS				
Local species diversity	Unaffected	Maintained	Maintained	Greatly reduced
Successional changes	Unaffected or utilized	Utilized	Utilized	Opposed
Environmental manipulation	Little to none	Great, temporary	Moderate, temporary	Great, permanent

*Allows for fallow areas.

herbicides requires about 3100 kJ ha^{-1} in human labor while a tractor and sprayer consumes 200,000 kJ ha^{-1} in petroleum alone (Terhune 1977). Fossil energy subsidies devoted to weeding can be reduced yet further by hand or mechanical hoeing instead of herbicides. However, in situations where insufficient labor is available at critical phases of the crop cycle, mechanization may be the only way to increase food production. Many of the suggestions in subsequent sections also reduce the amount of agrochemicals needed. The production, distribution, and application of these chemicals constitute the biggest energy subsidy by far in industrial agriculture (Table 9-3).

Climate.   Tropical climates present major problems for sustained industrial agriculture. In drier climates, rainfall is so unpredictable that feeding people from the land every year is impossible even at relatively moderate population densities (Figure 9-11; see also Figure 1-4). This is an acute problem in parts of Africa and India but is not experienced or understood by many people from temperate or moist tropical crop growing areas. Developments in agroclimatology can increasingly optimize water use by crops in more predictable climates and help to determine when crops should be planted. However, forecasts are based mainly on probability of adequate rainfall for a given crop, rather than on accurate prediction for a coming season. Many subsistence cultures,

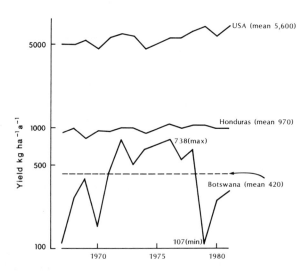

Figure 9-11   Maize yield in countries from different climatic zones; United States (temperate) and Honduras (moist, tropical) with relatively predictable precipitation show little annual variation, while Botswana (semiarid, tropical) with unpredictable rainfall shows enormous variation. The yield axis is logarithmic to show proportional variation. (Data from Food and Agriculture Organization of the United Nations [1967–1983].)

through their lengthy association with their climates, are able to predict weather patterns better than professional meteorologists, and it is important that such information be assimilated before it is lost (see Antúnez de Mayolo 1983).

Another approach is to intensify research into drought-resistant crops. Although maize is favored by many cultivators, it is much more susceptible to drought than cereals such as sorghum, millets, and the root crop cassava. With just a fraction of the effort devoted to maize breeding, these crops could increase the predictability of crop yields and thereby reduce the frequency of famines. Cassava has a higher productivity than cereals on infertile acid soils and much potential for breeding varieties with more protein than current strains. (Cock 1982).

Soils and Nutrients. The unpredictability of rainfall is a major and direct effect of climate in drier tropical agriculture, but "too much" rain has a detrimental effect on soils in the humid tropics. Tropical soils are often much older than those in temperate regions and as a consequence are highly leached and low in nutrients in many parts of the humid tropics. Even when similar types of tropical and temperate soils are compared, apparently tropical soils experience much larger nutrient reductions after deforestation than do the temperate soils (Allen 1985). In natural ecosystems on soils of low-nutrient content, a large proportion of most elements is locked up in the vegetation (Chapter 3). Once this nutrient capital has been used up after felling and burning of forest vegetation, the soils are generally poorer than temperate agricultural soils. Nutrients removed in crops or by erosion are not replaced quickly enough to maintain fertility (Table 9-2). It seems, therefore, that the use of organic (section 9.2.5) or inorganic fertilizers is unavoidable. For example, approximately 80% of soils in the Amazon basin cannot support continuous agriculture without such additions. To minimize energy and cash costs, applications need to be carefully tailored to the requirements of specific soils. Experimental programs judiciously using crop rotations and inorganic fertilizers have produced sustained yields for more than five years on some of the poorest soils of Amazonia and Indonesia (Sanchez et al. 1982, Conway et al. 1983, respectively). In the Amazon Basin of Peru, rotations of rice, maize, and soybean, or rice, peanuts, and soybean require lime to reduce acidity, nitrogen for the nonlegume crops, phosphorus, potassium, and magnesium with every crop and less frequent applications of the trace nutrients copper, zinc, boron, and molybdenum. Using only human labor, this carefully planned system has increased the income of small holders and gradually improved many aspects of soil quality.

Erosion is insignificant on the flat lands of the Peruvian program, but different techniques are called for when farming on slopes. Using no-tillage methods on slopes up to 15% in Nigeria, soil losses were reduced to less than 1% of those experienced on ploughed fields

(Greenland 1975). Such techniques also enhanced moisture retention and nutrient cycling because of the mulch of crop residues and weeds that cover the soil. Herbicides are necessary to kill the weeds while the crops are growing, but the soil surface is left undisturbed.

The experimental agroecosystems in Peru, Indonesia, and Nigeria just mentioned include legumes in crop rotations because, in their root nodules, these plants have symbiotic bacteria that fix atmospheric nitrogen. The use of nitrogen fixers has great potential for reducing the reliance on inorganic nitrogen fertilizers. Besides some bacteria, many cyanobacteria can fix nitrogen, and they contribute substantially to the nitrogen economy of paddy rice both as free-living forms and in symbiosis with the floating aquatic fern *Azolla*. This fern has been widely used for hundreds of years as a green manure in China in both wet and dry farming and has distinct possibilities for use in the tropics.

Removal of natural forest vegetation leads to **laterization** in some tropical soils. Exposure to the sun causes soft clays to become brick-hard soils that support little plant growth, including crops. This process was thought to be widespread among rain forest soils, but in fact is more typical of dry forest environments with a marked dry season. Consequently, laterization only affects a small proportion of the soils of Amazonia and Southeast Asia but is much more common in Indian and African forest soils. In the Amazon Basin, for example, only 4% of soils are susceptible to laterite formation (Sanchez et al. 1982).

Biotic Interactions. Competitors of crops (weeds) and predators (pests) often pose greater problems in tropical agroecosystems than they do in temperate regions. Although previous chapters have suggested that there is no conclusive evidence that biotic interactions are more intense in natural ecosystems of the tropics, in simple agroecosystems, both weeds and pests seem to greatly reduce crop yield. In temperate areas the cold winter suppresses these organisms such that populations have to build up from low numbers each year. However, in the humid tropics the climate is favorable and food is available throughout the year so that weeds and pests are a perpetual problem. The high diversity of tropical communities also means that there are many potential weed and pest species, and where continuous cultivation is practiced, these species have every opportunity to establish.

Some biotic interactions can, however, be beneficial to crop production. While sequential crop rotations help to maintain soil fertility, simultaneous rotations or intercropping often has the additional benefit of increasing total yield from a given field. In Nigeria, a cowpea monoculture yielded 1185 kg ha^{-1}, but when interspersed with maize, the total production was 2370 kg ha^{-1} (665 cowpeas, 1705 maize). Use of no-tillage methods further increased the combined harvest to 3750 kg ha^{-1} (Greenland 1975). All the reasons for enhanced yield are not known, but reduced susceptibility to pests and diseases is a contributory factor (see section 9.3.2). Intercropping also reduces soil erosion because at least one crop is usually in the ground at any particular time, and its canopy reduces the impact of rain drops and its roots help to bind the soil.

Another biotic process that continuous cultivation has to fight against is the rapid secondary succession found in the tropics (Figure 9-4b). Much of the effort put into ploughing, weeding, fertilizing, and weed and pest control is devoted to preventing succession. In sharp contrast, shifting cultivation actually uses succession to maintain productivity and minimize energy subsidies. Hart (1980) has suggested that agroecosystems could mimic secondary succession, although such ideas have yet to be tested. He proposes a scheme for Central America that first plants maize and legumes as the equivalents of herbaceous pioneers, interspersed with slower growing shrub and tree crops such as cassava and bananas. Larger trees including coconut, cocoa, and rubber would be planted next, followed by timber species to give a mixed forest climax of food and fiber tree crops. This proposal is one of many agroforestry schemes in which herbaceous and woody crops are grown together to provide food, fuel, and building materials for rural populations. Combined with some of the improved cultivation methods outlined, agroforestry has much to recommend it for producing sustainable and relatively self-contained communities.

It should be clear from the preceding pages that there is no panacea for tropical agriculture. Rather there are many possible avenues that need to be assessed in terms of their applicability to any particular situation. Although the discussion has been subdivided into energy subsidy, climate, soils, and biotic factors, there are many areas of overlap. For example, climate affects soil fertility and the biotic interactions between nitrogen-fixers and crops is often mediated through soils. This stresses the need to consider agroecosystems as a whole, choosing farming techniques that harmonize with local environments and cultures, and that simultaneously produce sustainable yields of food and other crops. Nowhere is this holistic approach more necessary than in the control of pests, as described in the next section.

9.3.2   The Ecology and Control of Pests

Much effort has been applied to trying to understand the ecology of pests with a view to controlling them. Pests have no special ecological quality but are creations of the human imagination. They can only be defined in terms of economic damage, itself an abstract human concept that has no ecological meaning. Outbreaks of population do occur in natural communities, whether of insects or ungulates, but such eruptions either level off or crash because further increase cannot be sustained by the resources available. To allow this natural course in agroecosystems would mean greatly reduced food production and that is why control is desirable.

Pest species may arise in several ways. In many cases they are recruited from the local fauna when natural communities are converted to agroecosystems. Uvarov (1964) describes the changes in insect faunas that accompanied the conversion of natural grasslands to wheat fields in the USSR. Total density increased almost twofold, while species richness was halved. Of the remaining species, a few became extremely abundant, comprising more than 90% of the fauna, and these become the potential pests. Predators and parasites of the pests were also greatly reduced. For example, the parasitization rate of a noctuid caterpillar dropped from 90% to less than 10%. Other pest species originate by the introduction of organisms from different geographical areas (see section 5.5.3) and the use of new crops to which local species are by chance preadapted. Whatever the origin of pest species, the disruption and simplification of vegetation in agroecosystems lead to the rapid development of pest faunas.

Initially the diversity of a pest community increases with time as shown by the history of tea pests (Figure 9-12). When the number of pest species stabilizes (after about 100–200 years), the area under cultivation becomes the major determinant of pest species richness. Such a situation is found in the pest fauna of sugar cane worldwide. For example, in the West Indies, Antigua has only 5000 ha under cane and 16 pest species, whereas Cuba has 999,000 ha and 81 species. Both islands began cultivation of sugar cane early in the 16th century (Strong et al. 1977). Latitude has much less influence on the diversity of both tea and sugar pests than in natural ecosystems, reflecting the likely importance of plant species richness and structural diversity in producing high herbivore diversity. While some pests are geographically widespread, most species develop locally. Only 3% of tea pests and less than 1% of sugar cane pests are common to all cultivated areas. The effect of

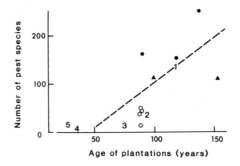

Figure 9-12   The number of insect pest species on tea crops in different regions is related to the age of plantations. Data from Banerjee (1981). •, Indian subcontinent; ▲, Southeast Asia; ○, Africa; 1, Sri Lanka; 2, Taiwan; 3, Japan; 4, Turkey; 5, Papua New Guinea.

pests on crop yields is often insignificant at low density because plants grow more to compensate for losses. At high densities, damage becomes devastating, and this is when control is necessary (Figure 9-13).

In industrial agroecosystems, **chemical control** has been highly effective until recently. However, the manufacture of these chemicals requires large energy subsidies averaging 200,000 kJ for each kilogram produced (Pimentel and Pimentel 1979). A number of other problems with these chemicals have also become apparent following their widespread use. First, because they are poisons, pesticides kill beneficial organisms as well as pests. Pest populations may even rise following spraying, if the chemicals more effectively kill the natural enemies rather than the pest itself. Such a situation arose following the use of dieldrin in Sri Lankan tea plantations to control a beetle pest. Another pest, a tortix caterpillar, increased because its natural enemy was killed by the pesticide (Danthanarayana and Kathirauetpillai 1969). Soil animals are also often reduced, thereby affecting soil fertility by disruption of decomposition and nutrient cycles. Reduction in the soil fauna seems to have been an important factor in the decline of maize yields in Nigerian plots sprayed with DDT (Perfect et al. 1979).

Second, many pesticides are persistent and cannot be broken down by natural processes. The resulting pollution accumulates in higher trophic levels, including humans (section 3.4.2). The effects of most of these chemicals on human physiology are unknown, but their increasing concentrations and ubiquity in people, even those far from where pesticides are used, are causing increasing concern (see Ehrlich et al. 1977).

Finally, the application of toxic chemicals imposes a large artificial selection pressure on the target organisms to evolve resistance to the

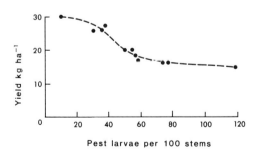

Figure 9-13   Light pest infestations have little effect on yield in many crops because the plants compensate with extra growth. As pest density increases above a threshold, the loss of yield increases markedly. This example shows the reduction in rice yield in relation to the density of African white rice borers. Reproduced from Brenier, J.; Walker, P. T. 1971. *FAO Manual on the Evaluation and Prevention of Losses by Pests, Diseases and Weeds.* Rome: Food and Agriculture Organization of the United Nations.

effects of those substances. Since 1930 the number of insect and mite pests resistant to one or more pesticides has been increasing exponentially to its 1977 level of 364 species.

Pests are not a significant problem to shifting cultivators because they exercise **cultural control**. When pests (and weeds) begin to build up to damaging levels, the farmer simply abandons the field and moves elsewhere. Because each field is only cultivated for a short period, pest populations (particularly highly specialized species) do not have the opportunity to build up. Cultural control methods, which generally involve maintaining a more complex agroecosystem, are also available to continuous cultivators. Both crop rotations and intercropping often reduce pest damage as well as improving soil fertility. In Costa Rican fields, monophagous pest populations in particular are much reduced on their host plants if other species of crop are interplanted. Natural predators and parasites of pests are also more abundant in such situations (Risch 1979, 1980).

Detailed knowledge of pest ecology may facilitate cultural control. *Quelea* birds, which form large migratory flocks, are a serious pest in much of Africa. By timing the rice harvest to coincide with the seasonal absence of the birds in mid-May to mid-June, in many years, damage can be kept below 1% of the crop in parts of Chad and Cameroon (Elliott 1979).

Where naturally occurring enemies are unable to control pests, it is sometimes possible to introduce potential enemies from elsewhere to reduce pest populations. This type of **biological control** can be spectacularly successful but in other cases has been totally ineffective (see Conway 1981). A typical example of a successful program is that directed against the potato tuber moth in Zambia. The moth attacks tubers in the soil and in storage, and insecticides have proved ineffective (Cruikshank and Ahmed 1973). Ten exotic species of possible parasites were introduced, and one of these, a chalcid wasp, attacked 56% of the moth larvae. Other species, including native parasites that reestablished in the absence of insecticides, played a minor role, with a maximum parasitization rate of only 4%. Yield of potatoes increased by 22% in two years, and the financial gains outweighed expenditure fifteenfold.

Although cultural and biological control are attractive because of their cheapness, low fossil energy subsidy, and lack of pollution, they remain very hit or miss in application. No sound theoretical basis exists that predicts which organisms or practices will effect control. Trial and error remains a major method of selection.

Integrated control is the best approach, using a battery of methods including chemical, biological, and cultural techniques in combination with a program to breed crop strains resistant to pests. If necessary, selective pesticides (which only kill harmful species) are used, together with other methods. For example, in the oil palm plantations of Malaysia, the pesticide Tetradifon EC was selected to control the spider mite

Oligonychus because it kills the mite but has little effect on its beetle predator *Stethorus*. In the same plantations, herbaceous vegetation is allowed to grow around the trees because it greatly reduces the density of rhinoceros beetles, which kill young palm leaves, although the reasons for the effectiveness of control are not clear (Wood 1971).

Summary

For most of human history, populations have subsisted as one forager among many species. Because of cultural biases and a degree of physiological diversity, the essentials of human nutrition are not well understood. Over the last 10,000 years human populations have increasingly exercised control over their environments and, in many cases, have created agroecosystems. In tropical countries the entire gamut of human feeding strategies coexist. Included are foraging, shifting cultivation, pastoralism, intensive "organic" agriculture, and intensive fossil fuel-based agriculture. Food production models based on those found in temperate developed countries cannot be transferred wholesale to the tropics because of ecological and cultural differences. An understanding of natural ecosystems and subsistence economies in the tropics can help toward developing sustainable methods of food production and crop protection.

Study Questions

Review

1. What are the ecological reasons for human mobility in many traditional subsistence systems?
2. Discuss the role of succession in human subsistence systems.
3. Why is the temperate, developed-country, agricultural model difficult to apply in poor tropical countries?
4. Compare and contrast nutrient cycling in natural ecosystems, shifting cultivation, mixed subsistence systems, and modern intensive agriculture.

Related Topics

1. Investigate the problems involved in determining human dietary standards.
2. How can tropical agriculture be intensified without becoming wholly dependent on fossil fuels?
3. Review cultural and ecological interactions between different types of traditional subsistence populations.
4. Report on the potential for biological control of agricultural pests in the tropics.

Further Reading ———————————————————————

Unesco (1978) and Unesco (1979) contain useful data on human food procurement in tropical forests and savannas, respectively (see "Further Reading" for Chapter 8).

Conway, G. (1981). Man versus pest. In: May, R. M. editor. *Theoretical Ecology*. 2d ed. Oxford: Blackwell, pp. 356–386. A good introductory account of the ecology and economics of pest control.

Harris, D. R. editor (1980). *Human Ecology in Savanna Environments*. London: Academic Press. A wealth of information about subsistence systems throughout the tropics, plus some discussion of human nutrition.

Nye, P. H.; Greenland, D. J. (1960). *The Soil Under Shifting Cultivation*. Tech. Comm. 51. Farnham Royal, Buckinghamshire, England: Comonwealth Agricultural Bureaux. An excellent ecological account of the reduction and subsequent restoration of soil fertility under shifting cultivation in Africa.

Pimentel, D.; Pimentel, M. (1979). *Food, Energy and Society*. London: Arnold. An excellent comparative account of energy subsidies in food procurement.

Scientific American. (1976). *Food and Agriculture*. New York: Freeman. Considers human nutrition, domestication and agricultural systems in developed and developing regions.

Age structure pyramids in developing countries usually have a broad base. This family from Malacca, Malaysia, has one grandparent, two parents, and nine young children. (Photograph by James N. Anderson.)

CHAPTER 10

Human Populations

This chapter is an ecological view of human populations that is comparable to the approach adopted in Chapters 4 and 5 toward populations of other organisms. Socioeconomic and political factors have important effects on human populations, particularly when comparing tropical and temperate regions. However, an ecological approach allows these factors to be put into a biological perspective. Cultural factors sometimes obscure the basic biology that needs to be understood before rational population policies can be devised. From an ecological perspective, human populations differ from those of most other animals in two important respects. First, they are distributed in all the major terrestrial biomes. No other land animals, except some of those intimately associated with people (parasites, pests, and domestic animals), are so widespread. Second, this distribution results to a large degree from human adaptability to different environments, which is itself enhanced by the ability to manipulate those environments to make them habitable.

Sections 10.1 and 10.2 deal with population parameters, models of growth, and carrying capacity in a similar way to those discussed in Chapter 5 dealing with prey-predator interactions. A word of caution is needed concerning the accuracy of many of the population statistics. Some are based on total censuses, others on sample censuses, and some are estimates based on predictions of models. None of these methods gives completely accurate results, and the degree of uncertainty varies greatly between different countries and regions.

10.1   Human Population Statistics _____

10.1.1   Density

On a national basis, human population densities range from 0.02 km^{-2} in Greenland to 18,300 km^{-2} in Macau, but most regions fall within the range 10–100 km^{-2}. The wide range of densities is best analyzed in terms of the following aspects of human environments:

1. *Abiotic environment.* The lowest locally sustainable densities are found in the most hostile climates such as cold or aridity. For example, the density of aboriginal foragers in Australia is closely correlated with rainfall, ranging from 0.04 km^{-2} in regions of 250 mm annual rainfall to 0.43 km^{-2} at 2500 mm (Birdsell 1978). Countries with predominantly semiarid climates have population densities below 5 km^{-2}, but those in wetter environments may be as low or much higher, depending on the following biotic and cultural factors.

2. *Biotic environment.* The availability of food or the prevalence of diseases may control population density. Even in the most productive natural ecosystems, rain forests, there is little food suitable for people, and consequently densities of foragers are low (Table 9-4). The effects of diseases are considered in section 10.3.

3. *Cultural environment.* As the intensity of human manipulation of ecosystems increases, so does the possibility of supporting higher population densities (Table 9-4). Historical factors such as war, colonization, epidemics, and famine also have major effects on population density.

Population densities cannot be separated in terms of tropical versus temperate nor between developing versus developed regions (Table 10-1). Europe, China, and Southeast Asia have similar high densities while Africa, North America, South America, and the USSR have comparable, lower densities. Regional and national statistics also hide the enormous variation found within a country. For example, rural districts in Kenya range from less than 1 km^{-2} to more than 300 km^{-2}, according to the 1969 census.

10.1.2   Age Structure

While density reveals little about the differences between populations in developed and developing countries, age structure highlights some of the causes of rapid population growth in the latter. Figure 10-1 compares the age structure of three developing countries with that of the United States. The developing countries have a much greater preponderance of young people, and the sides of the pyramids are concave. Such a shape means high population growth, not only now,

Table 10-1 Comparison of population statistics for tropical and temperate regions (or countries). Vital rates are annual averages for the period 1975–1980; total population and density are for 1981. Data from United Nations (1981).

	TOTAL POPULATION (MILLIONS)	DENSITY (km^{-2})	VITAL RATES PER YEAR		
			BIRTHS (⁰/oo)	DEATHS (⁰/oo)	GROWTH (%)
WORLD	4508	33	29	11	1.8
TROPICAL REGIONS					
Central America	95	38			
Tropical South America	204	14	35	9	2.6
Caribbean Islands	31	131	28	9	1.9
Africa	484	16	46	17	2.9
Southeast Asia	319	82	35	14	2.1
Melanesia, Micronesia, and Polynesia	1.4	48	34	7	2.7
TEMPERATE REGIONS AND COUNTRIES					
Temperate South America	42	11	22	9	1.3
North America	250	12	16	9	0.7
Europe	485	98	14	11	0.3
USSR	268	12	18	9	0.9
China	1008	105	21	7	1.4

but most likely long into the future, because of the large number of girls who will join the reproductive sector of the population in the near future. The age structure of the United States population in 1900 is intermediate between the current situation in developing and developed nations and gives some idea of the length of time needed for substantial change. To attain a stable population, the United States would need an almost vertical-sided pyramid between the ages of 0 and 30 years, given current mortality rates (Figure 10-1).

10.1.3 Natality and Mortality

Demographers usually express human birth and death rates per thousand members of the population per year; that is, a birth (or death) rate of 40 per thousand (⁰/oo) means that 40 individuals are born (or die) each year for every thousand people present in that population. The difference between these two rates is the exponential growth rate of the population per thousand per year. (If this figure is negative, it is

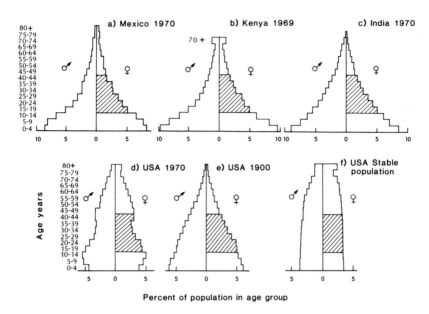

Figure 10-1   Age structure pyramids for various nations around 1970 and for the United States in 1900 and the projected pattern for a stable population in the United States, assuming current age-specific mortality rates prevail. Shaded portions are women of reproductive age. Note the large proportion of girls about to enter the reproductive age group in developing countries. (Data from Ehrlich et al. [1977] and Kenya census [1969].)

the exponential rate of decline; see 4.2.2 for a precise account of assumptions). Population growth rates are, however, usually expressed as percent per year. For example,

$$\text{Annual birth rate} - \text{Annual death rate} = \text{Annual growth rate}$$
$$40\,^o/oo \quad - \quad 20\,^o/oo \quad = \quad 20\,^o/oo \; or$$
$$2\% \text{ per year}$$

One of the striking discontinuities revealed in Table 10-1 is that between birth and growth rates in developing and developed regions. This divergence is even clearer in Figure 10-2*a* and *c*, where individual countries are shown. Most developing countries, whether tropical or temperate, have birth rates above 25°/oo and rates of growth greater than 1.5%. However, there is no similar clear distinction in death rates. All the developed countries have less than 10°/oo deaths annually, but so do a wide geographical range of developing countries (Figure 10-2*b*).

It is presumed that primitive human populations had high birth and death rates, perhaps around 40°/oo and consequently slow growth. In developed countries, both rates are much lower (below 20°/oo)

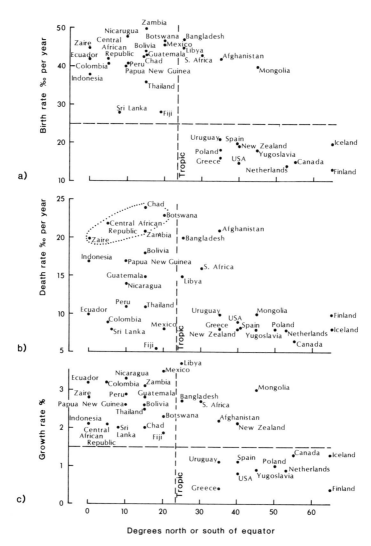

Figure 10-2 Annual vital rates of a variety of countries in relationship to latitude. Note that developing countries have higher birth and growth rates than developed countries (*a* and *c*, divergences illustrated by dashed lines), but there is no clear relationship for death rates (*b*). Highest death rates are mostly in African countries surrounded by dotted line. (Data from Ehrlich et al. [1977].)

and again growth rates are low. The historical change in the developed countries from the first state to the second is called the **demographic transition** (Figure 10-3*a*). This model of demographic change as economic development proceeds is often assumed to be applicable to developing countries also. However, in most developing countries there

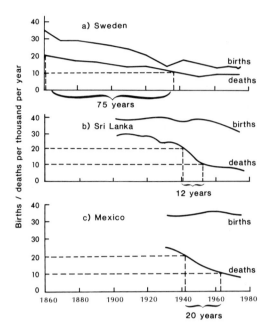

Figure 10-3   Demographic transition. Changes in birth and death rates with time in *(a)* Sweden, *(b)* Sri Lanka, and *(c)* Mexico. Dashed lines incorporate the time for death rates to decline from 20 °/oo to 10 °/oo. (Modified from Ehrlich et al. [1977].)

has been no such orderly progression as shown by Sri Lanka and Mexico in Figure 10-3. In these two countries, death rates have fallen more than three times as rapidly as in Sweden and other developed countries, but birth rates have declined only slowly or not at all. This has led to a more rapid increase in the rate of population growth in developing countries than was ever experienced in the developed regions.

Why has the demographic transition failed to happen in so many tropical countries? It seems that the idea of a causal relationship between economic development and vital rates is too simplistic. While death rates do appear to be less in richer countries, birth rate is not connected with national wealth (Figure 10-4). Money can "buy" improved health and reduced mortality as a medical product, but birth rate is a much more complex cultural phenomenon. Also the reduction in death rate can be effected extremely quickly, even within one generation, whereas changes in birth rate occur much more slowly with only small adjustments in any one generation.

Fertility rates of women of all age groups are much greater in developing countries. Enormous changes in the pattern of childbearing are required to substantially reduce the birth rate of the population as

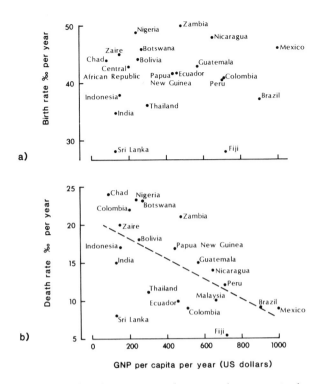

Figure 10-4   Relationship between vital rates and economic development (measured as per capita gross national product or GNP) in a variety of tropical countries. *a,* There is no relationship between birth rate and GNP. *b,* Death rate tends to be lower in countries with a higher GNP. (Data from Ehrlich et al. [1977].)

a whole (Figure 10-5*b*). In contrast, survivorship patterns are already similar to those in the developed countries and different to those of European cultures prior to the demographic transition (Figure 10-5*b*). Whether a transition to low birth rates will occur in all developing countries is open to question. The pattern will certainly be different to that typified by Sweden (Figure 10-3). Birth rates have dropped in some developing countries such as Sri Lanka but not in others (for example, Mexico) (Figure 10-3). The cultural diversity of developing countries is much greater than that in developed countries. As a result, it is unlikely that all of developing countries will respond demographically in the same way, or at the same rate, to economic development.

10.1.4   Immigration and Emigration

Movements of individuals or groups of individuals have had profound effects on the structure of human populations. From hominid origins, presumed to be in eastern Africa, human populations have spread to

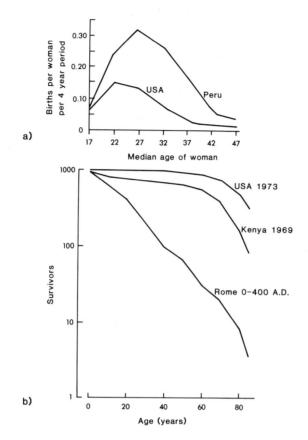

Figure 10-5   Fertility and survivorship in human populations. *a,* Comparison of fertility in women from Peru and the United States. (Modified from Ehrlich et al. [1977].) *b,* Comparison of survivorship in recent history in Kenya and the United States and in ancient Rome (Data from Kenya census [1969], Ehrlich et al. [1977], Hutchinson [1978].)

most of the earth as *Homo sapiens,* which has diversified into a handful of races and a multitude of cultures. Imperial cultures, from those of ancient Asia and the Mediterranean to the recent global empires based in Europe, have made a practice of colonization. The consequent migrations, murders, diseases, medicines, and new socioeconomic systems have had profound effects on population structure. Three examples of such large scale migrations follow:

1. The European slave trade between 1500 and 1800 took more than 12 million Africans from their own continent, most of them going to the Americas. In the same period, Arab slavers took around 1 million as a continuation of a trade that they had practiced for thousands of years. Assuming a population of around 50 million in sub-Saharan Africa during that period, and a maximum removal rate of around 75,000 per

year, there was probably little demographic impact on the African population as a whole. Rather, this rate of removal would have constituted an inhuman form of sustainable yield (McEvedy and Jones 1978). However, there was an enormous impact on the American lands to which they were taken.

2. Colonization of the Americas has led to the virtual replacement of the native populations by Europeans and Africans. Only in Mexico, Central America, and northwestern South America (Colombia, Venezuela, Ecuador, Peru, and Bolivia) do the Amerindians constitute more than one half of the population. Four million Africans were taken to the Caribbean, one-half million to North America, and three and one-half million to Brazil. These areas now have 18 million, 20 million, and 50 million descendants of slaves, respectively. Between 1500 and 1975, 65 million Europeans voluntarily moved to the Americas: mainly northern Europeans to what are now the United States and Canada and Spanish and Portugese to South America. Overall, the descent of present American peoples is 15% African, 60% European, 20% of mixed descent, and only 5% original inhabitants.

3. Colonization of tropical Africa by Europeans has been very different to that of the Americas. European administrations occupied most of the continent by the end of the 19th century but only remained until the middle of the 20th century. Relatively few people immigrated to Africa during this period, and many of these left at or shortly after independence. The British also encouraged Indians to settle, but even in one of the most favored colonies, Kenya, Europeans and Indians only consituted 0.4% and 1.3%, respectively, of the total population in 1969. The establishment of national boundaries by colonial administrations has had another effect in that it has largely stopped the intermittent but large-scale migrations of many African populations.

These three examples illustrate some of the very different effects that migrations have had on the ethnic structure of populations. Another effect has been on growth patterns, which are discussed further in section 10.2. Another type of migration is that from rural to urban areas. More than 60% of the population of developed countries now live in urban areas compared with around 40% in Latin America and 10% in other developing regions. If present trends continue, more than one half of the world population will live in urban areas by 2000, with all the social and economic consequences that that will entail.

10.2   Population Growth

The current growth of the global population is often likened to an explosion. In fact, world population has been growing, with only occasional respite and with increasing momentum for many thousands of years (Figure 10-6). However, the present rate of growth of around 2% is alarming because the present population of 4.5 billion (4×10^9) already presses on food resources in many countries, and yet will double in only 35 years. **Doubling time** is a useful concept because its effects are much easier to visualize than an exponential percentage rate of increase (see Appendix D for method of calculation). In many developing countries the doubling time is less than 25 years. The fastest current growth rate, 4% per anum in Kenya, yields a doubling time of 17 years.

Figure 10-6 shows the progress of global population growth over the last 12,000 years and identifies three phases of rapid growth preceded and interspersed with brief periods of no growth or even decline. More detailed patterns of change are shown for individual countries in Figure 10-7. The patterns for Brazil, the Philippines, and Tanzania, all of which had largely subsistence economies prior to European colonization, are similar. The cultural changes wrought by the colonizers, of which the Christian religions, continuous agriculture, and western medicine are

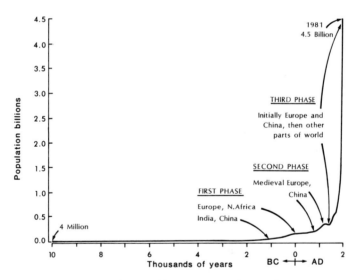

Figure 10-6   The growth of global human population from 10,000 BC to the present. The steady growth from 10,000–2000 BC reflects lack of information rather than a real absence of irregularities. The first phase of growth coincides with the Iron Age in Europe and the northeastern Mediterranean and the ancient empires of Egypt, Greece, Rome, India, and China. The second phase reflects medieval agricultural development in Europe and in China, the dynasties prior to the Mongol invasion. The third phase corresponds to the almost global colonization by Europeans and their effects on local cultures and the Manchurian dynasty in China. (Data from McEvedy and Jones [1978].)

arguably the most disruptive, have led to greatly increased rates of population growth, which have continued beyond independence and show little sign of moderating. Population growth in India has been rather different (Figure 10-7). The long history of agriculture, empires, and bureaucracies meant that population density was already fairly high before European colonization, with substantial growth over the previous 300 years. Nevertheless, colonization led to a new phase of growth that has resulted in India having the densest self-supporting population of

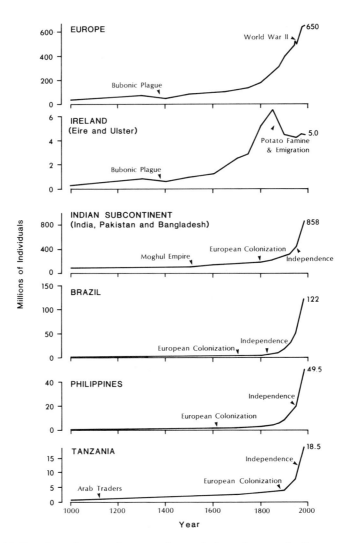

Figure 10-7   One thousand years of population changes in Europe and Ireland contrasted with the Indian subcontinent, Brazil, the Philippines, and Tanzania. See text. (Data from McEvedy and Jones [1978].)

any large land mass outside Europe. Much more detailed information is available for Europe where the effects of specific events such as bubonic plague and major wars can be illustrated (Figure 10-7). Population growth is leveling off markedly in most European countries, with a doubling time of 173 years overall and the populations of East Germany and Austria actually declining. The most dramatic well-documented reversal of population growth was that experienced in Ireland following a disease that ruined the staple potato crop in 1846 and 1847. More than three fourths of a million people died as a result of famine and a further million left for North America.

Overall, the global population shows an approximately exponential pattern of growth in contrast to the logistical model proposed in Chapter 4 as a more usual form of growth. This is because carrying capacity for human populations has increased with technological development, and it has not had the normal inhibitory effect much of the time (section 10.3; Table 9-4). However, population growth shows signs of outstripping food production. Such a situation can lead to catastrophic mortality in populations of other organisms that grow exponentially (Figure 4-12). There is no reason to suppose that the human population is immune from such an outcome.

10.3   Carrying Capacity and Human Population _____

There is no doubt that local and global carrying capacities have increased immensely during human history. As more advanced technologies have been applied to food procurement, so has the number of people supported per unit area of land. In the past and present, local populations have been constrained by prevailing carrying capacity as discussed previously and in Chapter 9. Perhaps the biggest difference between population growth in this century and previous phases has been the enormous improvements in medicine and hygiene, which have rapidly reduced death rates, leaving them markedly out of balance with birth rates. Combined with rapid increases in food production and distribution (increases in carrying capacity), this has led to the unstable situation now facing the populations of many tropical countries. The situation is unstable because food production cannot continue to increase indefinitely; indeed it currently shows signs of leveling off. But population growth has enormous momentum as a result of the age structure in developing countries (section 10.4). As described in Chapter 9, recent increases in food production are fragile because of the heavy dependence on fossil fuels and the use of demanding and delicate strains of crops.

To continue the present trends of population growth and resource use may have catastrophic consequences as illustrated in Figure 10-8. This "limits-to-growth" model has received much criticism, but its crude revelations of the dynamic interrelatedness of the many factors determining the carrying capacity for industrial populations are undeniable, even though the detailed predictions are inaccurate. In the first half of the period shown in Figure 10-8, population, food production, and industrial output all continue to rise, following their present trends. However, the latter two activities use up nonrenewable resources (minerals and fossil energy) increasingly rapidly and increase pollution. Eventual shortages of the nonrenewable necessities of industry (including agriculture) and increasing pollution lead to a rapid decline in carrying capacity followed by catastrophic human mortality. Faults of the model include its pessimistic view of the discovery of new resources and technologies, its treatment of the world as one homogeneous unit, and the inaccuracy of some of the data inputs. More specific and more refined models are less pessimistic, but nevertheless indicate the importance of curtailing population growth if a decent standard of living for most people is to be obtained (see Biswas 1979).

In many regions, population already exceeds the demand for food. These areas have become increasingly dependent on cereals imported mostly from North America (Table 10-2). Besides any political considerations, such dependence is also dangerous for ecological reasons. The grain production of the United States and Canada is based on a few strains of a few species. These crops are reliant on industrial methods that are

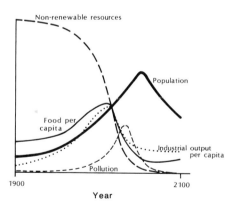

Figure 10-8   Limits-to-growth "standard model run" showing the interactions between population growth and factors such as food and other industrial production, nonrenewal resources, and pollution, which determine the global carrying capacity for human population. See text. (Modified from Meadows at al. [1972].)

Table 10-2 Changes in the pattern of international trade in cereals. Eastern Europe and the developing regions have become increasingly dependent on grain imports from the United States, Canada, Australia, and New Zealand to feed their people. Data from Brown (1981).

	1960	1970	1980
Total world grain production	863*	1137	1432
EXPORTERS			
North America	39	56	131
Australia and New Zealand	6	12	19
IMPORTERS			
Asia	17	37	63
Eastern Europe	0	0	46
Western Europe	25	30	16
Africa	2	5	15
Latin America	0	0†	10

*Figures are millions of tonnes.

†Latin America exported 4×10^6 t in 1970.

not sustainable indefinitely in their present forms (Chapter 9). In addition, the massive areas planted are potentially susceptible to a new disease that might arise and not be immediately controllable (note the Irish potato blight in Figure 10-7). Finally, any change in climate of this region, either a temporary drought or a long-term trend to drier climate as some predict (Chapter 3), could curtail the export of these cereals to the dependent regions.

Global food production is likely to continue to rise in the future but not as quickly as population growth. In the 1950s and 1960s, grain production per capita increased by 14% and 8%, respectively. But the increase was only 5% in the 1970s, and there was a decline of 14% in Africa (Brown 1981). Against this background the desirability of reducing growth and of eventually stabilizing population at sustainable levels is clear.

10.3.1   Stabilization of Population

To avoid starvation the human population needs to be stabilized at the local or global carrying capacity that is ecologically and economically sustainable. In fact, sporadic famine can only be avoided if the population is maintained below the average carrying capacity because of the vagaries of climate, particularly in drier areas (Chapters 8 and 9). A stable population has matching birth and death rates such that the net reproductive rate, $R_o = 1$ (Chapter 4). This situation is rare in human history, but during periods of slow growth it has been obtained approximately as it is now in some foraging cultures and some European countries. However, the way in which stable populations have been achieved differ greatly.

In societies without modern medicine, high birth rates have been roughly matched with high death rates resulting from famine or disease or by culturally induced deaths. The second phase of world population growth in Figure 10-6 seems to have been halted by the level of technology then available for food production. The population stabilized at or oscillated around 350 million (80% of whom lived in Asia and Europe) because of famines, wars, and diseases (McEvedy and Jones 1978).

Many human foragers seem to match their populations to the resources available, and one of the mechanisms employed is the killing of newborn children. Infanticide is sometimes used to remove children with congenital defects, but of greater demographic consequence is the killing of normal babies (see Dickeman 1975 for a review). Australian aborigines living on the desert fringes selectively kill baby girls, thereby keeping the reproductive potential of their populations in check (Birdsel 1978). The reason for killing babies rather than attempting abortion is to minimize risk to the mother as primitive methods of abortion carry a high maternal death rate. Whether human infanticide is a result of group selection, kin selection, or individual selection is open to question. In Chapter 2 group selection was rejected as a likely method of regulation in nonhuman populations. *H. Sapiens* has the ability for abstract thought, making possible adaptations that benefit the whole population. However, most traditional cultures have relatively small local bands or clans whose members are often closely related. In such circumstances, kin selection or individual selection could favor infanticide if survival of those already present is reduced by adding to the population .

The alternative to matching high birth rate with a high death rate is to match the low death rate made possible by modern food distribution, hygiene, and medicine, with a low birth rate. This balance is the cause of low rates of population growth in developed countries. Most European women choose to have few babies because cultural conditions make small families desirable.

The need for, and indeed inevitability of, a stable or reduced population in the future is clear. There are two ways of achieving this, each with two options. The first is increased death rates, either through deliberate killings or famine resulting from neglect. The second is a reduction in birth rates, either forcibly or by individual choice. Most people reject increased death rates or forcibly reduced birth rates for reasons of humanity, which leaves the fourth alternative as an obligation. Effective methods of birth control and family planning exist from which individuals can choose. National programs can provide the means, education, and economic incentives to reduce average family sizes to the replacement level. If such a committment is made, specific policies are needed in addition to generalized improvements in education and economic conditions as shown by the following example. Among African countries, Kenya has a high rate of economic development and educational advance. Yet, at the time of writing (mid-1985), Kenya has the highest rate of population growth in the world. Fertility among women who have attended school is higher than among those who did not, probably because of increased awareness of nutritional and hygiene needs in the former group (Myers 1980). General education has not led to a reduction in birth rate. The current rise in fertility may be a prelude to a subsequent reduction, a pattern that occurred in England and Wales between 1850 and 1900. However, such a course is neither inevitable nor predictable.

The immediacy of the need for population control programs is revealed by projecting present populations into the near future (Figure 10-9). Even if each pair of parents alive today left only two survivors ($R_0 = 1$), the world population will grow to more than six billion because of an age structure in which almost half of the people are below reproductive age now but will have children later (Figure 10-1). This problem is most acute in developing countries, most of which will more than double their present populations even if a net reproductive rate of unity is achieved within the next 20 years (Figure 10-9).

10.4   Ecology and Human Diseases

A sharp difference between developed and developing countries is the type of diseases that prevail and their demographic effects. Three types of disease can be distinguished with respect to their ecological and cultural environments:

1. Infectious diseases caused by poor hygiene and lack of basic health care. These diseases are not peculiar to tropical countries, although are often prevalent there. They include malnutrition, dysenteries, cholera, and typhoid.

2. Diseases of great concern in developed countries are those that often become manifest in middle- or old-age, such as obesity, diabetes, hypertension, heart diseases, and cancers. They are not infectious and may be related to urban and industrial cultures. The onset of most of them usually occurs beyond the reproductive age group. This has two biological consequences. First, they are not affected by natural selection, and second, they only have minor demographic effects.

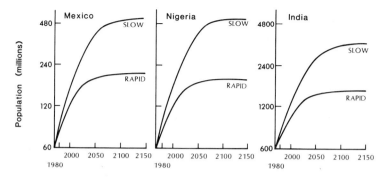

Figure 10-9   Population projections calculated by Frejka (1973), which assume that net reproductive rates will achieve unity in the future for three developing nations. In the rapid path, this is achieved around 2005 and in the slow path, by 2045, which mean more than doubling and trebling the 1970 populations, respectively. In 1981, Mexico had a population of 71 million, Nigeria 80 million, and India 684 million, respectively, indicating the limited progress toward stabilization since Frejka's computations.

3. The major infectious diseases of high incidence in tropical populations (Table 10-3). Many of these such as malaria, schistosomiasis, and the filariases are caused by animal parasites; others such as yellow fever and Lassa fever are viral.

The focus of this section will be on the ecological factors related to the third group of diseases. This is not to deny the importance of the first group in tropical countries. Their control or eradication in terms

Table 10-3   Comparison of the occurrence of some helminthic diseases in different parts of the world. Note the high prevalence in Africa. Data are very approximate from Peters (1978).

PARASITIC ORGANISM	AFRICA	CENTRAL AND SOUTH AMERICA	SOUTH ASIA	NORTH AMERICA	EUROPE	USSR
Roundworms (*Ascaris*)	40	32	41	2	8	12
Hookworms	33	30	32	1	Negligible	2
Filaria worms	44	12	13	0	0	0
Schistosomes	42	5	4	0	0	0
Tapeworms	8	<1	2	Negligible	Negligible	11

of public health is well understood and relates mainly to socioeconomic issues that are outside the scope of this book.

10.4.1   Types and Ecology of Tropical Parasitic Diseases

Table 10-4 surveys some of the common parasitic diseases of tropical human populations. Two groups of animals are the main contributors: the Protozoa and the helminths. The helminths include two phyla: the Nematoda (round worms) and the Platyhelminthes (flat worms such as the cestode tapeworms and the trematode flukes). Many of these diseases are **endemic** to particular populations; that is, they are present all of the time at roughly constant levels of infection. In contrast, **epidemic** diseases arise suddenly, sweep through the population, and then disappear again, as happened with the bubonic plagues in Medieval Europe (Figure 10-7).

Another important distinction is between diseases that are transmitted directly from one human host to another, such as the dysenteries and intestinal nematodes, and those that have a second host or **vector,** which transmits the disease from one person to another. The vectors are also infected but by a different stage in the parasite's life cycle. Many important tropical diseases such as malaria, onchocerciasis, and the filariases have blood-sucking flies as vectors, while schistosomiasis is transmitted by snails. These invertebrate vectors all have important aquatic stages in their life cycles, an important consideration in the ecology of control measures. A notable exception to this pattern are the sleeping sicknesses (trypanosomiases), which are transmitted by terrestrial insects. Several qualifications apply to the data in Table 10-4. First, many of the estimates are low because some stages of some diseases may be present but not show up in tests. Second, many individuals carry several diseases simultaneously. For example, giardiasis is often associated with amoebic dysentry. Third, the presence of a disease organism does not mean that there are serious detrimental effects on the host. Where parasites and people have existed together for many generations, adaptations in one or both populations may lead to coexistence with subclinical effects on the human hosts (see p. 308). Finally, it seems that some parasites may reduce susceptibility to other diseases. Infection with hookworms causes anemia but also leads to resistance to many bacterial infections (Desowitz 1981).

Ecology can contribute to the understanding of diseases in several ways. The effects of ecological changes in human environments seem to increase the incidence of many diseases as discussed in section 10.4.2. The remainder of this section is a brief account of some other facets of the interactions between human and parasite populations.

Mathematical models of infectious diseases are rooted in the prey-predator models introduced in Chapter 5. Two-species models suffice for directly transmitted diseases, but where a vector is involved,

Table 10-4   Rural health surveys of animal parasites in tropical human populations from forest or forest farming environments. In the top portion, three tropical regions are compared. Only a few populations are included, and true ranges are therefore likely to be greater. Nevertheless the variation between populations is enormous. The bottom half of the table compares the incidence of diseases in two adjacent forest populations of the Central African Republic. The pygmies are seminomadic but also work on the farms; the Bantu group are forest farmers. Data from Unesco (1978).

	% OF PEOPLE INFECTED		
	MALAYSIA	SOUTH AMERICA	AFRICA
PROTOZOAL DISEASES			
Amoebic dysentery (*Entamoeba hystolica*)	1–6	0–30	12–36
Giardiasis diarrhea (*Giardia*)*	9–12	12	7–10
Malaria (*Plasmodium*)	2–20	14–95	12–70
HELMINTHIC DISEASES			
Hookworm (e.g., *Ancylostoma*)	3–93	40–95	40–86
Roundworms (e.g., *Ascaris*)	0–90	15–100	11–73
Filariasis (*Wucheria bancrofti*)	0–17	—	—
Onchocerciasis (*Onchocerca*) (river blindness)	—	4–60	4–100

	% OF PEOPLE INFECTED	
	PYGMIES	BANTU
INTESTINAL DISEASES		
One or more species of pathogenic Protozoa	86	77
Hookworms	74	80
Roundworms	5–24	12–53
SKIN DISEASES		
Onchocerciasis	1	5
BLOOD DISEASES		
Malaria	42	61
Calabar filariasis	2	11
Schistosomiasis	0	41

*Giardiasis is now a significant disease in North America and the USSR as a result of contamination of drinking water (see Desowitz 1981).

the dynamics of three interacting populations must be considered, as shown below (see also section 5.1):

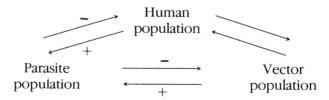

No signs are given to the human-vector interaction because they are different for different diseases. If the vector is eaten, as in the case of tapeworms, the signs are $\xrightleftharpoons[+]{-}$; for bloodsucking vectors, it is $\xrightleftharpoons[-]{+}$; in the case of schistosomiases, people and snails have no direct interactions $\xrightleftharpoons[0]{0}$.

From the perspective of control or elimination of diseases, the population dynamics of the parasite and the vector are most important. Indeed, few of these endemic parasitic diseases are lethal to humans with access to modern medicines. The effect on human populations is more often one of chronic recurrent illness, with its social and economic consequences. Mortality is often due to secondary infections. However, it is possible that human population growth has been curtailed by diseases in the past, as many parasites are capable of regulating the density of their hosts. Models of disease population dynamics produce some predictions that are borne out by available evidence. For example, in diseases transmitted by vectors, models predict a low death rate, a high prevalence of infection, and a long life span in human populations, with the converse effect on vector populations (May and Anderson 1979). These qualities are found in many tropical diseases, including malaria and schistosomiasis. For a review of these and other features of ecological models of parasite populations, see Anderson (1981).

An important feature of host-parasite interactions is coevolution. Most parasites are highly specialized, with specific adaptations to life on or within their host and for dispersal from one host to another. They are detrimental to their host either through obvious illness or because of diversion of energy from the host's growth and reproduction. Natural selection places a premium upon host individuals that can evade or reduce the effects of parasites. Immunological responses are powerful means of suppressing many parasites, although coevolution sometimes turns these to the parasite's advantage. For example, in schistosomiasis, the adult worms live in blood vessels. They avoid attack by the host's antibodies because they can acquire antigens from the blood, which effectively hide them from the immune system.

It is often suggested that coevolution leads to reduced host mortality so that there is a long-term home for the parasites. There are many examples that support this conclusion such as the insignificant effects of trypanosomiasis on wild African ungulates with which the parasites evolved, compared with the virulent effect on exotic domestic ungulates more recently introduced. However, such a course is not inevitable. Very high virulence can produce a stable coexistence, as in rabies. An important method of transmission of this virus is by saliva-infected bites of potential hosts. Biting behavior is induced by the final stages of the disease, which then kills the host.

Many examples of genetic traits that confer enhanced resistance to diseases are known. One of the best documented is the sickle cell trait. This is a simple genetic polymorphism, which is maintained in areas where malaria is endemic because the heterozygote (with one "normal" and one sickle cell gene) has a better chance of survival than either homozygote (Figure 10-10). People homozygous for the sickle cell gene are anemic, and most die because their red blood cells are deficient in oxygen-carrying ability. However, the sickle cell gene confers resistance to malaria. The heterozygote combines the advantages of malaria resistance and good oxygen transport. The sickle cell gene therefore survives despite its great disadvantage in the homozygous state.

10.4.2   Ecological Change and Diseases

The ecological consequences of the cultural changes from being mobile foragers to settled farmers or urban dwellers may increase the prevalence of some diseases. Direct evidence from hunter-gatherer populations before their contact with other cultures is scanty, although surveys of parasite burdens hint at such a possibility (Table 10-4). From the ecological standpoint, there are good reasons to predict that disease burdens will be higher among settled populations. For instance, to be successfully transmitted, all diseases need a threshold level of host population density. The spread of epidemics in particular requires very

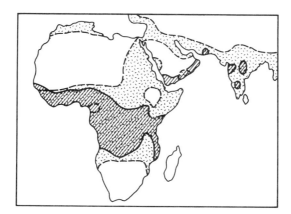

Figure 10-10   The sickle cell gene (in the heterozygous state) is advantageous in areas where malaria occurs but disadvantageous elsewhere. Hatched areas indicate where the gene frequency is more than 5%, and these are almost all within the stippled areas, which indicate the distribution of malignant malaria caused by *Plasmodium falciparum*. (Modified from Dobzhansky et al. [1977].)

dense populations. Transmission is also enhanced by permanent settle-
ment because the infective stages or the vectors are in constant contact
with their human hosts. Ecological disruption and the setting up of
agroecosystems means that many species are eliminated locally, but the
few that survive, including pests, and diseases often become more
successful. Diseases that were formerly restricted to nonhuman populations sometimes infect the workers involved in forest clearance. If these
diseases can be transmitted by other domestic vectors, they may spread
to urban human populations. Cooper and Tinsley (1978) document how
monkey yellow fever may enter human populations in this way in Africa
and South America.

There is more direct evidence to show that changes to more
industrialized forms of agriculture in developing countries have caused
much greater incidence of some of the debilitating tropical diseases. Two
of the many examples of the interrelatedness of ecological and cultural
changes with the prevalence of diseases are described next.

Malaria.   The malarial parasites (species of *Plasmodium*) are spread
by bloodsucking female mosquitos of the genus *Anopheles*. Because
these flies have an aquatic larval stage, standing water is needed to
complete the life-cycle. Relatively few such environments exist in mature
tropical forests. Where they do, such as in tree holes or water-holding
epiphytes, they are often in the canopy. For example, *Anopheles cruzi*
populations in South America are often restricted to the canopy where
they bite monkeys but do not encounter people (Unesco 1978). The
felling of forest for agriculture produces many permanent or transient
bodies of water at ground level, ranging from footprints and wheel ruts
to fish ponds and irrigation systems. Even in deforested areas, changes
in agricultural procedures can dramatically change the potential for
malarial mosquitos. In western Kenya when irrigated rice fields replaced
a diverse environment of scattered maize cultivation, seasonal swamps,
and water holes with profuse plant growth, the ratio of individuals of
malarial to nonmalarial species of mosquitos changed from 1:99 to 2:1
(Desowitz 1980).

Industrial agroecosystems using large pesticide applications have
led to resurgence of malaria in Central America and the Indian
subcontinent (Chapin and Wasserstrom 1981). In the late 1960s and early
1970s, malaria was close to eradication in these regions. For example,
the number of cases in Honduras was reduced from 100 million to 50,000
between 1952 and 1962. However, by the late 1970s the number of
people infected had returned or was higher than previous levels in Central America and up to several millions in India. During the same period,
land holdings in both areas were consolidated and turned over to
larger-scale cash cropping of species such as cotton, which demand high
pesticide applications. This rapidly increasing use of pesticides to

maintain agricultural production has caused the malarial mosquitos to evolve resistance to the chemical control measures that had previously reduced malaria. Indeed, since 1969, the number of cases of malaria in India has increased exponentially in direct relation to increased applications of DDT to agroecosystems and the concurrent changes in agricultural practices (Figure 10-11). An irony of this situation is the conflicting policies within the United Nations. The World Health Organization was striving to eliminate malaria by judicious use of insecticide. At the same time the Food and Agriculture Organization was advocating the use of high-yield crops, which demand heavy pesticide applications.

Trypanosomiasis.   The trypanosomes, which cause sleeping sickness in Africa and Chagas' disease in Latin America, are transmitted by biting flies and bugs, respectively, which have wholly terrestrial life cycles (see Molyneux and Ashford 1983 for a detailed account). In Africa, trypanosomiasis affects both people and ungulates (plus a few other wild vertebrates). All species are transmitted by tsetse flies of many species, which are ecologically segregated to some extent by environment and by the species bitten (Ford 1971). The two most important strains of human sleeping sickness *(Trypanosoma brucei)* are chronic diseases that

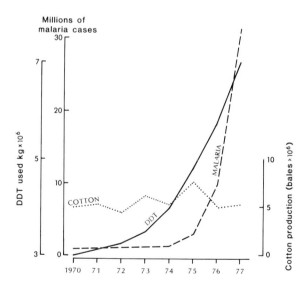

Figure 10-11   The incidence of malaria in India has increased dramatically as the application of DDT to agroecosystems has increased. Heavier pesticide applications have been necessary to maintain crop production (cotton in this example) but have had the side-effect of causing the mosquite vectors to evolve resistance to DDT used in disease control. (Modified from Chapin and Wasserstrom [1981].)

usually lead to death. This fate also awaits many breeds of cattle, which are not kept in large parts of Africa as a result. However, some strains of trypanosomiasis are not so drastic in effects, and some breeds of cattle and many wild ungulates carry the disease without obvious ill effects. Chagas' disease *(T. cruzi)* is also chronic and incurable. The trypanosomes usually invade muscle tissues, including the heart, and may cause death. The triamotomid bug vectors pick up the disease when taking a blood meal. Transmission between humans (and other mammals) is by way of the feces of these insects. The infective stage enters the vertebrate body either through wounds or by burrowing through mucous membranes. The trypanosomes may even pass from mother to child either in utero or in milk.

Both the African and Latin American diseases have spread because of the increase of favorable habitats for the vectors. Tsetse flies carrying sleeping sickness are widespread in Africa, *Glossina morsitans* occupying the vast savanna woodlands and bushlands of eastern and southern central Africa. *G. palpalis* is an inhabitant of forest edges and the perimeters of streams and lakes in west Africa and the Congo basin. The way in which human activities can, unintentionally, induce epidemics of sleeping sickness can be illustrated by an East African case study (see Ford 1971). A sequence of cultivators and pastoralists replaced the forests with savanna vegetation around Lake Victoria. Originally only *G. palpalis* had been present, but the new savanna environments led to the invasion of *G. morsitans* as well. However, up until the late 19th century, neither fly carried trypanosomes in this area. The disease was most likely introduced by the African porters of the European explorers in the 1880s. The result was an epidemic of sleeping sickness that killed perhaps a third of the local populations in the next 20 years. The strain of trypanosomiasis that took hold was that transmitted by *G. Morsitans* (which had been absent from the area until the forests were cleared). Since this species of tsetse fly also transmits trypanosomes that infect cattle, the area eventually became unsuitable for stock-rearing, although another epidemic, rinderpest (introduced from Europe), had already killed 95% of the cattle in the area before the arrival of cattle trypanosomiasis. Rinderpest subsequently died out, but trypanosomiasis remains a major problem.

The story of the spread of sleeping sickness shows how the mobility of human populations can introduce diseases to new geographical areas (see also section 5.4.3). Much of the decline in the populations of native Americans (by perhaps 2.5 million people) following the arrival of Europeans in the 16th century can be attributed to the introduced epidemic diseases such as smallpox and measles. Cultural factors are also important in Chagas' disease. The bug vectors rest in the crevices of the rough roofing, wall, and livestock pens typical of poor rural communities. Mammals associated with such dwellings, particularly rodents and domestic cats and dogs, probably act as reservoirs for the human disease. In natural communities as many as 150 species of forest mammals harbor trypanosomes. Even if Chagas' disease could be eliminated from farming populations by improved economic conditions and public health measures, new infections would probably arise. Workers involved in deforestation become infected with the forest form and would then spread it back to other rural people (Molyneux and Ashford 1983).

10.4.3   Ecology and the Control of Diseases

Mathematical models of disease populations can assist in devising control strategies. They may be able to identify threshholds of parasite density below which the disease will die out naturally or to indicate vulnerable stages in the life cycle. For example, Anderson (1979) has calculated a threshhold mean density for hookworms in India of 0.3 worms per person, suggesting that almost complete eradication is necessary to effect control. Otherwise the population will return to its equilibrium mean density of around 50 worms per person.

As with pests, integrated control is the best way of attacking diseases in many cases. Schad and Rozeboom (1976) give a review of integrated programs to control helminth diseases. Some of the control methods are analogous to or the same as in pest control, such as pesticides or the maintenance of environments unsuitable for vectors. Many of the methods, such as chemotherapy and vaccination, are in the realm of clinical medicine or public health and are outside the scope of this book. Unlike pest control, which results in higher crop yields, there is usually no immediate economic benefit to defray the costs of disease control. However, the increase in human well-being is priceless, and reduced illness leads to a potential increase in economic activity. It is certainly no coincidence that the high incidence of diseases is closely connected with individual or national poverty. Cheap programs of integrated control are therefore a high priority.

Many directly transmitted diseases such as hookworms reach hosts by shedding eggs with the host's feces. An active larval stage then searches for a new host, which it enters by ingestion or by penetration through the skin. The control of such diseases is best effected by public health and hygiene education and the use of sanitary latrines. Combined with drug therapy, such methods reduced the incidence of hookworms from 39% to 7% of the Korean population between 1949 and 1971 (Schad and Rozeboom 1976).

Vector-borne diseases need similar control efforts, but in addition, control of the vector population can greatly reduce and occasionally eliminate the disease. Schistosomiasis, like the diseases discussed in section 10.4.2, has spread enormously because of environmental changes caused by human endeavors. Dams and irrigation schemes have proved ideal environments for the aquatic snail vectors that inhabit still or gently flowing water. Prior to damming in 1966, the incidence of the disease in Ghanaians living around the fast flowing Black Volta river was less than 1%. By 1969 all children in the area harbored schistosomes (Desowitz 1981).

Although an Old World disease in origin, schistosomiasis is now a problem in South America, where it was introduced by African slaves. The African snails did not make the journey, but local snails proved to be suitable vectors. Biological control has made a significant and cost-effective contribution to the control of this disease in Puerto Rico. Virtually no one in that country was infected early in this century. By the time a control program was introduced in 1953, 20%–30% of the population was infected. Integrated control reduced this proportion to 5% by 1976. Methods used included a mixture of chemical and biological control of vectors, clearing of their plant food from irrigation and drainage ditches, and health education and drug therapy. The biological control agent is another herbivorous snail that competes for food with and eats the eggs of the vector. Together with small amounts of molluskicide, the per capita cost of vector control was only $1 compared with up to $10 when chemicals alone were used (Nergón-Aponte and Jobin 1979).

Summary

The biggest long-term problem facing many tropical countries is the unprecedented rate at which their populations are increasing. The rate of population growth and its consequences for social change is much greater than that experienced in developed countries during their transitions from subsistence to fossil fuel–based economies. Further, the global relationship between population and resources means that poor countries cannot reach a high standard of living in the same way as rich countries. The history of population growth shows great diversity between regions and countries. Therefore, each country needs to determine its own informed population policies, which are sensitive to cultural mores. Debilitating diseases are endemic in many tropical regions. Ecological changes caused by population growth and economic development often increase the incidence and the virulence of these diseases. Careful consideration of ecological factors, together with improved public health measures, can reduce the effects of these diseases.

Study Questions

Review

1. Discuss the concept of carrying capacity as applied to human populations.

2. Compare recent population changes (and their causes) in Europe with those in a tropical region.

3. How have recent human activities increased the incidence of many diseases in the tropics?

4. How can applied ecology contribute to the control of human diseases and agricultural pests (Chapter 9)?

Related Topics

1. Trace the history of population growth in a named country.

2. Compare demographic transitions in Europe with recent changes in birth and death rates in developing countries.

3. Describe in detail the interactions among populations of disease, vector, and human in a named protozoal disease.

4. Report on an integrated control effort that has succeeded in reducing the incidence of a named tropical disease.

Further Reading

Biswas, A. K. (1979). World models, resources and environment. *Env. Conserv.* 6:3–1. Critical comparisons of the structure and predictions of several world models of population and resources.

Ehrlich, P. R.; Ehrlich, A. H.; Holdren, J. P. (1977). *Ecoscience.* San Francisco: Freeman. A compendium of information on many aspects of human ecology and particularly strong on population. The source of much information in this chapter.

Frejka, T. (1973). *The Future of Population Growth: Alternative Pathways to Equilibrium.* New York: Wiley. Although many of the statistics are dated, the conclusions are not. Illustrates momentum of population growth graphically.

McEvedy, C.; Jones, R. (1978). *Atlas of World Population History.* Harmondsworth, Middlesex: Penguin. A fascinating region-by-region and country-by-country account of the history of human population growth.

Scientific American (1974). *The Human Population.* New York: Freeman. Eleven chapters on different aspects of population, including sociological, physiological, and genetic factors not considered here.

United Nations. *Demographic Yearbook.* New York: UN. An annual update of population statistics from all countries.

Finally, two very different accounts of the affects of diseases on populations.

Anderson, R. M. (1981). Population ecology of infectious disease agents. In: May, R. M. editor: *Theoretical Ecology.* 2d ed. Oxford: Blackwell, pp. 318–355. Discusses how theoretical models of populations can be used to clarify and control diseases.

Desowitz, R. S. (1981). *New Guinea Tapeworms and Jewish Grandmothers. Tales of Parasites and People.* New York: Avon Books. An illuminating entertaining account of the interactions between diseases, culture, and ecology.

Tropical forests are rapidly disappearing as illustrated by this clear-cut area in Malacca, Malaysia. (Photograph by James N. Anderson.)

CHAPTER 11

Ecology and Conservation

11.1 What Is Biological Conservation?

The broad term **conservation** refers to the sustainable use or preservation of natural resources. Biological conservation deals more specifically with biological resources, and ecologists play a key role in such endeavors. Three distinct but often intertwined threads underly the motivations of conservationists:

1. Design and protection of productive and sustainable land-use systems

2. Preservation of species with actual or potential economic benefit to humankind

3. Conservation of communities (or species) for noneconomic reasons

The latter two constitute motivations for the conservation of genetic resources. However, the two are distinct, even though the preservation of species for potential use is often raised to support aesthetic or ethical points of view. For example, large amounts of time and money are being spent in an attempt to preserve viable populations of the Chinese panda. The rationale is largely aesthetic, but there always remains the remote possibility that this species may be of unique use in some unsuspected way. Such a situation has arisen with the armadillo, which is the only nonhuman animal known to contract leprosy. As a result, it has become of use to medical researchers.

The biggest practical problem facing any kind of conservation effort is enforcement of regulations. There is little point in banning tree cutting on steep slopes or fighting a verbal battle against poaching if there is no effective machinery to deal with transgressors. It follows, therefore, that conservation programs are much more likely to succeed if they have the support of rural people and are compatible with local traditions.

Ecologically sustainable land-use systems adapted to tropical environments have been discussed in the previous two chapters and will be developed elsewhere in this chapter. Many of these are, coincidentally, better for the conservation of genetic resources than are the agricultural, forestry, and domestic land-use practices commonly used in developed countries. The emphasis is often on sustainable multiple use of land in rural areas to supply food, fuel, money, and a healthy life for the local population.

Four somewhat different topics are dealt with in this chapter. First, the conservation of genetic resources is discussed in terms of rationale and methods. That is followed by two land-use problems that are particularly acute in the tropics: the depletion of forests and the desertification of savannas. Finally, the application of ecology to the management of nature reserves is considered.

11.2   The Conservation of Genetic Resources

There are three interrelated facets to genetic conservation:

1. The preservation of species richness in communities

2. Maintainance of genetic variation within a population

3. The prevention of extinction of particular species

Because of their high diversity, tropical regions are vital to the first of these, and because of rapid changes in human activity, very important to the second and third.

11.2.1   Why Conserve Genetic Resources?

There are often good economic reasons to preserve genetic diversity within populations and between species. However, a coherent and unifying philosophical answer to this question remains elusive because different peoples react differently to the other species around them. Indeed, for many tropical people the unhappy imperative is food today, rather than some less tangible speculation about the future. Perhaps most profound are the biological implications of ignoring genetic conservation, since we are altering the very processes in the biosphere from which we arose.

Economic Benefits.   Only a minute proportion of species are used directly by people in their agroecosystems, fisheries, forests, other industries, or for medical purposes. It is probable that many more species and local varieties have potential for human use, either in their wild state or through domestication. Myers (1979) lists many such species. For example, he estimates that around 80,000 species of plants are likely to be edible, but only 3000 are used as human food and merely 150 are cultivated to any extent.

A well-proven use of wild species and varieties is in the breeding of improved or specialized varieties of closely related crop species. Particularly important is the production of crop strains that are resistant to pests and diseases. When a resistant wild form is found, it may be possible to crossbreed it with the appropriate cultigen to produce a resistant crop. For example, grassy stunt virus is a serious disease of cultivated Asian rice *(Oryza sativa)*. One of the twenty wild species (*O. nivara* from India) has been used in breeding programs to produce resistant crop varieties (Chang 1984). Knowledge of many wild species and varieties of potential use is scanty, and many species will inevitably become extinct before they are discovered because of the pressure of human populations on natural ecosystems. Species of possible benefit usually need other species with no direct economic use if they are to survive. The disruption of complex tropical foodwebs may lead to unpredicted extinctions, particularly when mutualisms are involved (see Gilbert 1980).

An immediate economic benefit of genetic conservation to some tropical countries is the foreign exchange earned by the promotion of tourism. People from developed countries are attracted to the biological diversity and can afford to visit tropical forests, savannas, and coral reefs. Loss of species from these communities reduces this attraction and the revenues that go with it. Tourism in Kenya earned more than $80 million in 1976 and employed 35,000 people at above average earnings. Returns from Amboseli National Park are equivalent to an income of $40 ha^{-1} a^{-1}, 50 times its potential agricultural value. In the same Park, one lion is worth $25,000 and an elephant herd $600,000 as tourist attractions (Western and Henry 1979). Dead animals are worthless by comparison, ivory notwithstanding.

In other situations, organisms or their products may be sold. Careful cropping or "farming" of such species can increase the probability of their survival in the wild. Notable success stories include farming of rare spectacular butterflies for sale to collectors and of crocodiles for their skins in Papua New Guinea. These commercial enterprises turning out "luxury" products reduce the pressure on wild populations and provide rural employment and revenue if properly organized (see Cherfas 1979). As with tourism, the contact with rich people from elsewhere can have detrimental effects on local cultures. Local interest, consent, and control are essential prerequisites for operations of this kind.

Philosophical Issues.   It is arrogant for a person from one culture to tell those from another how they should regard other species. Most subsistence peoples do not need to search for abstract conservation ideals, since their day-to-day existence depends on the persistence of natural communities (see Chapter 9). In contrast, developed countries have an appalling record of genetic conservation. The Judeo-Christian traditions and the economic systems promoted by colonial countries and the current superpowers, which have been exported to much of the world, exploit unsustainably rather than conserve natural resources. Conservation as an explicit ethic has a muddled history in Europe and North America, and it is only recently that clearer ideas rooted in ecological understanding have emerged. The economic viewpoint, given previously, is the practical strand to this philosophy, but the emotional underpinning is partly aesthetic. People from highly industrialized European cultures are divorced from natural biotic communities, yet many obtain great pleasure from visiting them. This has led to a popular movement for the development of and access to wilderness. The more abstract ethical concept of stewardship—the passing on of similar (or enhanced) opportunities to future generations—has also developed. Some people stress, in addition, the ethical importance of a sense of responsibility for all other organisms with which we share the biosphere.

Biological Issues.   Ecology and evolutionary biology are important components of any coherent philosophy of conservation. Contrasts have been made between the interactions within and the persistance of natural communities compared with the instabilities of simplified agroecosystems. There remains much to learn about natural systems that could be useful in the development of self-maintaining agroecosystems better adapted to tropical environments. On a larger scale, some natural ecosystems are needed to maintain the equilibria that human activities upset. The importance of forests in local and global ecological processes is stressed in section 11.3, but aquatic systems are also vital in decomposing the waste products of human populations. When they become overloaded with pollution, they are unable to do this job and may even become poisonous or reservoirs of diseases.

Evolutionary arguments are more subtle but the most profound of all. At the species level, extinctions far outweigh speciation rates because of human activities. Within many populations, genetic changes are dictated by environmental changes wrought by humans. In other words, we have unconsciously assumed control of much of the evolutionary process without establishing any guiding principles for the future. To take such an enormous step thoughtlessly is highly irresponsible. Conservation of as much of the natural environment as possible is a means of obtaining time to assess this development.

Tropical countries are ecologically the best placed but economically poorly equipped to be in the forefront of genetic conservation. Much of the economic responsibility for this global concern clearly lies with the rich countries, many of which have irrevocably impoverished their own genetic resources. These countries should provide money and expertise to assist poor countries in the development of their own conservation programs. Such programs must reflect the aspirations and cultural background of the recipient countries rather than the priorities of the donor, if they are to be effective.

A problem facing many developing countries is the need to earn foreign exchange by the sale of cash crops. These crops often take the best agricultural land, thereby pushing food production into more marginal environments. Sustained food production is difficult in such areas and often demands yet more land, some of which might otherwise be protected for its conservation value.

Local educational and lobbying groups of conservationists exist and have government support in many developing countries. The following examples from different continents illustrate the diverse ways in which a conservation movement may develop in different countries. Tanzania, despite being a very poor country, has the idea of stewardship of natural resources as part of the creed of its ruling party and has a good record in wildlife conservation. In India, an independent group of scientists and writers (the Center for Science and Environment) published a report on the state of the natural environment in 1982. Finally, Costa Rica protects 9% of its land surface for conservation. Government backing and international finance help, but the general populace contributed almost $2 million voluntarily to the establishment of Corcovado National Park (Myers 1979).

11.2.2   Preservation of Species Richness

It is impossible to preserve more than a few individual species by the intensive approaches described in this chapter. A far better and more cost-effective method is to establish land-use practices that allow many species to maintain themselves. Complete protection in nature reserves is the best way of preserving species-rich communities, but only limited areas can be demarcated in this way. However, as stressed elsewhere in this book, other areas occupied and used by people can be conserved if less disruptive technologies than many of those currently in use are employed.

11.2.3   Maintenance of Genetic Variation Within Populations

As human encroachment spreads, the ranges of many other populations are reduced. Some of these local populations become extinct, thereby reducing geographical variation within a species. Others are reduced in size, which leads to a reduction in intrapopulation variation and a potential loss of more genes by chance rather than by natural selection (see section 6.1.5). The overall effect of these factors is to decrease the probability of evolutionary adaptation in response to environmental changes and to lose varieties that have potential for human use.

Based in part on criteria from evolutionary genetics, a breeding population size of 500 has been advocated as the minimum necessary to maintain intrapopulation genetic diversity. In most populations many individuals such as juveniles or those socially excluded do not breed. In most cases, therefore, a total population much higher than 500 is required (see Frankel and Soulé 1981).

11.2.4   Prevention of Extinction

Extinction is a natural process and probably the ultimate fate of all species. However, the rate at which species become extinct has increased enormously as human populations have intensified their management of natural resources. Most temperate biomes were relatively poor in species before such developments, but the elimination of natural ecosystems has greatly increased this poverty. For example, in the last 100 years, 10% of native vascular plant species and 30% of bryophytes and lichens have been eliminated in many European countries. Many more are threatened by the greatly reduced area of natural communities (Maarel 1975).

Many factors contribute to extinction, including competitive exclusion, predation, and disease. On a larger scale, climatic and geological upheavals have produced periods of "mass extinction," but even during these periods the supposed rates at which species were lost is several orders of magnitude less than that induced by modern human activities. It is impossible to estimate accurately the magnitude of extinction caused by human activities. Informed guesses range from 1000 species per year currently to 40,000 species per year in the near future (Myers 1979). Such rates are unprecedented in global history.

There are three distinct approaches to species preservation which are appropriate to different situations:

1. *Nature reserves.* To conserve a piece (or pieces) of natural environment of a species is by far the best method because many other species are maintained simultaneously at low cost. The prospects for particular species can be enhanced by the translocation of individuals

from areas in which they are threatened to protected areas, although to be successful such efforts must be based on a sound understanding of the ecology and behavior of the organisms.

2. *Botanical and zoological gardens.* Captive propagation is a useful research and education tool but is a method of last resort for conservation. If survival in natural environments becomes permanently impossible, then the organism is extinct as a wild species, and its preservation ceases to be an ecological issue. However, gardens are useful holding grounds from which locally extinct species may be reintroduced to their remaining natural environments. For example, the Arabian oryx, a desert ungulate, was eliminated in the wild in 1972, but was successfully reintroduced to Oman in 1980 from captive herds in the Phoenix and San Diego zoos in the United States.

3. *Germplasm banks.* Concern about the loss of genetic diversity in many crop species has led to the establishment of gene banks to preserve varieties that are not widely used (or not used at all). The rationale is to preserve genetic resources, both species and their varieties, for use in breeding programs to produce crops resistant to specific pests or diseases or suitable for particular environmental conditions. Until the 1920s such conservation was performed by small-scale cultivation in botanic gardens, but technological developments now allow the preservation of seeds or other propagules at low temperatures for long periods. Initially, gene banks were concentrated in the developed countries, but developing countries are increasingly preserving their own crop resources in nationally and internationally organized gene banks (see Plucknett et al. 1983). However, most of the stored germplasm is of field crops, particularly cereals. There is a great need to establish or extend comparable facilities for the storage of other types of crop (actual and potential), including trees and medicinal plants.

Much effort is occasionally put into attempts to prevent the extinction of particular species. This raises a very real problem of priorities in conservation. For example, the substantial resources currently employed in efforts to maintain a small, declining population of California condors in the United States would have immeasurably greater impact, in terms of genetic conservation, if directed toward the preservation of an appropriate piece of tropical rain forest. Such rational options ignore the politics, economics, and emotionalism attached to the conservation of large homeothermic animals. The availability of public and private funds is often in response to issues of high local visibility and low conservation value beyond the aesthetic and anthropomorphic. Unfortunately, such funds usually would not be redirected to more worthwhile conservation projects. Also, such campaigns can play a more useful role if used to educate people in the broader issues of conservation.

11.3   Deforestation ————————————————————

Because they have the greatest species richness of all terrestrial biomes, tropical rain forests are a key feature of attempts to conserve genetic resources. Forests also have certain roles that are important to sustained, productive land-use at global and local levels as follows:

1. *Global climate.* Tropical forests contain almost one half of the carbon in terrestrial biomass. Deforestation releases much of this into the atmosphere as carbon dioxide, with possible drastic effects on climate (see section 3.4.1 for details). Even without the carbon dioxide effect, it has been predicted that removal of tropical forests will change atmospheric temperatures and circulation, causing a decrease in global precipitation and a change in its latitudinal distribution (Potter et al. 1975).

2. *Regional hydrological cycles.* Forest vegetation moderates seasonal fluctuations in stream flow and river discharge by absorbing much of the rainfall. Water supply to communities downstream is more predictable and easier to manage. Forests also affect local temperature, evaporation, and transpiration regimes, which may in turn affect local rainfall in some areas (see section 3.4.1).

3. *Soil degradation.* Closely connected with hydrological changes is the loss of soil and its nutrients when exposed to erosion following the removal of forest vegetation (see Table 9-2).

Despite the important roles listed, the rate of forest removal for agriculture, fuel and building materials, and timber is high overall and alarmingly so in some tropical countries (Table 11-1). Most of this deforestation in the rain forest zone is for farming or for commercial timber rather than for the immediate needs of local populations for forest products. However, the felling of trees for fuel as firewood or for charcoal manufacture is a major and increasing cause of reduced tree cover, not only in heavily wooded areas but also in savannas. Myers (1979) estimates that 25,000 km^2 of moist forest is lost annually to supply local fuel-wood needs. Between 1950 and 1973, global fuel-wood cutting doubled overall, with a threefold increase in tropical countries (Bolin et al. 1979). In the countries of southern Africa, 50%–90% of national energy consumption is in the forms of wood and charcoal. Rural populations in the moister countries of the region obtain their needs without seriously depleting forests, but this is not true in the drier countries such as Botswana and Swaziland. Throughout the region, urban populations, which constitute between 7% and 42% of the different nations, are causing widespread destruction of woody vegetation (Bhagavan 1984). Much forest land in Latin America is being converted to savanna for cattle rearing and subsequent beef export. Brazil has 66,000 km^2

Table 11-1 Status of tropical moist forests (global, regional, and some representative countries) in 1980. There is no connection between rate of deforestation and human population density, indicating crudely, that removal is not related directly to human needs for wood in the biome as a whole. Data from Grainger (1983) based upon the FAO/UNEP report of Lanly.

	TOTAL FOREST AREA (ha × 10^3)	DEFORESTATION PER ANNUM (ha × 10^3)	YEARS TO COMPLETE DEFORESTATION (CURRENT RATE)	HUMAN DENSITY (km^{-2})
ALL REGIONS	1,081,372	6113*	117*	
AFRICA	204,622	1204	170	
Ivory Coast	4458	310	14	24
Uganda	750	10	75	54
Zaire	105,650	165	640	12
Nigeria	5950	285	21	78
ASIA AND PACIFIC	263,647	1608	164	
Indonesia	113,575	550	206	72
Thailand	8135	325	25	88
Sri Lanka	1659	25	66	217
Papua New Guinea	331,750	21	1605	6
LATIN AMERICA	613,103	3301	186	
Brazil	331,750	1360	244	14
Colombia	46,400	800	58	23
Costa Rica	1638	60	27	41
Jamaica	67	1.5	45	194

*Some estimates are much higher, such as 24,500 × 10^3 ha a^{-1}) leading to a demise of forest within 50 years (Myers 1979); for an explanation of such discrepancies see Melillo et al. (1985).

of former forest devoted to cattle, most of it owned by multinational companies. However, it appears that sustained beef production at economic levels will be impossible in some of these areas because of poor soils, low productivity, and the invasion of noxious weeds (Myers 1979). Although a relatively small proportion of tropical forest land has been used in this way to date, the lack of sustainability means a continued demand for new land.

11.3.1 Conservation and Forest Management

The benefits accruing from leaving forests undisturbed on steep slopes and alongside rivers to minimize soil erosion should be obvious. In addition, exploited forests can be managed to maximize conservation of hydrological systems, soils, and genetic resources.

Timber is an increasingly important resource for tropical countries because of its contribution to foreign exchange earnings. The harvest of tropical hardwood tripled between 1950 and 1973 and was probably half as much again by 1980, while the proportion exported has risen from 13% to 50% (Myers 1979). During a similar period, foreign exchange earnings from this source have increased from $250 million to $4 billion. Much of the timber extraction is licensed to multinational companies, with the intention of obtaining a single crop. In such cases, companies have only a short-term interest in the logged areas and are not encouraged to promote sustainable use and conservation. To encourage the development of better policies, an Organization of Timber Exporting Countries (analogous to OPEC, the oil cartel) has been advocated. Such an organization would assign higher prices to tropical timber, with some of the extra profit used for sustainable forest management (Guppy 1983).

Selective logging, in which a portion of the trees in a stand is harvested, is one practice that reduces damage. However, considerable disruption is still possible. In some Malaysian sites, more than 50% of trees are destroyed to obtain the 3% that is salable (Johns 1983). The method of timber extraction is also important. Heavy machinery may deflect and delay secondary successions much more than other human activities (Figure 11-1). Most of the unused timber is left to rot, although it could be used for fuel or pulpwood. Even with more careful logging, it is questionable whether this form of management is economically sustainable under present conditions. For example, government regulations in Kalimantan, Indonesia, require that only commercial trees above 50 cm in diameter are removed and that 25 large individuals of such species are left standing per hectare. This policy limits the ecological impact (Figure 11-2) but means that a period of 80 or more years is required before the economically attractive dipterocarps regain their former dominance. In the same area, plantations are much more attractive economically, with potential harvests of up to 110,000 kg ha^{-1} of wood after only 15 years. Such single species plantations of uniform age are of low value in terms of genetic conservation but can take the pressure off other areas designated as nature reserves.

Rural wood needs, for fuel and construction, are best supplied by tree growing that is integrated into local multiple land-use practices such as agroforestry (see Chapter 9). In South Korea, government-backed but locally managed village forestry has been highly successful. More than 640,000 ha have been planted, often on land with degraded soils and denuded of trees (Eckholm 1979). Some Costa Rican coffee plantations have two additional canopy layers: one of nitrogen-fixing *Erythrina* trees that are kept low by pruning and a second of *Cordia* for timber. In the same country, another nitrogen-fixing tree, a native alder, is grown on

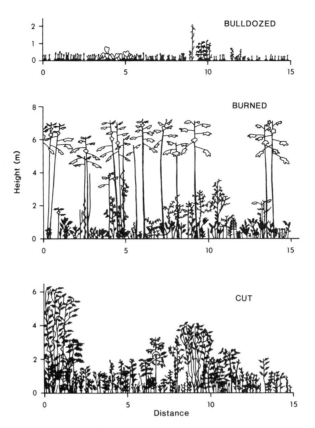

Figure 11-1   Comparison of the progress of secondary succession in an Amazonian rainforest of Venezuela one year after tree removal by different methods. Careful cutting or cutting and burning (similar to shifting cultivation) lead to rapid succession, in the former predominantly from coppice shoots and in the latter from seeds. Restoration of the former biomass takes about 100 years. In contrast, bulldozing disrupts soil, seed banks, and local hydrology, and recovery may take 1000 years. (From Uhl, C.; et al. 1982. *Oikos* 38:313–320.)

a 15–20 year rotation in some cattle pastures for fuel wood and other uses (Myers 1979). Livestock grazing in forest plantations is a common practice in India, where 13% of the national herd graze in these forests. Herbaceous forage production in teak forests ranges from 400–1200 kg ha^{-1} a^{-1}. Grazing is not detrimental to the trees if supplementary fodder is provided in the dry seasons (Pathak et al. 1978). Such intensive rural land use not only makes the local people more self-sufficient but also reduces disturbance to other areas, which can be designated for low-impact use or protected as nature reserves.

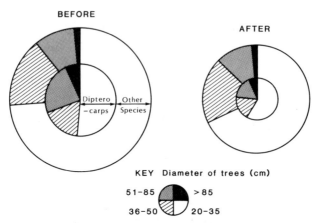

Figure 11-2   Comparison of rain forests in Kalimantan, Indonesia (formerly Borneo) prior to and one year after selective logging. Following government regulations (see text), the density of trees (proportional to the total area of the circles) is reduced by about 45%, but 60% of dipterocarps are removed (inner circle) compared with only 35% of other species (outer ring). The size structure is not greatly altered in most categories, but 70% of larger dipterocarps (> 50 cm diameter) are removed. (Data from Long and Johnson [1981].)

11.4   Desertification

In Chapter 8, deserts were characterized primarily in terms of vegetation structure and secondarily in terms of climate and soils. **Desertification** is a set of processes that changes savanna vegetation to that of a desert by reducing ground cover. Specific vegetation changes often include the replacement of trees and perennial grasses with a sparser cover of small shrubs, annual grasses, and ephemeral forbs. Loss of vegetation also leaves soil exposed to wind and water erosion, causing degradation to mobile sands or rocky substrates of low organic content. Areas prone to desertification are dry and semiarid savannas and semideserts adjacent to extreme deserts (Figure 11-3). The biggest threat in terms of area and number of people affected is in northern and southern Africa and in northwest India. The interactions between climatic fluctuations and cultural change are the causes of desertification.

11.4.1   Climatic Factors

Between 10,000 and 20,000 years ago, the extent of the Sahara and Kalahari deserts was much greater, with mobile dunes in West Africa stretching 500 km further south than at present. Subsequently, there was a humid phase for about 5000 years during which these former deserts carried savanna vegetation, followed by a variable but drier period that continues today (Hamilton 1982).

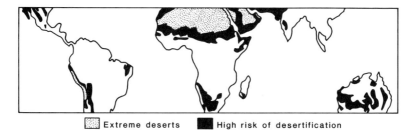

Extreme deserts   High risk of desertification

Figure 11-3   Tropical areas threatened by desertification are semiarid and arid lands, most of which are adjacent to the extreme deserts (annual rainfall < 100 mm) at the northern and southern edges of the tropical zone. (Modified from International Union of Nature and Natural Resources [1980].)

Has a trend to increasing aridity caused the desertification observed during the last 30 years? Because of the enormous year-to-year variation in rainfall, the answer must remain equivocal. What is clear is that the devastating famine and southward spread of the Sahara Desert into the African Sahel in the late 1960s and early 1970s coincided with a long and deep depression in regional rainfall that was unprecedented during the previous 45 years (Figure 11-4). Indeed, rainfall throughout the Sahel has shown a net decline in the middle of this century. A decline in rainfall has also occurred in the central Australian desert over the same period. However, no similar trend is discernible in northwest India nor in the Sonoran Desert of North America (Hare 1977). Whether these different patterns represent long-term trends or medium-term fluctuations is unknown.

11.4.2   Human Factors

Human cultures are never static, but recent changes in land use and population density (Chapters 9 and 10) have had greater ecological effects than those of previous periods. Agriculture in dry areas was generally restricted to the vicinity of rivers and lakes or to localities with unusually reliable rainfall. With increased numbers of cultivators, rain-fed agriculture is now attempted in semiarid and arid areas where harvests can be gathered in only one year out of every three or four in many places. The disturbance caused by such unproductive and unpredictable cultivation is a significant cause of local desertification (Le Houérou 1976a).

The most widespread human activity in tropical lands subjected to desertification is nomadic pastoralism. In Chapter 9 it was argued that traditional nomadic cultures are unlikely to cause prolonged degradation of vegetation, but there have been the following changes to traditional practices in recent times:

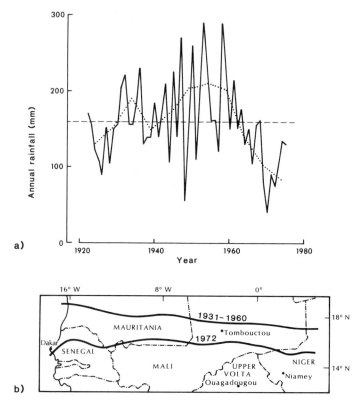

Figure 11-4   Changes in the temporal and geographical pattern of rainfall in the Sahel region of West Africa. *a,* Fifty-four years of rainfall at Agadez, Niger. Note the higher than average (dashed line) rainfall in the 1950s and the lower than average rainfall during the subsequent famine in the 1970s. Dotted line is average rainfall for five-year periods, which accentuates these medium-term trends. (Data from Unesco [1980].) *b,* The 200 mm annual rainfall isohyet (heavy line) was 200 km further south in the early 1970s compared with its average postion in the previous 30 years. Upper volta is now Burkina Faso. (Modified from Rapp [1976].)

1. *Population growth and increased density.* While cultivators have been able to secure new land by deforestation to partially offset increases in population, the areas of traditional grazing land have decreased, causing even greater human and livestock densities than would be dictated by population growth alone. More livestock are needed to support this population and, therefore, grazing intensity has increased. Furthermore, food aid on a scale not previously available has decreased human mortality during droughts. In 1973, 7 million people in the Sahel were supported in this way while 100,000 died (Wade 1974).

2. *Reduced nomadism.* Traditionally, desert vegetation is used during wet seasons and the pastoralists move to moister areas during dry seasons. Such migrations have been curtailed for several reasons. Active encouragement to settle is practiced by some governments that regard nomadism as being out of place in a modern state with fixed administrative boundaries. Indirect pressures are also important. For example, the drilling of wells means that permanent water is available in places that had temporary water supplies in the past. In parts of Sahelian Niger, individual bore holes may supply water to more than 10,000 head of livestock. This is several times the number supportable within feasible daily grazing movements even under good rainfall and pasture conditions (Unesco 1980).

3. *Loss of grazing lands.* The traditional dry season grazing areas, particularly those used during prolonged dry periods, extended into areas that can be cultivated. As populations increase, these marginal areas are used for permanent rain-fed or irrigated agriculture. Sometimes such developments have led to mutualistic relationships between pastoralist and cultivator, with the latter providing water and temporary pasture in return for livestock dung as fertilizer. However, in most cases, grazing by nomadic livestock is actively discouraged.

11.4.3   The Interactions Between Climate and Culture

Climatic or cultural changes are each capable of causing desertification, but the alarming rapidity of the process results from the two in combination. It is possible that loss of vegetation cover and increased aridity are linked by positive feedback mechanisms. For example, Charney et al. (1975) suggest that removal of vegetation increases radiative air cooling because of increased reflection. This cooler air sinks and thereby reduces the formation of cumulus rain clouds by convection. From a simulation model, they predict a reduction of rainfall over the Sahara of 40%. Other models predict similar effects, sometimes for slightly different reasons, but an empirical study in the Sonoran Desert in the United States suggests that loss of vegetation cover can increase rainfall in some situations (Jackson and Idso 1975). The applicability or generality of these studies is unclear, but even without such effects the changing interaction between climate and culture is pivotal to an understanding of recent desertification in Africa. The breakdown of traditional systems coincided with a wetter than average period in the 1950s and early 1960s. This disguised the medium- and long-term effects of increasing human and livestock density, decreasing nomadism, and losses of grazing land. A prolonged dry period followed that has revealed the instability of these changes, with such tragic results as the Sahelian droughts of the early 1970s and mid 1980s.

11.4.4   Ecology and Rangeland Management

Are Grazing Lands Overstocked?   It is impossible to define a constant
optimal stocking rate for communities subjected to large fluctuations in
rainfall and primary production. Moderate grazing and browsing intensities maximize both primary and secondary production; low livestock
density allows the accumulation of dead and less nutritious grasses and
woody plants; high densities lead to overgrazing. The term **overgrazing**
is widely used, although loosely defined. It is applicable where palatable
vegetation continues to decline through successive wet and dry periods
(Figure 11-5). Unfortunately it is often used to describe temporary
denudation of vegetation during a series of dry years, which subsequently recovers during a wet period. The ability of herbaceous and woody
vegetation to recover is demonstrated circumstantially by the continuation of herbivory by livestock and wild ungulates over thousands of
years and by simulation models of savanna vegetation (Table 11-2). More
direct evidence comes from observations of vegetation recovery
following temporary drought and very heavy grazing. Provided soil
degradation is not advanced, unmanaged vegetation has been observed
to increase in net production by 300%–500% in two to three years in
arid areas. However, where soils are greatly disturbed, recovery may not
occur even after 50 years (Le Houérou 1976b).

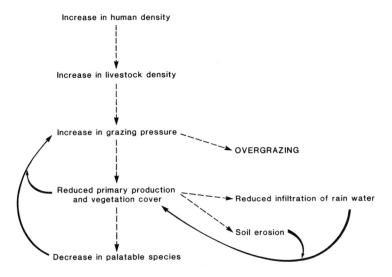

Figure 11-5   Overgrazing can be identified only in the medium-to-
long-term because it involves a continued reduction in palatable vegetation
through successive dry and wet periods over 5–10 years or more. Once that
point is reached degradation is self-reinforcing through several positive
feedback loops (solid arrows).

Table 11-2   Some properties of managed grazing systems in semiarid savannas of southern Africa predicted by a simulation model of water use by woody and herbaceous vegetation under different types of grazing regime. Nomadic pastoralism best fits the properties of the natural system (resilient and variable), while ranching produces the highest but most fragile livestock yield. Settled subsistence systems work against the natural system and have low productivity. Data from Walker et al. (1981).

| | ECOSYSTEM PROPERTIES | | |
MANAGEMENT REGIME	RESILIENCE OF ECOSYSTEM	GRASS AND HERBIVORE PRODUCTION	TEMPORAL VARIATION OF ECOSYSTEM
Nomadic pastoralism	High	Intermediate	High
Settled subsistence mixed agriculture	Intermediate	Low	Intermediate
Commercial ranching	Low	High	Low

Figure 8-9*b* shows stocking rates of pastoral populations in arid and semiarid parts of Kenya in relation to rainfall. It was suggested that these do not cause permanent land degradation where traditional practices are maintained. A country-by-country survey of cattle populations throughout Africa paints a rather different picture (Bourn 1979). Cattle biomass alone falls above the stocking rates of Figure 8-9*b* in Sudan, Somalia, and Mauritania. When other livestock and wild animals are taken into account, it seems obvious that periodic and widespread overgrazing is likely. Crude nationwide averages do not give the full picture, as Bourn showed within Ethiopia. In that country about one third of rangelands were seriously overstocked, and, significantly, these were concentrated in the driest and most fragile areas. The situation is particularly acute in much of the Sahel where primary production per millimeter of rainfall is much less than in other regions of Africa (Figure 8-9*a*).

In conclusion, overgrazing is rarely spread throughout a region. Rather, it is concentrated in particularly sensitive and poorly managed areas from which desertification may then radiate. Many other areas are undergrazed or not grazed at all. Political conflicts, banditry, and lack of water all contribute to the absence of grazing in such areas.

What Can Be Done?   The short answer is local destocking, but the cultural implications of enforcing such action are enormous. Most attempts to change the patterns of traditional pastoralism have been highly inappropriate. It seems that some form of nomadism remains the best way of utilizing semiarid and arid rangelands (Table 11-2). Settled mixed agriculture ignores well-known ecological factors, while commercial ranching seeks to overcome them by using expensive energy subsidies. Seeding, fertilizers, grazing rotations, and water management can reduce environmental variability and supply more livestock to market in the short-term. However, fewer people are supported economically than by nomadism in a similar area, and net economic returns may be smaller because of the financial investments needed (Pratt and Gwynne 1977).

Alternative management systems have been proposed that combine some of the ecological features of nomadism and ranching and that are also thought to be compatible with local cultures. An example for the Sahelian region is shown in Figure 11-6. For eastern Africa, Hjort (1976) has suggested "cattle insurance blocks" in which cattle in excess of subsistence needs are grazed in areas of more reliable rainfall. Meanwhile, nomadic herds are grazed, as before, in less reliable areas, but stock is replenished following a drought from the insurance blocks. This scheme avoids the necessity of building up very large herds in wet periods to ensure the survival of a few animals through a dry period. With such insurance, it is argued, pastoralists may be more willing to sell stock in times of plenty and thereby have money to buy supplementary food during droughts.

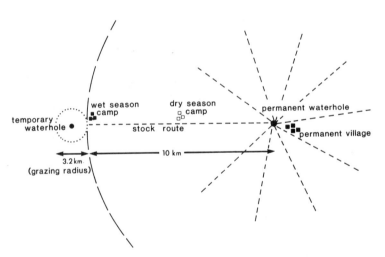

Figure 11-6 Proposed scheme for sustained use of Sahelian pastures. An area of 500–600 km² is centered on a permanent source of water. At the begining of the wet season, livestock and herders radiate to graze around temporary rain pools (one radius is shown in detail). As the dry season progresses, grazing gradually moves back to permanent water. A rotation system means that all radii are not used every year, allowing recovery of some areas and avoidance of those without rain. A few cattle may te taken on longer migrations to maintain traditional grazing rights. (Modified from Boudet [1976].)

The choice of specific systems will vary, depending on local conditions. There should be two guiding principles: the ecological conditions and the cultural milieu. Idealistic and alien preconceptions that ignore these factors are bound to fail.

The use of novel species adapted to arid conditions as a means of increasing economic returns is often advocated. For example, jojoba and guayule shrubs of the North American deserts are suggested as plantation crops. Jojoba produces a unique liquid wax with many actual and potential applications, while guayule contains a natural rubber, a resin, and a hard wax. Similarly, wild herbivores such as the eland and oryx of semiarid African savannas are promoted as candidates for domestication. Many other species could be cropped for meat in their wild state. However, none of these species, plant or animal, has yet demonstrated sufficient commercial viability and cultural acceptance to justify widespread adoption in the arid areas of developing countries. The unpredictable productivity of these environments reduces the attraction for commercial investment in new crops. The large financial and energy subsidies necessary to overcome this unpredictability could support more conventional crops and livestock equally well.

11.5   Ecology and Conservation in Nature Reserves

To be successful, biological conservation should be a feature of most land uses. The agricultural developments suggested in section 9.3.1 and the sustainable forestry and rangeland practices outlined previously all meet this end to a greater extent than the more industrial land use systems so often pursued. In addition, it is desirable to establish protected areas purely for biological conservation, which contain natural communities and are devoid of human populations. For convenience, such areas are called **nature reserves** here, although they are often given other designations such as national park, game reserve, or wildlife sanctuary. Ecology makes a vital contribution to the planning of nature reserves. There are two somewhat different steps involved in such planning: the selection of sites and the design of subsequent management practices.

11.5.1   Site Selection

There are three distinct reasons for the designation of nature reserves in a particular area:

1. Sites of high diversity

2. Sites containing rare organisms

3. Sites containing representative communities

Ideally, a given reserve can encompass the three objectives simultaneously. The first two address the maximization of genetic conservation, but all three are directed toward the conservation of national (or international) ecological heritage.

On a global scale, conservation is best served by establishing reserves in as many of the world's biotic communities as possible. There are important biogeographical areas in which no reserves exist, particularly within tropical latitudes (Figure 11-7). Ecology can also help to select the areas of highest diversity. The greatest species richness is found where former forest refuges persisted (Chapter 7). Reserves established in these areas will protect more species than those established elsewhere in the same biome. However, the designation of reserves is a political process that is subject to local, national and international pressures of various kinds. As a consequence, reserves are usually established in areas of low agricultural potential because of climate, soil, or remoteness from major population centers. In practice such sites are more easily protected because they are less threatened by the agricultural needs of burgeoning populations or by commercial pressures. This does not necessarily mean that areas of low conservation value are chosen. Several national parks already exist, and more are planned in the forest refugia of Brazil, Peru, Venezuela, Surinam, and Colombia, most of which are coincidentally in remote areas. Many of the savanna reserves of Africa are in areas that could support very few people, but because of their environmental heterogeneity they support a rich community of large mammals. After site selection, size is the most important factor in designation of a nature reserve. In general, large reserves contain more species than smaller ones, as indicated by species-area curves such as Figure 7-10. Particularly sensitive are many species in tropical rain forests that occur at low densities. For example, in the dipterocarp forests of Sarawak an area greater than 2000 ha is needed to contain 200 individuals of most tree species (Ashton 1976). Large predatory animals need much larger areas; 10,000 km^2 is recommended by Frankel and Soulé (1981). Unfortunately, less than 5% of existing national parks are this big, and almost 80% are less than 1000 km^2 in size.

Theories of island biogeography have been applied and extended to indicate preferred numbers, sizes, and shapes of reserves. Such ideas have been canonized in the World Conservation Strategy (IUCN 1980), although some applications of these theories have been questioned (for example, McCoy 1983). Reserves are visualized as being similar to land-bridge islands. Equilibrium theory is then used to predict rates of extinction and final species richness in protected areas (see section 7.6.1). Such predictions have been confirmed in several cases, particularly with respect to bird communities. As an example, an 87 ha plot of undisturbed forest surrounded by exploited land in Ecuador lost 44 of its 170 bird species in the five years following isolation, and another 15 species have precariously small populations (Leck 1979).

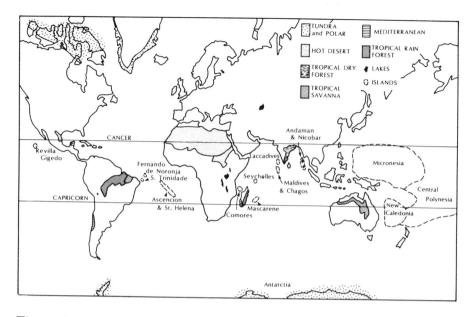

Figure 11-7   Outlined areas are biogeographical provinces (geographical units separable in terms of species composition; see Udvardy 1975) that contain no national parks or areas of equivalent conservation status. Note the predominance of tropical areas, particularly islands and lakes, and polar areas. An additional 19 tropical biogeographical provinces have total areas of less than 1000 km² (100 km² for islands) fully protected. Shading shows the type of vegetation in each of the provinces with no fully protected areas. After IUCN (1980) and Udvardy (1975). (Modified from Udvardy [1975] and International Union of Nature and Natural Resources [1980].)

There is little doubt that community fragmentation, isolation, and habitat reduction have profound effects on community structure and cause extinctions (Wilcox and Murphy 1985). However, mainland reserves need not be as isolated as true islands. A forest reserve surrounded by industrial agriculture may be equivalent, but other land-use practices are not as extreme. Low-intensity, shifting cultivation or even selective logging may allow many forest species to coexist with disturbance. In logged Indonesian forests with only 50% of trees left undamaged, a similar set of large animals remained as in undisturbed forest, although at lower densities (Wilson and Johns 1982; see also Figure 11-2). Most of the savanna reserves in eastern and southern Africa are not self-contained as migratory species spend wet seasons in surrounding rangelands that clearly are not hostile environments. Indeed, these reserves constitute only 10%–50% of the ecosystems used by the migratory mammals. A reserve of 200 km² is often adequate to preserve most ungulate species, provided surrounding land is managed appropriately (Western and Ssemakula 1981). Should these reserves become completely isolated (say by fences), many more species will be lost (East 1983).

Species-area curves suggest that one large reserve will contain more species than two smaller ones of equal total area. Such a consequence is not necessarily true, since no two communities are identical in species composition and these differences may compensate for the area effect. Also, a single reserve is much more vulnerable to political upheaval or to natural disruptions caused by climate or disease. In general, when more land becomes available, it is sensible to create additional reserves rather than add to the existing one, provided each is of adequate size.

Further development of and experiments on ecological concepts such as those discussed previously will undoubtedly increase ability to select and manage reserves. However, this basis for generalization should not be regarded as a substitute for a working knowledge of the interactions within a particular community. It is impossible to make allowance for all populations, but some of the species vulnerable to local extinction can usually be identified. If appropriate steps are taken to protect these key species, much of the community will also survive. Such species are often large (trees, large birds, and mammals), top carnivores (raptors, cats), or highly specialized (few food species or scarce environmental requirements). Important ecological factors to be taken in account (which are dealt with in more detail in other chapters) include the following.

1. *Population characteristics.* K-tendency populations are more vulnerable to unusual forms of environmental disturbance that might result from inadequate protection or increased insularity. However, r-tendency populations are more likely to become locally extinct, an argument favoring more than one reserve containing the species. Special consideration is needed for migratory populations as no amount of conservation in one part of their range will be effective if they are vulnerable elsewhere.

2. *Social organization.* Populations of solitary individuals need to be at sufficient density to enable successful mating. Black rhinoceros populations in Africa have been drastically reduced by poachers, and it is questionable whether viable populations persist in some areas where only a few individuals remain. Knowledge of social groupings is crucial to a determination of the size of the breeding populations, particulary where some adults are excluded from mating.

3. *Mutualisms.* Many plant populations depend on animals for pollination and seed dispersal. Where such mutualisms are closely evolved, viable populations of all participants need to be maintained. A classical example of mutualistic dependence is seen in the tree *Calvaria major* of Mauritius and its seed-dispersing bird, the now extinct dodo. The seeds only germinate after being eaten by the bird. As a conse-

quence, none have germinated for 300 years and the tree species faces extinction (Temple 1977).

4. *Phenology*. Even in the wettest environments, plants exhibit seasonal rhythms that affect other trophic levels in turn. Ignorance of these patterns can have serious consequences. For example, two species of fig fruit are crucial food for howler monkeys and other vertebrates during the drier season on Barro Colorado Island, Panama. During this period other fruit species are scarce and despite their low density (1 tree per 5 ha in some areas), loss of these fig species would have serious consequences for frugivore populations (Milton et al. 1982).

5. *Spatial heterogeneity*. An important cause of tropical species diversity is the pattern of secondary successions, which produce a varied environment in space and time. Maintaining an array of gap colonizers and high canopy trees in forests and natural fluctuations in herbaceous and woody vegetation in savannas requires insight into the temporal and spatial scales of these successions.

11.5.2   Management of Existing Nature Reserves

There are two divergent philosophies of reserve management. **Laissez-faire** is the ideal because it requires no interference beyond protection. It is applicable to well-defined ecosystems that are self-maintaining. In contrast, **active management** involves manipulating the environment to achieve attributes of high conservation value. Even where active management is practiced, minimum interference is the usual aim. The two approaches are appropriate to different situations. For savanna reserves, judicious and continuing active management may be needed to control the balance between herbivores, fire, herbaceous vegetation, and trees. These reserves are also major tourist attractions in Africa because of their combinations of open landscapes and large mammal populations. Tourism requires the planning of roads, lodgings, and other facilities. Laissez-faire is more appropriate to large forest reserves where protection and a surrounding buffer zone of low impact human activities may be adequate. Forest secondary successions are often endogenous and self-sustaining. Active management should be restricted to monitoring and possibly reintroduction of species excluded by prior human disturbance. Forests are not as superficially attractive to tourists except for more specialized groups such as ornithologists and botanists. However, there is much potential for forests to provide conservation education to nonspecialists.

Planning a system of reserves needs a clear definition of objectives. Ideally these would be dictated by international priorities for conservation such as those illustrated in Figure 11-7. In practice, most planning is at the national level, although regional cooperation is vital in some areas. For example, the Virunga volcanoes spread into Rwanda, Uganda, and Zaire, and their conservation value, which includes one of the few surviving gorilla populations, can only be maximized by international cooperation. Within a country, objectives such as conservation and tourism may conflict. Careful planning is needed to achieve a workable balance. This may involve different types of reserve, some of which prohibit and others which encourage tourists, or zoning within particular reserves to protect more sensitive areas. Careful planning can minimize the impact of tourists while maximizing their enjoyment.

Each nature reserve needs a **management plan** that fits into national priorities but takes account of the specific local situation. The first step in the production of such a plan is to establish the current status of the designated area. Prior literature is updated and gaps in information filled by field surveys. From these data, management options are determined and a plan drawn up to meet the specific objectives for the reserve in question. Such a plan should include an ecological monitoring program to identify changes that may call for future intervention. Monitoring of density and age structure of key populations, successional changes, and phenological patterns are particularly important. Specific research projects may be commissioned to clarify particular problems and to determine new management options.

The management plan also establishes the policy toward human access. For example, are independent research projects to be permitted? Many such projects improve management capabilities and should be encouraged. Some, however, such as those involving removal of native species or introduction of alien ones, or those involving large-scale disruption, are inappropriate in a nature reserve.

Another important role of nature reserves is in education. Guided tours, nature trails, and interpretation centers are important features where public access is encouraged. People educated by their visit are more likely to make worthwhile contributions to future conservation than tourists whose only objective is to photograph a tiger while learning nothing of the ecology of the reserve.

The preceding account of reserve planning is highly abbreviated, very generalized, and idealized. The politics, economics, and engineering involved in planning and managing reserves and in enforcing plans are outside the scope of this book but are vital components of the process. Some more specific types of active management stemming from ecology are outlined in the concluding section.

11.5.3   Ecology and Active Management in Savanna Reserves

Manipulation of secondary successions is the key to landscape and wildlife management in savannas. To avoid unnecessary intervention it is important to distinguish between short- and medium-term changes that are part of the natural dynamics from those brought about by changing human activities. For example, tree populations have declined recently in several eastern and southern African nature reserves. Is this due to climatic or edaphic changes, to fire, or to destruction by elephants or other herbivores? Any of these factors alone or in combination can be the cause. Research is needed to clarify management options. If edaphic or climatic changes are the ultimate cause, then shooting elephants will not solve the problem. In Amboseli National Park, Kenya, it seems that fluctuations in the water table, which is saline, are responsible. Between 1890 and 1960 the water table was relatively low and extensive *Acacia* woodlands developed, but prior to and since that period, a higher water table killed many trees (Western and van Praet 1973). In other situations, elephant population densities have greatly increased because of migration into reserves as surrounding rangelands have become peopled. Culling is an appropriate response in this case if the conservation value of the reserve is reduced by large numbers of elephants. In yet other places the elephant and tree populations may be components of a natural cycle driven by rainfall fluctuations; laissez-faire is the appropriate response (Phillipson 1975). Eltringham (1979) reviews this so-called "elephant problem," but it is important to realize that there is no universal explanation of the decline in tree densities, and only research can reveal the options relevant in a particular area.

The cheapest and easiest way of managing savanna successions is with fire. Woody vegetation can be reduced to improve the conditions for the grazing fauna or increased if landscape variety or browser populations are threatened. Fires need to be carefully controlled with respect to intensity, location, and frequency. Those set early in the dry season are cooler and less damaging to plants because the accumulation of dead, burnable vegetation is less. Grazing in specific areas is increased because the postburn sward is more nutritious than adjacent unburned areas. In this way, grazing rotations can be managed to attract grazers from unprotected areas or to provide good viewing opportunities for visitors. The frequency of fires is particularly important in controlling tree populations (Figure 11-8a), but interactions with browsers should not be ignored, as shown in Figure 11-8b. These results, which are from a simulation model, illustrate another powerful management tool. Such predictions should not, however, be accepted uncritically. Simulation models are only useful if two criteria are met. First, the baseline data input must be sufficiently accurate, and second, the model must be constructed with a clear understanding of (or good insight into) the most important interactions of the ecosystem. It is essential that monitoring continues to test the accuracy of model predictions and that management (and the model) be altered if unexpected changes occur.

The siting of water holes can also be used to control the distribution

a)

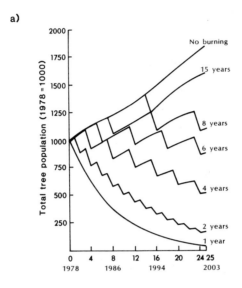

b)

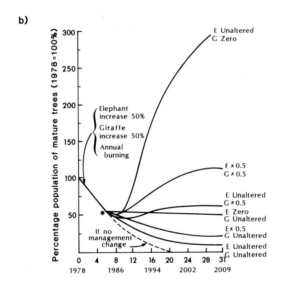

Figure 11-8 Simulation models and the management of Serengeti National Park, Tanzania. *a,* The effects of varying fire frequency upon the population of *Acacia tortilis* trees of the Seronera woodlands. *b,* The effects of various changes in elephant and/or giraffe populations and burning policy will alter the current decline of mature *A. tortilis* trees. E = elephants; G = giraffes (* denotes the onset of several possible alternative management schemes in 1984). In this particular simulation, burning is reduced from every year to every eight years beginning in 1984. Reproduced by permission of Blackwell Scientific Publications Ltd. (From Pellew, R. A. 1983. *Afr. J. Ecol.* 21:41–74.)

of water-dependent animals, particularly in the dry seasons. Feeding, trampling, and dung inevitably alter the surrounding vegetation. Such effects may spread for several kilometers around a water hole. Nevertheless, the provision of extra water holes in Wankie National Park, Zimbabwe, has helped to promote management objectives. Tourist viewing is enhanced and large herbivores are attracted to the Park, reducing their damage to surrounding farms (Weir 1971).

Summary

Conservation of biological resources requires the application of ecological principles. Motivations for conservation stem from both abstract philosophical and materialistic concerns. Two major conservation problems in the tropics are deforestation and desertification. The rapid rate of destruction of tropical forests may lead to some countries having no forests within two decades; overall, it is projected that all tropical forests could disappear within a century. The implications for global, regional, and local climate, hydrology, soils, and agriculture are not completely understood, but deforestation would be detrimental (probably severely) to most species including *Homo sapiens*. At the same time many livestock-rearing areas are being degraded from savanna to desert, with consequent loss of species and human food production. Ways to alleviate these changes present enormous ecological and cultural challenges. Conservation of nonhuman species and varieties is best carried out in nature reserves. Knowledge of ecological interactions is essential to the management of such reserves. Some guiding principles from ecology for reserve management are well understood, but others remain obscure.

Study Questions

Review

1. Do you think that genetic conservation is important? If so, why? If not, why not?

2. Why should poor countries conserve their forests when many rich countries have depleted their own?

3. Should nomadic pastoralism be stopped to prevent desertification? (See Chapter 9 also.)

4. Compare the roles of nature reserves and captive propagation as means of species conservation.

Related Topics

1. Collect information about a named nature reserve and suggest a management plan for it.

2. Investigate the role of gene banks as a means of conservation.

3. Discuss the importance of agroforestry as a conservation strategy.

4. Assess commercial ranching, game exploitation, and nomadic pastoralism as options for supporting the human populations of dry and semiarid savannas.

Further Reading

Black, J. (1970). *The Dominion of Man*. Edinburgh: Edinburgh University Press. A comparative historical account of conservation philosophy (mainly in nonsubsistence societies).

Duffey, E.; Watt, A. S. (1971). *The Scientific Management of Animal and Plant Communities*. Oxford: Blackwell. A symposium volume; mainly temperate, but useful information and includes a section on savanna management.

Frankel, O. H.; Soulé, M. E. (1981). *Conservation and Evolution*. Cambridge: Cambridge University Press. A good complement to *Conservation Biology* listed below; deals with genetic aspects of nature reserve management and captive propagation.

Myers, N. (1979). *The Sinking Ark*. Oxford: Pergamon Press. A lively, polemical account of the need for genetic conservation, with particular emphasis on tropical deforestation.

Rapp, A.; Le Houérou, H. N.; Lundholm, B. editors. (1976). Can desert encroachment be stopped? *Ecol. Bull.* Vol. 24. A description of arid ecosystems, their use by humans, desertification, and prescriptions for sustainable land use.

Sinclair, A. R. E.; Norton-Griffiths, M. editors. (1979). *Serengeti: Dynamics of an Ecosystem*. Chicago: Chicago University Press. Chapters 12 and 13 illustrate the usefulness of simulation models in savanna management.

Sutton, S. L.; Whitmore, T. C.; Chadwick, A. C. editors. (1983). *Tropical Rain Forest: Ecology and Management*. Oxford: Blackwell. Section IV deals with conservation and management of rain forests, including the need for international cooperation.

Soulé, M. E.; Wilcox, B. A. editors. (1980). *Conservation Biology*. Sunderland, MA: Sinauer. Strong on relationship between ecological theory and conservation, captive propagation and the relationship between exploitation and preservation.

Unesco (1980). *Case Studies on Desertification*. Paris: Unesco/UNEP/FAO.

Calculation of the Intrinsic Rate of Natural Increase (r) from Life Table Data

The information from sections A and C of Table 4-3 are used. First R_0 is calculated as follows:

$$R_0 = \frac{\Sigma l_x m_m}{1000} = \frac{1325}{1000} = 1.325$$

The generation time (T) is defined as the mean time from birth of parent to birth of offspring and is estimated as:

$$T = \frac{\Sigma x l_x m_x}{\Sigma l_x m_x} = \frac{8512}{1325} = 6.4 \text{ years}$$

r is now calculated as the natural logarithm of net reproductive rate (R_0) divided by the generation time:

$$r = \frac{\ln R_0}{T} = \frac{0.281}{6.4} = 0.044$$

Introduction to Mathematical Models of Interspecific Interactions

Interspecific competition _____

The Lotka-Volterra competition equations are derived from the logistic growth equation (section 4.2.2). For species 1:

$$\frac{dN_1}{dt} = r_1 N_1 \left(\frac{K_1 - N_1}{K_1} \right)$$

When together in scramble competition, growth of both species is reduced. This effect is included by the use of **competition coefficients**, where α is the effect of one individual of species 2 on population growth of species 1, and β is the same for the effect of species 1 on species 2. Thus, a pair of simultaneous population growth equations can be derived. For species 1:

$$\frac{dN_1}{dt} = r_1 N_1 \left(\frac{K_1 - N_1 - \alpha N_2}{K_1} \right)$$

For species 2:

$$\frac{dN_2}{dt} = r_2 N_2 \left(\frac{K_2 - N_2 - \beta N_1}{K_2} \right)$$

The magnitude of α and β determine the outcome of interspecific competition. For example, if α is greater than unity, interspecific effects of species 2 on species 1 are greater than intraspecific effects within species 1. The model predicts:

1. Species 1 alone survives if $\alpha < K_1/K_2$; $\beta > K_2/K_1$ (and vice versa for species 2).

2. Both species coexist if $\alpha < K_1/K_2$; $\beta < K_2/K_1$.

3. The species beginning at higher density survives if $\alpha > K_1/K_2$; $\beta > K_2/K_1$.

Specific outcomes of these predictions using numerical values are given in Figure 5-1.

Predation

The Lotka-Volterra equations for prey-predator systems are based on the exponential growth equation (section 4.3.2). For the prey population (N_1):

$$\frac{dN_1}{dt} = r_1N_1 - \gamma N_1N_2$$

Exponential growth is reduced by the predator population (N_2) with γ being the rate of feeding contacts between predator and prey. γN_1N_2 is the functional response (assumed to be linear) of the predator population on the growth of the prey population. For the predator population (N_2):

$$\frac{dN_2}{dt} - d_2N_2 + \delta N_1N_2$$

$-d_2$ is the intrinsic death rate of the predator (negative because it dies in the absence of the prey), and δ is the effect of acquiring one prey individual upon the predator population. δN_1N_2 is the numerical response (assumed to be linear) of the predator population to the prey population.

Most other models are structurally similar but substitute different algebraic formulations for the functional and numerical responses (see May 1981b).

Mutualisms

The simplest model involves changing the signs of α and β in the competition equations to make them positive mutualism coeffecients. However, this leads to infinite growth of both populations in most formulations. One approach to overcoming this problem is as follows. Suppose species 1 is wholly dependent on species 2; then in the absence of species 2, $K_1 = 0$, and carrying capacity of species 1 can be defined simply in terms of abundance of species 2. Thus:

$$\frac{dN_1}{dt} = r_1N_1 \left(\frac{\alpha N_2 - N_1}{\alpha N_2} \right)$$

To produce stable populations, it is necessary to set an upper limit to the benefit that each individual of species 2 obtains from individuals of species 2 (the saturation value referred to in section 5.3). Thus:

$$\frac{dN_2}{dt} = r_2N_2 \left(\frac{K_2 - N_2 + N_1Q}{K_2} \right)$$

where Q describes the saturation value and is formulated in such a way that its value decreases as N_2 decreases. See Whittaker (1975a) and May (1981b) for further discussion.

Indices of
Species Diversity

Several indices have been devised to combine both species richness and relative abundance into a single statistic. The most commonly used is the Shannon-Wiener index (H), which is derived from information theory and is calculated as:

$$H = -\Sigma p_i \ln p_i$$

where p_i is the proportional abundance (or importance measured in some other way) of the ith species such that $p_i = N_i/N$ (N_i = number of individuals in the ith species and N = the total number of individuals of all species in the community).

Similar proportionality calculations are used in Simpson's index (D), which uses probability theory and sums the probabilities of two successive individuals being of the same species:

$$D = 1 - \Sigma p_i^2$$

Specimen calculations are given in Table C-1.

H is strongly advocated by Pielou (1975), but recent work by Routledge (1979, 1980) and May (1975a, 1981c) suggests that D is better for both theoretical and practical reasons.

A third index, α, assumes the log series model of relative abundance (Figure 7-7b), but it is difficult to calculate as it involves trial-and-error substitutions in the formula:

$$S_T = \alpha \ln(1 + N/\alpha)$$

where S_T = total number of species and N = total number of individuals in the community. Alternatively, estimates of α can be obtained from the nomograph reproduced in Southwood (1978). This index (more accurately a statistic of the log series model) is strongly advocated by Taylor (1978).

The indices are used to compare diversity in different communities and usually give the same ranking sequence in a series of communities from most to least diverse. Unfortunately, this is not always the case and so caution must be exercised when interpreting information derived from these indices. For a fuller discussion of these and other indices, see the references just listed and Peet (1974), Hurlbert (1971), and Hutchinson (1978).

Table C-1 Computation of the diversity indices H and D.

SPECIES	NUMBER OF INDIVIDUALS (I)	PROPORTION OF INDIVIDUALS IN EACH SPECIES ($p_i = I/\Sigma I$)	(p_i^2)	$p_i \ln p_i$
A	30	0.36	0.130	− 0.265
B	26	0.32	0.102	− 0.233
C	18	0.22	0.048	− 0.146
D	5	0.06	0.004	− 0.022
E	3	0.04	0.002	− 0.012
ΣI 82		Σp_i 1.00	Σp_i^2 0.286	$\Sigma p_i \ln p_i$ − 0.678

$$D = 1 - \Sigma p_i^2 = 0.714$$

$$H = -\Sigma p_i \ln p_i = 0.678$$

Computation of Doubling Time from Exponential Growth Rates

Doubling time is that required for the number of individuals in a population to double. If geometric growth is assumed, this period can be computed simply. The exponential growth equation is (see Chapter 4):

$$N_t = N_0 \cdot e^{rt}$$

For a population to double, then $N_t/N_0 = 2$, reducing the equation to $2 = e^{rt}$. To solve this equation for t, first take the natural logarithm of each side ($\ln e = 1$; $e = 2.718$) to give:

$$\ln 2 = rt \ \ or \ \ \frac{\ln 2}{r} = t = \frac{0.693}{r}$$

Suppose the annual growth rate is 2%, then $e^r = 1.02$ (1 + the percentage increase as a decimal), and $r = 0.198$ (that is, $\ln 1.02$). Therefore:

$$t = \frac{0.693}{0.0198} = 35 \text{ years.}$$

Note that if a population is declining (a negative growth rate), this method will give halving time.

References

Abe, T. (1978). Studies on the distribution and ecological role of termites in a lowland rain forest of West Malaysia, I. Faunal composition, size, coloration and nest of termites in Pasoh Forest Reserve. *Kontyû* 46:273–290.

Abe, T.; Matsumoto, T. (1979). Studies on the distribution and ecological role of termites in a lowland rain forest of West Malaysia. (3) Distribution and abundance of termites in Pasoh Forest Reserve. *Jap. J. Ecol.* 29:337–351.

Ajtay, G. L.; Ketner, P.; Duvineaud, P. (1979). Terrestrial primary production and phytomass. In: Bolin, B.; Degens, E. T.; Kempe, S.; Ketner, P. editors *The Global Carbon Cycle*. New York: Wiley, pp. 129–181.

Allen, J. C. (1985). Soil response to forest clearing in the United States and the tropics: geological and biological factors. *Biotropica* 17:15–27.

Amarasinghe, L.; Pemadasa, M. A. (1982). The ecology of a montane grassland in Sri Lanka. II The pattern of four major species. *J. Ecol.* 70:17–23.

Anderson, J. M.; Coe, M. J. (1974). Decomposition of elephant dung in an arid tropical environment. *Oecologia* 14:111-125.

Anderson, J. M.; Proctor, J.; Vallack, H. W. (1983). Ecological studies in four contrasting lowland rain forests in Gunung Mulu National Park, Sarawak. III. Decomposition processes and nutrient losses from leaf litter, *J. Ecol.* 71:503–527.

Anderson, J. M.; Swift, M. J. (1983). Decomposition in tropical forests. In: Sutton, S. L.; Whitmore, T. C.; Chadwick, A. C. editors. *Tropical Rain Forest: Ecology and Management*. Oxford: Blackwell, pp. 287–309.

Anderson, R. M. (1979). The influence of parasitic infections on the dynamics of host population growth. In: Anderson, R. M.; Turner, B. D.; Taylor, L. R. editors. *Population Dynamics*. Oxford: Blackwell, pp. 245–281.

Anderson, R. M. (1981). Population ecology of infectious disease agents. In: May, R. M. editor. *Theoretical Ecology*. Oxford: Blackwell pp. 318–355.

Andersson, M. (1982). Female choice selects for extreme tail length in a widow bird. *Nature* 299:818–820.

Andrews, R.; Coleman, D. C.; Ellis, J.; Singh, J. S. (1974). Energy flow relationships in a shortgrass prairie ecosystem. *First Int. Congr. Ecol.* pp. 22–28.

Antúnez de Mayolo, R. S. E. (1983). Climate prediction and agriculture in pre-Colombian Peru. In: Cusack, D. F. editor. *Agroclimate Information for Development*. Boulder, CO: Westview Press, pp. 25–33.

Anway, J. C. (1978). A mammalian consumer model for grasslands. In: Innis, G. S. editor. *Grassland Simulation Model*. New York: Springer–Verlag, pp. 89–125.

Armstrong, J. T. (1965). Breeding home range in the nighthawk and other birds: its evolutionary and ecological significance. *Ecology* 46:619–629.

Ashton, P. S. (1969). Speciation among tropical forest trees: some deductions in the light of recent evidence. *Biol. J. Linn. Soc.* 1:155–196.

Ashton, P. S. (1976). Factors affecting the development and conservation of tree genetic resources. In: Burley, J.; Styles, B. T., editors. *Tropical Trees: Variation, Breeding and Conservation. Symposium of the Linnaean Society Series*. No. 2. London: Academic Press; pp. 189–198.

Attwell, C. A. M. (1982). Population ecology of the blue wildebeest *Connochaetes taurinus taurinus* in Zululand, South Africa. *Afr. J. Ecol.* 20:147–168.

Augspurger, C. K. (1983). Seed dispersal of the tropical tree, *Platypodium elegans,* and the escape of its seedlings from fungal pathogens. *J. Ecol.* 71:759–771.

Austin, M. P.; Ashton, P. S.; Greig-Smith, P. (1972). The application of quantitative methods to vegetation survey. III. A reexamination of rainforest data from Brunei. *J. Ecol.* 60:305–324.

Aweto, A. O. (1981a). Secondary succession and soil fertility restoration in southwestern Nigeria. I. Succession. *J. Ecol.* 69:601–607.

Aweto, A. O. (1981b). Secondary succession and soil fertility restoration in southwestern Nigeria. II. Soil fertility restoration. *J. Ecol.* 69:609–614.

Bagine, R. K. N. (1982). *The Role of Termites in Litter Decomposition and Soil Translocation with Special Reference to* Odontotermes *in Arid Lands of Northern Kenya.* University of Nairobi. Thesis.

Baillie, I. C.; Ashton, P. S. (1983). Some soil aspects of the nutrient cycle in mixed dipterocarp forests, in Sarawak. In: Sutton, S. L.; Whitmore, T. C.; Chadwick, A. C. editors. *Tropical Rain Forest: Ecology and Management.* Oxford: Blackwell, pp. 347–356.

Baker, H. G. (1970). Evolution in the tropics. *Biotropica* 2:101–111.

Baker, H. G. (1973). Evolutionary relationships between flowering plants and animals in American and African tropical forests. In: Meggers, B. J.; Ayensu, E. S.; Duckworth, W. D. editors. *Tropical Forest Ecosystems in Africa and South America: A Comparative Review.* Washington, DC: Smithsonian, pp. 145–159.

Baker, H. G. (1978). Chemical aspects of the pollination biology of woody plants in the tropics. In: Tomlinson, P. B.; Zimmerman, M. H. editors. *Tropical Trees as Living Systems.* London: Cambridge University Press, pp. 57–82.

Baker, H .G., Baker, I. (1983). A brief historical review of the chemistry of floral nectar. In: Bently, B.; Elias, T. editors. *The Biology of Nectaries.* New York: Columbia University Press; pp. 126–152.

Banerjee, B. (1979). A key-factor analysis of the population fluctuations in *Andraca bipunctata* Walker (Lepidoptera: Bombycidae). *Bull. Ent. Res.* 69:195–201.

Banerjee, B. (1981). An analysis of the effect of latitude, age and area on the number of arthropod pest species of tea. *J. Appl. Ecol.* 18:339–342.

Banse, K.; Mosher, S. (1980). Adult body mass and annual production/biomass relationships of field populations. *Ecol. Mon.* 50:355–379.

Barbour, M. G. (1981). Plant-plant interactions. In: Goodall, D. W.; Perry, R. A. editors. *Arid-land Ecosystems: Structure, Functioning and Management.* Vol. 2. Cambridge: Cambridge University Press, pp. 33–49.

Baroni-Urbani, C.; Josens, G.; Peakin, G. J. (1978). Empirical data and demographic parameters. In: Brian, M. V. editor. *Production Ecology of Ants and Termites.* Cambridge: Cambridge University Press, pp. 5–44.

Beals, E. W. (1969). Vegetational change along altitudinal gradients. *Science* 165:981–985.

Beaver, R. A. (1979). Host specificity of temperate and tropical animals. *Nature* 281:139–141.

Beddington, J. R. (1979). Harvesting and population dynamics. In: Anderson, R. M.; Turner, B. D.; Taylor, L. R. editors. *Population Dynamics.* Oxford: Blackwell, pp. 307–320.

Bell, E. A. (1978). Toxins in seeds. In: Harborne J. B. editor. *Biochemical Aspects of Plant and Animal Coevolution.* London:Academic Press, pp. 143–161.

Bell, R. H. V. (1970). The use of the herb layer by grazing ungulates in the Serengeti. In: Watson, A. editor: *Animal Populations in Relation to Their Food Resources.* Oxford: Blackwell, pp. 111–124.

Bennett, A. F.; Gorman, G. C. (1979). Population density and energetics of lizards on a tropical island. *Oecologia* 42:339–358.

Benson, W. W.; Brown, K. S., Jr.; Gilbert, L. E. (1975). Coevolution of plants and herbivores: Passionflower butterflies. *Evolution* 29:659–680.

Bentley, B. L. (1977). Extrafloral nectaries and protection by pugnacious body guards. *Ann. Rev. Ecol. Syst.* 8:407–427.

Bernays, E. A.; Chapman, R. F. (1978). Plant chemistry and acridoid feeding behaviour. In Harborne, J. B. editor. *Biochemical Aspects of Plant and Animal Coevolution.* London: Academic Press, pp. 99–141.

Berry, R. J. (1982). *Neo-Darwinism*. London: Arnold.

Bess, H. A.; Bosch, R. van den; Haramoto, F. H. (1961). Fruit fly parasites and their activities in Hawaii. *Proc. Hawaii Ent. Soc.* 17:367–378.

Bhagavan, M. R. (1984). The woodfuel crisis in the SADCC countries. *Ambio* 13: 25–27.

Bhatia, R. (1977). Energy and rural development in India. In Lockeretz, W. editor. *Agriculture and Energy*. New York: Academic Press, pp. 559–579.

Birch, L. C.; Dobzhansky, T.; Elliott, P. O.; Lewontin, R. C. (1963). Relative fitness of geographic races of *Drosophila serrata*. *Evolution* 17:72–83.

Birdsell, J. B. (1978). Spacing mechanisms and adaptive behaviour of Australian Aborigines. In: Ebling, F. J.; Stoddart, D. M. editors. *Population Control by Social Behaviour*. London: Institute of Biology, pp. 213–244.

Biswas, A. K. (1979). World models, resources and environment. *Env. Conserv.* 6:3–11.

Black, C. C. (1971). Ecological implications of dividing plants into groups with distinct photosynthesis production capacities. *Adv. Ecol Res.* 7:87–114.

Blaxter, K. L. (1973). The purpose of protein production. In: Jones, J. G. W. editor. *The Biological Efficiency of Protein Production*. Cambridge: Cambridge University Press, pp. 3–11.

Blueweiss, L.; Fox, H.: Kudzma, V.; Nakashima, D.; Peters, R.; Sams, S. (1978). Relationships between body size and some life history parameters. *Oecologia* 37:257–272.

Boag, P. T.; Grant, P. R. (1981). Intense natural selection in a population of Darwin's finches (Geospizinae) in the Galápagos. *Science* 214:82–85.

Bolin, B.; Degins, E. T.; Kempe, S.; Ketner, P. (1979). *The Global Carbon Cycle*. Chichester, England: Wiley.

Borchert, R. (1983). Phenology and control of flowering in tropical trees. *Biotropica* 15:81–89.

Boudet, G. (1976). Mali. *Ecol. Bull.* 24:137–153.

Bourn, D. (1978). Cattle, rainfall and tsetse in African countries. *J. Arid Environ.* 1:49–61.

Boyce, M. S. (1984). Restitution of r- and K-selection as a model of density-dependent natural selection. *Ann. Rev. Ecol. Syst.* 15:427–447.

Brassel, H. M.; Sinclair, D. F. (1983). Elements returned to forest floor in two rain forest and three plantation plots in tropical Australia. *J. Ecol.* 71:367–378.

Britton, P. L.; Zimmerman, D. A. (1979). The avifauna of Sokoke Forest, Kenya. *J. E. Afr. Nat. Hist. Soc. Nat. Mus.* 169:1–15.

Broadhead, E. (1983). The assessment of faunal diversity and guild size in tropical forests with particular reference to the Psocoptera. In: Sutton, S. L.; Whitmore, T. C.; Chadwick, A. C. editors. *Tropical Rain Forest: Ecology and Management*. Oxford: Blackwell, pp. 107–119.

Brokaw, N. V. L. (1982). The definition of treefall gap and its effect on measures of forest dynamics. *Biotropica* 14:158–160.

Brosset, A. (1971). Recherches sur la biologie des Pycnonotidés du Gabon. *Biol. Gabon.* 7:424–460.

Brosset, A. (1981). La periodicitie de la reproduction chez un bulbul de foret equatoriale Africaine *Andropadus latirostris* ses incidences demographiques. *Rev. Ecol. (Terre et Vie)* 35:109–129.

Brower, J. V. Z. (1960). Experimental studies of mimicry. IV. The reactions of starlings to different proportions of models and mimics. *Am. Nat.* 94:271–282.

Brower, L. P.; Alcock, J.; Brower, J. V. Z. (1971). Avian feeding behaviour and the selective advantage of incipient mimicry. In: Creed, R. editor. *Ecological Genetics and Evolution*. Oxford: Blackwell, pp. 261–274.

Browman, D. L. (1974). Pastoral nomadism in the Andes. *Cur. Anthropol.* 15:188–196.

Brown, L. H. (1971). The biology of pastoral man as a factor in conservation. *Biol. Conserv.* 3:93–100.

Brown, L. R. (1981). World population growth, soil erosion and food security. *Science* 214:995–1002.

Brown, S.; Lugo, A. E. (1984). Biomass of tropical forests: a new estimate based on forest volumes. *Science* 223:1290–1293.

Bullock, S. H. (1982). Population structure and reproduction in the neotropical dioecious tree *Compsoneura sprucei*. *Oecologia* 55:238–242.

Burns, R. C.; Hardy, R. W. F. (1975). *Nitrogen fixation in Bacteria and Higher Plants*. Berlin: Springer-Verlag.

Caldwell, M. (1975). Primary production of grazing lands. In: Cooper, J.P. editor. *Photosynthesis and Productivity in Different Environments*. Cambridge: Cambridge University Press, pp. 41–73.

Calvert, W. H.; Zuchowski, W.; Brower, L. P. (1983). The effect of rain, snow and freezing temperatures on overwintering monarch butterflies in Mexico. *Biotropica* 15:42–47.

Campbell, K. E., Jr.; Frailey, D. (1984). Holocene flooding and species diversity in southwestern Amazonia. *Quaternary Res.* 21:369–375.

Caswell, H. (1976). Community structure: a neutral model analysis. *Ecol Mon.* 46:327–354.

Caughley, G. (1976). The elephant problem—an alternative hypothesis. *E. Afr. Wildl. J.* 14:265–283.

Caughley, G. (1977). *Analysis of Vertebrate Populations*. Chichester, England: Wiley.

Chang, T. T. (1984). Conservation of rice genetic resources: luxury or necessity? *Science* 224:251–256.

Chapin, G.; Wasserstrom, R. (1981). Agricultural production and malaria resurgence in Central America and India. *Nature* 293:181–185.

Chapman, R. F.; Page, W. W. (1979). Factors affecting the mortality of the grasshopper *Zonoceros variegatus* in southern Nigeria. *J. Anim. Ecol.* 48:271–288.

Chapman, S. B. editor. (1976). *Methods in Plant Ecology*. Oxford: Blackwell.

Charney, J.; Stone, P. M.; Quirk, W. J. (1975). Drought in the Sahara: a biogeophysical feedback mechanism. *Science* 187:434–435.

Cherfas, J. (1979). How to raise protection money. *New Sci.* 84: 786–788.

Clark, D. A.; Clark, D. B. (1984). Spacing dynamics of tropical rain forest trees: evaluation of the Janzen-Connell model. *Am. Nat.* 124:769–788.

Clark, D. B.; Guayasamín, C.; Pazmiño, O.; Donoso, C.; de Villacís, Y. P. (1982). The tramp ant *Wassmania auropunctata:* autecology and the effects on ant diversity and distribution on Santa Cruz Island, Galápagos. *Biotropica* 14:196–207.

Clarke, C. A.; Sheppard, P. M. (1962). Disruptive selection and its effect on a metrical character in the butterfly *Papilio dardanus*. *Evolution* 16:214–226.

Clarke, C. A.; Sheppard, P. M. (1971). Further studies on the mimetic butterfly *Papilio memnon* L. *Phil. Trans. Roy. Soc. Lond.* B 263:35–70.

Clarke, G. L. (1954). *Elements of Ecology*. New York: Wiley.

Cloudsley-Thompson, J. L. (1969). *The Zoology of Tropical Africa*. London: Weidenfield & Nicholson.

Cloudsley-Thompson, J. L.; Chadwick, M. J. (1964). *Life in Deserts*. London: Foulis.

Clutton-Brock, T. H.; Harvey, P. H. (1977). Primate ecology and social organization. *J. Zool. Lond.* 183:1–39.

Cock, J. H. (1982). Cassava: a basic energy source in the tropics. *Science* 218:755-762.

Coe, M. J.; Cumming, D. H.; Phillipson, J. (1976). Biomass and production of large African herbivores in relation to rainfall and primary production. *Oecologia* 22:341–354.

Cole, D. W.; Rapp, M. (1981). Elemental cycling in forest ecosystems. In: Reichle, D. E. editor. *Dynamic Properties of Forest Ecosystems*. Cambridge: Cambridge University Press, pp. 341–409.

Cole, N. H. A. (1984) Tropical ecology research. *Nature* 309:204.

Coleman, D. C. (1976). A review of root production processes and their influence on soil biota in terrestrial ecosystems. In: Anderson, J. M.; MacFadyen, A. editors. *The Role of Terrestrial and Aquatic Organisms in Decomposition Processes*. Oxford:Blackwell, pp. 417–434.

Colinvaux, P. A.; Miller, M. C.; Liu, K-B; Steinitz-Kannan, M.; Frost, I. (1985). Discovery of permanent Amazonian lakes and hydraulic disturbances in the upper Amazon basin. *Nature* 313:42–45.

Collins, N. M. (1981). Populations, age structure and survivorship of colonies of *Macrotermes bellicosus* (Isoptera: Macrotermitinae). *J. Anim. Ecol.* 50:293–311.

Condominas, G. (1980). Agricultural ecology in the southeastern Asian savanna region: The Mnong Gar of Vietnam and their social space. In: Harris, D. R. editor: *Human Ecology in Savanna Environments.* London: Academic Press, pp. 209–251.

Connell, J. H. (1975). Some mechanisms producing structure in natural communities: a model and evidence from field experiments. In: Cody, M. L.; Diamond J. M. editors. *Ecology and Evolution of Communities.* Cambridge, MA: Belknap, pp. 460–490.

Connell, J. H. (1979). Tropical rain forests and coral reefs as open nonequilibrium systems. In: Anderson, R. M.; Turner, B. D.; Taylor, L. R. editors. *Population Dynamics.* Oxford: Blackwell, pp. 141–163.

Connell, J. H. (1980). Diversity and the coevolution of competitors, or the ghost of competition past. *Oikos* 35:131–138.

Connell, J. H.; Sousa, W. P. (1983). On the evidence needed to judge ecological stability or persistence. *Am. Nat.* 121:789–814.

Conway, G. (1981). Man versus pests. In: May, R. M. editor. *Theoretical Ecology.* Oxford: Blackwell, pp. 356–386.

Conway, G. R.; Manwan, I.; McCauley, D. S. (1983). The development of marginal lands in the tropics. *Nature* 304:392.

Cooper, J. J.; Tinsley, T. W. (1978). Some epidemiological consequences of drastic ecosystem changes accompanying exploitation of tropical rain forest. *Terre et Vie* 32:221–240.

Cornaby, B. W. (1974). Carrion reduction by animals in contrasting tropical habitats. *Biotropica* 6:51–63.

Cornforth, I. S. (1970). Reafforestation and nutrient reserves in the humid tropics. *J. Appl. Ecol* 7:609–615.

Coupland, R. T. (1979). Conclusion. In: Coupland, R. T. editor *Grassland Ecosystems of the World.* Cambridge: Cambridge University Press, pp. 335–355.

Cox, C. B.; Healey, I. N.; Moore, P. D. (1976). *Biogeography.* Oxford: Blackwell.

Crawford, C. S. (1981). *Biology of Desert Invertebrates.* Berlin: Springer-Verlag.

Crombie, A. C. (1946). Further experiments on insect competition. *Proc. Roy. Soc. B* 133:76–109.

Cruden, R. W.; Hermann-Parker, S. M. (1979). Butterfly pollination of *Caesulpinia pulcherrima,* with observations on the psychophilus syndrome. *J. Ecol.* 67:155–168.

Cruikshank, S.; Ahmed, F. (1973). Biological control of potato tuber moth *Phthorimaea operculellia* (Zell.) (Lep.: Geleschiidae) in Zambia. *Commonwealth Inst. Biol. Control Tech. Bull.* 16:147–162.

Culver, D. C. (1974). Species packing in Caribbean and north temperate ant communities. *Ecology* 55:974–988.

Cummins, K. W.; Wuycheck, J. C. (1971). Caloric equivalents for investigations in ecological energetics. *Mitt. Int. Ver. Limnol.* No. 18.

Cunningham, W. J. (1954). A nonlinear differential-difference equation of growth. *Proc. Natl. Acad. Sci.* 40:708–713.

Danthanarayana, W.; Kathirauetpillai, A. (1969). Studies on the ecology and causes of outbreaks of *Ectropis bhurmitra* (Geometridae), the twig caterpillar of tea in Ceylon. *J. Appl. Ecol* 6:311–322.

Darwin, C. (1859). *On the Origin of Species by Means of Natural Selection.* London: John Murray. (1st edition reprinted by Penguin 1968).

Dash, M. C. (1979). Tropical grasslands: consumers. In: Coupland, R. T. editor. *Grassland Ecosystems of the World.* Cambridge: Cambridge University Press, pp. 219–225.

David, J. R.; Bocquet, C. (1975). Evolution in a cosmopolitan species: genetic latitudinal clines in *Drosophila melanogaster* wild populations. *Experientia* 31:164–166.

Davidson, D. W. (1977). Species diversity and community organization in desert seed-eating ants. *Ecology* 58:711–724.

Davies, N. B.; Houston, A. I. (1984). Territory economics. In: Krebs, J. R.; Davies, N. B. editors. *Behavioural Ecology*. 2d ed. Oxford: Blackwell.

Delany, M. J. (1974). *The Ecology of Small Mammals*. London: Arnold.

Delany, M. J.; Happold, D. C. D. (1979). *Ecology of African Mammals*. London: Longman.

Delvi, M. R.; Pandian, T. J. (1979). Ecological energetics of the grasshopper *Poecilocerus pictus* in Bangalore fields. *Proc. Ind. Acad. Sci.* 88B:241–256.

Denslow, J. S. (1980). Gap partitioning among tropical rain forest trees. *Biotropica* (Suppl). 12:47–55.

Deshmukh, I. K. (1974). *Primary production and leaf litter breakdown in a sand dune succession*. University of Dundee, Thesis.

Deshmukh, I. K. (1984). A common relationship between precipitation and grassland peak biomass for East and southern Africa. *Afr. J. Ecol.* 22:181–186.

Deshmukh, I. K. (1985). Decomposition of grasses in Nairobi National Park, Kenya. *Oecologia*. 67:147–149.

Desowitz, R. S. (1980). Epidemiological-ecological interactions in savanna environments. In: Harris, D. R. editor. *Human Ecology in Savanna Environments*. London: Academic Press. pp.457–477.

Desowitz, R. S. (1981). *New Guinea Tapeworms and Jewish Grandmothers. Tales of Parasites and People*. New York: Avon Books.

Diamond, A. W. (1980). Seasonality, population structure and breeding ecology of the Sychelles brush warbler *Acrocephalus sechellensis*. *Proc. Fourth Pan-Afr. orn. Congr.* pp. 253–266.

Diamond, A. W.; Hamilton, A. C. (1980). The distribution of forest passerine birds and quaternary climatic changes in tropical Africa. *J. Zool. Lond.* 191:379–402.

Diamond, J. M. (1973). Distributional ecology of New Guinea birds. *Science* 179:759–769.

Diamond, J. M. (1975). Assembly of species communities. In: Cody, M. L., Diamond, J. M. editors. *Ecology and Evolution of Communities*. Cambridge, MA: Belknap, pp. 342–444.

Diamond, J. M. (1979). Community structure: is it random, or is it shaped by species differences and competition? In: Anderson, R. M.; Turner, B. D., Taylor, L. R. editors. *Population Dynamics*. Oxford: Blackwell; pp. 165–181.

Diamond, J. M.; May, R. M. (1981). Island biogeography and the design of nature reserves. In: May, R. M. editor. *Theoretical Ecology*. Oxford: Blackwell, pp. 228–252.

Dickeman, M. (1975). Demographic consequences of infanticide in man. *Ann. Rev. Ecol. Syst.* 6:107–137.

Dobzhansky, T. (1950). Evolution in the tropics. *Am. Sci.* 38:209–221.

Dobzhansky, T.; Ayala, F. J.; Stebbins, G. L.; Valentine, J. W. (1977). *Evolution*. San Francisco: Freeman.

Dodson, C. H. (1975). Coevolution of orchids and bees. In: Gilbert, L. E.; Raven, P. H. editors. *Coevolution of Animals and Plants*. Austin, TX: University of Texas Press, pp. 91–99.

Dougall, H. W.; Drysdale, V. M.; Glover, F. E. (1964). The chemical composition of Kenya browse and pasture herbage. *E. Afr. Wildl. J.* 2:86–121.

Doyle, T. W. (1981). The role of disturbance in the gap dynamics of a montane rain forest: an application of a tropical forest succession model. In West, D. C.; Shugart, H. H.; Botkin, D. B. editors: *Forest Succession: Concepts and Applications*. New York: Springer-Verlag, pp. 56–73.

Dunbar, R. I. M. (1980). Demographic and life history variables of a population of gelada baboons *(Theropithecus gelada)*. *J. Anim. Ecol.* 49:485–506.

Dwyer, P. D. (1982). Prey switching: a case study from New Guinea. *J. Anim. Ecol.* 51:529–542.

Dyson-Hudson, N. (1980). Strategies of resource exploitation among East African pastoralists. In: Harris, D. R. editor: *Human Ecology in Savanna Environments*. London: Academic Press, pp. 171–184.

East, R. (1983). Application of species-area curves to African savanna reserves. *Afr. J. Ecol.* 21:123–128.

Eckholm, E. (1979). Forestry success of the seventies. *New Sci.* 81:764–765.

Edroma, E. L. (1974). Copper pollution in Rwenzori National Park, Uganda. *J. Appl. Ecol.* 11:1043–1056.

Edwards, P. J. (1977). Studies of mineral cycling in a montane rain forest in New Guinea. III. The production and decomposition of litter. *J. Ecol.* 65:971–992.

Edwards, P. J. (1982). Studies of mineral cycling in a montane forest in New Guinea. V. Rate of cycling in throughfall and litter fall. *J. Ecol.* 70:807–827.

Edwards, P. J.; Grubb, P. J. (1982). Studies of mineral cycling in a montane rain forest in New Guinea. IV. Soil characteristics and the division of mineral elements between the vegetation and soil. *J. Ecol.* 70:649–666.

Eggler, W. A. (1971). Quantitative studies of vegetation on sixteen young lava flows on the island of Hawaii. *Trop. Ecol.* 12:66–100.

Egunjobi, J. K.; Bada, S. O. (1979). Biomass and nutrient distribution in stands of *Pinus caribea* L. in the dry forest zone of Nigeria. *Biotropica* 11:130–135.

Egunjobi, J. K.; Onweluzo, B. S. (1979). Litter fall, mineral turnover and litter accumulation in *Pinus caribea* L. stands in Nigeria. *Biotropica* 11:251–255.

Ehrlich, P. R.; Ehrlich, A. H.; Holdren, J. P. (1977). *Ecoscience.* San Francisco: Freeman.

Ehrlich, P. R.; Gilbert, L. E. (1973). Population structure and dynamics of the tropical butterfly *Heliconius ethilla. Biotropica* 5:69–82.

Ehrlich, P. R.; Raven, P. H. (1964). Butterflies and plants: a study in coevolution. *Evolution* 18:586–608.

El-Din, A. S. (1971). Ecological studies of the vegetation of the Sudan. The effect of simulated grazing on the growth of *Acacia senegal* (L.) Willd. seedlings. *J. Appl. Ecol.* 8:211–216.

Elias, T. S. (1983). Extrafloral nectaries: their structure and distribution. In: Bentley, B.; Elias, T. editors. *The Biology of Nectaries.* New York: Columbia University Press, pp. 174–203.

Elliott, C. C. H. (1979). The harvest time method as a means of avoiding *Quelea* damage to irrigated rice in Chad/Cameroon. *J. Appl. Ecol.* 16:23–35.

Elton, C. C. (1927). *Animal Ecology.* London: Sidgwick & Jackson.

Elton, C. S. (1958). *The Ecology of Invasions by Animals and Plants.* London: Chapman & Hall.

Elton, C. S. (1966). *The Pattern of Animal Communities.* London: Methuen.

Elton, C. S. (1973). The structure of invertebrate populations inside neotropical rain forest. *J. Anim. Ecol.* 42:55–104.

Eltringham, S. K. (1979). *The Ecology and Conservation of Large African Mammals.* London: Macmillan.

Emmons, L. H. (1979). Observations on litter size and development of some African rain forest squirrels. *Biotropica* 11:207–213.

Ernst, A. (1908). *The New Flora of the Volcanic Island of Krakatau.* (Translated by A. C. Steward). Cambridge: Cambridge University Press.

Ewel, J. J. (1976). Litter fall and leaf decomposition in a tropical forest succession in eastern Guatemala. *J. Ecol.* 64:293–308.

Farlow, J. O. (1976). A consideration of the trophic dynamics of a late Cretaceous large-dinosaur community (Oldman Formation). *Ecology* 57:841–857.

Farnworth, E. G.; Golley, F. B. editors. (1974). *Fragile Ecosystems.* New York: Springer-Verlag.

Feeny, P. (1976). Plant apparency and chemical defense. *Rec. Adv. Phytochem.* 10:1–40.

Feinsinger, P. (1983). Coevolution and pollination, In: Futuyma, D. J.; Slatkin, M., editors. *Coevolution.* Sunderland, MA: Sinauer. pp. 282–310.

Fischer, A. (1960). Latitudinal variation in organic diversity. *Evolution* 14:64–81.

Flannery, K. V. editor (1982). *Maya Subsistence.* New York: Academic Press.

Fleming, T. H. (1971). Population ecology of three species of neotropical rodents. *Misc. Publ. Mus. Zool. Univ. Michigan.* No. 143.

Fleming, T. H. (1973). Number of mammal species in north and central American forest communities. *Ecology* 54:555-563.

Fleming, T. H. (1974). The population ecology of two species of Costa Rican heteromyid rodents. *Ecology* 55:493-510.

Fogden, M. P. L. (1972). The seasonality and population dynamics of equatorial forest birds in Sarawak. *Ibis* 114:307-342.

Food and Agriculture Organization of the United Nations. (1967-1983). *Production Yearbooks*. Rome: FAO.

Ford, E. B. (1975). *Ecological Genetics*. London: Chapman & Hall.

Ford, J. (1971). *The Role of Trypanosomiasis in African Ecology*. Oxford: Oxford University Press.

Forman, R. T. T.; Hahn, D. C. (1980), Spatial patterns of trees in a Caribbean semi-evergreen forest. *Ecology* 61:1267-1274.

Foster, R. B. (1982). Famine on Barro Colorado Island. In: Leigh, E. G.; Rand, A. S.; Windsor, D. M. editors. *The Ecology of a Tropical Forest*. Washington, DC: Smithsonian, pp. 201-212.

Frankel, O. H.; Soulé, M. E. (1981) *Conservation and Evolution*. Cambridge: Cambridge University Press.

Frankie, G. W.; Baker, H. G.; Opler, P. A. (1974). Comparative phenological studies of the trees in tropical wet and dry forests in the lowlands of Costa Rica. *J. Ecol.* 62:881-919.

Franks, N. R.; Bossert, W. H. (1983). The influence of swarm raiding army ants on the patchiness and diversity of a tropical leaf litter ant community. In: Sutton, S. L.; Whitmore, T. C.; Chadwick, A. C. editors. *Tropical Rain Forest: Ecology and Management*. Oxford: Blackwell, pp. 151-163.

Freeman, B. E. (1980). A population study in Jamaica on adult *Sceliphron assimile* (Dahlbon) (Hymenoptera: Sphecidae). *Ecol. Entomol.* 5:19-30.

Freeman, B. E. (1981). The dynamics in Trinidad of the sphecid wasp *Trypoxylon palliditarsa*: a Thompsonian population? *J. Anim. Ecol.* 50:563-572.

Freeman, B. E.; Ittyeipe, K. (1976). Field studies on the cumulative response of *Melittobia* sp. (Hawaiiensis complex) (Eulophidae) to varying host densities. *J. Anim. Ecol.* 45:415-423.

Frejka, T. (1973). *The Future of Population Growth: Alternative Pathways to Equilibrium*. New York: Wiley.

Freson, P.; Goffinet, G.; Malaise, F. (1974). Ecological effects of the regressive succession muhulu-miombo-savannah in Upper-Shaba (Zaire). *Proc. First Int. Congr. Ecol.* pp. 365-371.

Fry, C. H. (1980). Survival and longevity among tropical land birds. *Proc. Fourth Pan-Afr. Orn. Congr.* pp. 333-343.

Fryer, G.; Iles, T. D. (1972). *The Cichlid Fishes of the Great Lakes of Africa: Their Biology and Evolution*. Edinburgh: Oliver & Boyd.

Futuyma, D. J.; Slatkin, M., editors. (1983). *Coevolution*. Sunderland, MA: Sinauer.

Gagné, W. C. (1979). Canopy associated arthropods in *Acacia koa* and *Metrosideros* tree communities along an altitudinal transect on Hawaii island. *Pac. Insects* 21:56-82.

Gandar, M. V. (1982). Trophic ecology and plant/herbivore energetics. In: Huntley, B. J.; Walker, B. H. editors. *Ecology of Tropical Savannas*. Berlin: Springer-Verlag, pp. 514-539.

Garwood, N. C.; Janos, D. P.; Brokaw, N. (1979). Earthquake caused landslides: a major disturbance to tropical forests. *Science* 205:997-999.

Gause, G. F. (1934). *The Struggle for Existence*. Baltimore: Williams & Wilkins. (Reprinted by New York: Dover, 1971).

Gautier-Hion, A. (1980). Seasonal variations in diet related to species and sex in a community of *Cercopithecus* monkeys. *J. Anim. Ecol.* 49:237-269.

Gentry, A. H. (1974). Flowering phenology in tropical Bignoniaceae. *Biotropica* 6:64-68.

Gentry, A. H. (1982). Patterns of neotropical plant species diversity. *Evol. Biol.* 15:1-84.

Gilbert, L. E. (1975). Ecological consequences of coevolved mutualism between butterflies and plants. In: Gilbert, L. E.; Raven, P. H. editors. *Coevolution of Plants and Animals*. Austin, TX: University of Texas Press, pp. 210–240.

Gilbert, L. E. (1980). Food web organization and the conservation of neotropical diversity. In: Soulé, M. E.; Wilcox, B. A. editors. *Conservation Biology*. Sunderland, MA: Sinauer, pp. 11–33.

Gilbert, L. E. (1983). *Coevolution and mimicry*. In: Futuyma, D. J.; Slatkin, M. editors. *Coevolution*. Sunderland, MA: Sinauer, pp. 263–281.

Gilbert, L. E.; Smiley, J. T. (1978). Determinants of local diversity in phytophagous insects: host specialists in tropical environments. In: Mound, L. A.; Waloff, N. editors: *Diversity of Insect Faunas*. Oxford: Blackwell, pp. 89–104.

Gill, D. E.; Halverson, T. G. (1984). Fitness variation among branches within trees. In: Shorrocks, B. editor. *Evolutionary Ecology*. Oxford: Blackwell, pp. 105–116.

Gill, F. B.; Wolf, L. L. (1975a). Economics of feeding territoriality in the golden-winged sunbird. *Ecology* 56:333–345.

Gill, F. B.; Wolf, L. L. (1975b). Foraging strategies and energetics of East African sunbirds at mistletoe flowers. *Am. Nat.* 109:491–510.

Gilpin, M. E.; Diamond, J. M. (1976). Calculation of immigration and extinction curves for the species-area-distance relations. *Proc. Natl. Acad. Sci.* 73:4130–4134.

Golley, F. B. (1969). Caloric values of wet tropical forest vegetation. *Ecology* 50:517–519.

Golley, F. B.; Leith, H. (1972). Bases of organic production in the tropics. In: Golley, P. M.; Golley, F. B. editors. *Tropical Ecology with an Emphasis on Organic Production*. Athens, GA: University of Georgia, pp. 1–26.

Golley, F.B.; McGinnis, J.T.; Clements, R. G.; Child, G. I.; Deuver, M. J. (1975). *Mineral Cycling in a Tropical Moist Forest Ecosystem*. Athens, GA: University of Georgia Press.

Gorham, E. (1979). Shoot height, weight and standing crop in relation to density of monospecific plant stands. *Nature* 279:148–150.

Gorman, M. (1979). *Island Ecology*. London: Chapman & Hall.

Grainger, A. (1983). Improving the monitoring of deforestation in the humid tropics. In: Sutton, S. L.; Whitmore, T. C.; Chadwick, A. C. editors. *Tropical Rain Forest: Ecology and Management*. Oxford: Blackwell, pp. 387–395.

Grant, V.; Grant, K. A. (1965). *Flower pollination in the Phlox Family*. New York: Columbia University Press.

Greenland, D. J. (1975). Bringing the green revolution to the shifting cultivator. *Science* 190:841–844.

Greenslade, P. J. M. (1983). Adversity selection and the habitat template. *Am. Nat.* 122:352–365.

Greenwood, J. J. D. (1984). The evolutionary ecology of predation. In: Shorrocks, B. editor. *Evolutionary Ecology*. Oxford: Blackwell, pp. 233–273.

Griffiths, D. (1981). Sub-optimal foraging in the ant-lion *Macroleon quinquemaculatus*. *J. Anim. Ecol.* 50:697–702.

Grubb, P. J. (1977). Control of forest growth and distribution on wet tropical mountains: with special reference to mineral nutrition. *Ann. Rev. Ecol Syst.* 8:83–107.

Guppy, N. (1983). The case for an organisation of timber exporting countries (OTEC). In: Sutton, S. L.; Whitmore, T. C.; Chadwick, A. C. editors: *Tropical Rain Forest: Ecology and Management*. Oxford: Blackwell, pp. 459–463.

Hamilton, A. C. (1982). *Environmental History of East Africa*. London: Academic Press.

Happold, D. C. D. (1977). A population study on small rodents in the tropical rain forest of Nigeria. *Terre et Vie* 31:385–457.

Harcombe, P. (1977). The influence of fertilization on some aspects of succession in a humid tropical forest. *Ecology* 58:1375–1383.

Hardin, G. (1960). The competitive exclusion principle. *Science* 131:1292–1297.

Hare, F. K. (1977). Connections between climate and desertification. *Env. Cons.* 4:81–90.

Harper, J. L. (1977). *Population Biology of Plants*. London: Academic Press.

Harper, J. L. (1977). The concept of population in modular organisms. In: May, R. M. editor: *Theoretical Ecology*. Oxford: Blackwell, pp. 53–77.

Harper, J. L. (1982). After description. In: Newman, E. I. editor. *The Plant Community as a Working Mechanism.* Oxford: Blackwell, pp. 11–25.

Harris, D. R. (1977). Subsistence strategies across Torres Strait. In: Allen, J.; Golson, J.; Jones, R. editors. *Sunda and Sahul.* New York: Academic Press, pp. 421–463.

Harris, D. R. (1980). Tropical savanna environments: definition, distribution, diversity and development. In: Harris, D. R. editor: *Human Ecology in Savanna Environments.* London: Academic Press, pp. 3–27.

Hart, R. D (1980). A natural ecosystem analog approach to the design of a successional crop system for tropical forest environments. *Biotropica* (Suppl.) 12:73–82.

Hartshorn, G. S. (1978). Tree falls and tropical forest dynamics. In: Tomlinson, P. B.; Zimmerman, M. H. editors. *Tropical Trees as Living Systems.* Cambridge: Cambridge Universtiy Press, pp. 617–638.

Hayden, B. (1981). Subsistence and ecological adaptations of modern hunter/gatherers. In: Harding, R. S. O.; Teleki, G. editors. *Omnivorous Primates: Gathering and Hunting in Human Evolution.* New York: Columbia University Press, pp. 344–421.

Heal, O. W.; McLean, S. F., Jr. (1975). Comparative productivity in ecosystems—secondary productivity. In: van Dobben, W. H.; Lowe-McConnell, R. H. editors. *Unifying Concepts in Ecology.* Hague: Junk, pp. 89–108.

Heaney, L. R. (1984). Mammalian species richness on islands on the Sunda shelf, Southeast Asia. *Oecologia* 61:11–17.

Heatwole, H.; Levins, R. (1973). Biogeography of the Puerto Rican bank: species-turnover on a small cay, Cayo Ahogado. *Ecology* 54:1042–1056.

Hebert, P. D. N. (1980). Moth communities in montane Papua New Guinea. *J. Anim. Ecol.* 49:593–602.

Hedberg, I; Hedberg, O. (1979). Tropical-alpine life-forms of vascular plants. *Oikos* 33:297–307.

Hedberg, O. (1964). Études écologiques de la flore Afroalpine. *Bull. Soc. Roy. Bot. Belg.* 97:5–18.

Hedberg, O. (1969). Evolution and speciation in a tropical high mountain flora. *Biol. J. Linn. Soc.* 1:135–148.

Heinrich, B. (1975). Energetics of pollination. *Ann. Rev. Ecol. Syst.* 6:139–170.

Heithaus, E. R. (1979). Flower-feeding specialization in wild bee and wasp communities in seasonal neotropical habitats. *Oecologia.* 42:179–194.

Heithaus, E. R.; Culver, D. C.; Beattie, A. J. (1980). Models of some ant-plant mutualisms. *Am. Nat.* 116:347–361.

Heithaus, E. R.; Stashko, E.; Anderson, P. K. (1982). Cumulative effects of plant-animal interactions on seed production by *Bauhinia ungulata,* a neotropical legume. *Ecology* 63:1294–1302.

Herrera, R.; Jordan, C. F.; Medina, E.; Klinge, H. (1981). How human activities disturb the nutrient cycles of a tropical rain forest in Amazonia. *Ambio* 10:109–114.

Hjort, A. (1976). Kenya. *Ecol. Bull.* 24:165–169.

Hladik, A.; Hladik, C. M. (1977). The occurence of alkaloids in rain forest plants and its ecological significance. Preliminary screening study in Gabon. *Terre et Vie* 31:515–555.

Hodkinson, I. D.; Hughes, M. K. (1982). *Insect Herbivory.* London: Chapman & Hall.

Holdridge, L. R.; Grenke, W. C.; Hatheway, W. H.: Liang, T.; Tosi, J. A., Jr. (1971). *Forest Environments In Tropical Life Zones: A Pilot Study.* Oxford: Pergamon.

Holling, C. S. (1959). Some characteristics of simple types of predation and parasitism. *Can. Entomol.* 91:385–398.

Hopkins, B. (1970). Vegetation of the Olokemenji Forest Reserve, Nigeria. VI. The plants on the forest site with special reference to their seasonal growth. *J. Ecol.* 58: 765–793.

Hopkins, B. (1974). *Forest and Savanna.* London: Heineman.

Houston, D. C. (1979). The adaptations of scavengers. In: Sinclair, A. R. E.; Norton-

Griffiths, M. editors. *Serengeti*. Chicago: University of Chicago Press, pp. 262–286.

Howarth, F. G. (1979). Neogcoaeolian habitats on new lava flows on Hawaii Island: an ecosystem supported by windborne debris. *Pac. Insects* 20:133–144.

Howe, H. F.; Smallwood, J. (1982). Ecology and seed dispersal. *Ann. Rev. Ecol. Syst.* 13:201–228.

Hubbell, S. P. (1979). Tree dispersion, abundance and diversity in a tropical dry forest. *Science* 203:1299–1309.

Hubbell, S. P. (1980). Seed predation and the coexistence of tree species in tropical forests. *Oikos* 35:214–229.

Huber, O. (1982). Significance of savanna vegetation in the Amazon territory of Venezuela. In: Prance, G. T. editor. *Biological Diversification in the Tropics*. New York: Columbia University Press, pp. 221–244.

Humphreys, W. F. (1979). Production and respiration in animal populations. *J. Anim. Ecol.* 48:427–453.

Humphreys, W. F. (1984). Production efficiency in small mammal populations. *Oecologia* 62:85–90.

Huntley, B. J.; Walker, B. H. editors (1982). *Ecology of Tropical Savannas*. Berlin: Springer-Verlag.

Hurlbert, S. H. (1971). The nonconcept of species diversity: a critique and alternative parameters. *Ecology* 52:577–586.

Hutchinson, G. E. (1959). Homage to Santa Rosalia or why are there so many kinds of animals? *Am. Nat.* 93:145–159.

Hutchinson, G. E. (1965). *The Ecological Theater and the Evolutionary Play*. New Haven, CN: Yale University Press.

Hutchinson, G. E. (1978). *An Introduction to Population Ecology*. New Haven, CN: Yale University Press.

Huxley, C. (1980). Symbioses between ants and epiphytes. *Biol. Rev.* 55:321–340.

Inger, R. F.; Colwell, R. K. (1977). Organisation of contiguous communities of amphibians and reptiles in Thailand. *Ecol. Mon.* 47:229–253.

International Union for the Conservation of Nature and Natural Resources. (1980). *World Conservation Strategy*. Morges, Switzerland: IUCN.

Iverson, J. B. (1978). The impact of feral cats and dogs on populations of the West Indian rock iguana *Cyclura carinata*. *Biol. Conserv.* 14:63–73.

Jackson, I. J. (1977). *Climate, Water and Agriculture in the Tropics*. London: Longman.

Jackson, R. D.; Idso, S. B. (1975). Surface albedo and desertification. *Science* 189:1012–1013.

Jacobs, A. H. (1975). Maasai pastoralism in historical perspective. In: Monod, T. editor. *Pastoralism in Tropical Africa*. London: International African Institute, pp. 406–425.

Janos, D. P. (1980). Mycorrhizae influence tropical succession. *Biotropica* (Suppl.) 12:56–64.

Janos, D. P. (1983). Tropical mycorrhizas, nutrient cycles and plant growth. In: Sutton, S. L.; Whitmore, T. C.; Chadwick, A. C. editors. *Tropical Rain Forest: Ecology and Management*. Oxford: Blackwell, pp. 327–345.

Janzen, D. H. (1966). Coevolution of mutualism between ants and acacias in Central America. *Evolution* 20:249–275.

Janzen, D. H. (1967). Why mountain passes are higher in the tropics. *Am. Nat.* 101:233–249.

Janzen, D. H. (1970). Herbivores and the number of tree species in tropical forests. *Am. Nat.* 104:501–528.

Janzen, D. H. (1973). Tropical agroecosystems. *Science* 182:1212–1219.

Janzen, D. H. (1977). The impact of tropical studies on ecology. In: Goulden, C. E. editor. *Changing Scenes in Natural Sciences*. Spec. Publ. No. 12. Philadelphia: Academy of Natural Science, pp. 159–187.

Janzen, D. H. (1979). How to be a fig. *Ann. Rev. Ecol. Syst.* 10:13–51.

Janzen, D. H. (1980a). When is it coevolution? *Evolution* 34:611–612.

Janzen, D. H. (1980b). Specificity of seed-attacking beetles in a Costa Rican deciduous forest. *J. Ecol.* 68:929–952.

Janzen, D. H.; Ataroff, M.; Fariñas, M.; Reyes, S.; Rincon, N.; Soler, A.; Soriano, P.; Vera, M. (1976). Changes in the arthropod community along an elevational transect in the Venezuelan Andes. *Biotropica* 8:193–203.

Jarman, P. J.; Sinclair, A. R. E. (1979). Feeding strategy and the pattern of resource partitioning in ungulates. In: Sinclair, A. R. E.; Norton-Griffiths, M. editors. *Serengeti.* Chicago: University of Chicago Press, pp. 130–163.

Jarvis, J. U. M. (1973). The structure of a population of the mole rats *Tachyoryctes splendens* (Rodentia: Rhizomyidae). *J. Zool. Lond.* 171:1–14.

Jayasingh, D. B.; Freeman, B. E. (1980). The comparative population dynamics of eight solitary bees and wasps (Aculeata: Hymenoptera) trap-nested in Jamaica. *Biotropica* 12:214–219.

Jeanne, R. L. (1979). A latitudinal gradient in rates of predation. *Oecologia* 55: 238–242.

Jeník, J. (1978). Roots and root systems in tropical trees: morphological and ecological aspects. In: Tomlinson, P. B.; Zimmerman, M. H. editors. *Tropical Trees as Living Systems.* Cambridge: Cambridge University Press, pp. 323–349.

Jenny, H.; Gessel, S. P.; Bingham, F. T. (1949). Comparative study of decomposition rates of organic matter in temperate and tropical regions. *Soil Sci.* 68:419–432.

Jewell, P. A. (1980). Ecology and management of game animals and domestic livestock in African savannas. In: Harris, D. R. editor. *Human Ecology in Savanna Environments.* London: Academic Press, pp. 353–381.

Johns, A. (1983). Wildlife can live with logging. *New Sci.* 99:206–209.

Johnstone, I. M. (1981). Consumption of leaves by herbivores in mixed mangrove stands. *Biotropica* 13:252–259.

Jolly, A.; Oliver, W. L. R.; O'Connor, S. M. (1982a). Population and troop ranges of *Lemur catta* and *Lemur fulvus* at Berenty, Madagascar: 1980 census. *Folia Primatol.* 39:115–123.

Jolly, A.; Gustafson, H.; Oliver, W. L. R. (1982b). *Propithecus verreauxi* population and ranging at Berenty, Madagascar, 1975 and 1980. *Folia Primatol.* 39:124–144.

Jones, R. (1980). Hunters in the Australian coastal savanna. In: Harris, D. R. editor. *Human Ecology in Savanna Environments.* London: Academic Press, pp. 107–146.

Jordan, C. F. (1982). The nutrient balance of an Amazonian rain forest. *Ecology* 63:647–654.

Jordano, P. (1983). Fig-seed predation and dispersal by birds. *Biotropica* 15:38–41.

Karn, M. N.; Penrose, L. S. (1951). Birth-weight and gestation time in relation to maternal age, parity and infant survival. *Ann. Eugen.* 16:147–164.

Karr, J. R. (1971). Structure of avian communities in selected Panama and Illinois habitats. *Ecol. Mon.* 41:207–238.

Karr, J. R.; Roth, R. R. (1971). Vegetation structure and avian diversity in several New World areas. *Am. Nat.* 105:423–435.

Keeler, K. H. (1981). A model of selection for facultative nonsymbiotic mutualism. *Am. Nat.* 118:488–498.

Keiyoro, P. P. N. (1982). *The effects of burning on grazing resources in Nairobi National Park.* University of Nairobi, Thesis.

Kellman, M. (1979). Soil enrichment by neotropical savanna trees. *J.Ecol.* 67:565–577.

Kelly, R. D.; Walker, B. H. (1976). The effects of different forms of land use on the ecology of a semiarid region in southeastern Rhodesia. *J. Ecol.* 64:553–576.

Kira, T. (1975). Primary production of forests. In: Cooper, J. P. editor. *Photosynthesis and Productivity in Different Environments.* Cambridge: Cambridge University Press. pp. 5–40.

Kjekhus, H. (1977). *Ecology Control and Economic Development in East African History.* London: Heinemann.

Kowal, J. M.; Kassam. A. H. (1978). *Agricultural Ecology of Savanna.* Oxford: Clarendon Press.

Krebs, C. J. (1978). *Ecology.* New York: Harper & Row.

Krebs, J. R. (1978). Optimal foraging: decision rules for predators. In: Krebs, J. R.; Davies, N. B. (eds.) *Behavioural Ecology.* Oxford: Blackwell, pp. 23–63.

Kruuk, H. (1972). *The Spotted Hyena.* Chicago: University of Chicago Press.

Kunkel-Westphal, I.; Kunkel, P. (1979). Litter fall in Guatemala primary forest, with details of leaf shedding by some common species of trees. *J. Ecol.* 67:665–686.

Kunstadter, P.; Chapman, E. C. (1978). Problems of shifting cultivation and economic development in Northern Thailand. In: Kunstadter, P.; Chapman, E. C.; Sabhasri, S. editors. *Farmers in the Forest.* Honolulu: University Press of Hawaii, pp. 3–23.

Kushawa, S. P. S.; Ramakrisnan, P. S.; Tripathi, R. S. (1981). Population dynamics of *Eupatorium odoratum* in successional environments. following slash and burn agriculture. *J. Appl. Ecol.* 18:529–535.

Lack, A. (1978). The ecology of the flowers of the savanna tree *Maranthes polyandra* and their visitors, with particular reference to bats. *J. Ecol.* 66:287–295.

Lack, D. (1947). *Darwin's Finches.* Cambridge: Cambridge University Press.

Lack, D. (1966). *Population Studies of Birds.* Oxford: Oxford University Press.

Lamoral, B. H. (1978). Soil hardness, an important limiting factor in burrowing scorpion of the genus *Opisthophthalmus* C. L. Koch, 1837 (Scorpionidae, Scorpionida) *Symp. Zool. Soc. Lond.* 42:171–181.

Lamotte, M. (1979). Structure and functioning of the savanna ecosystem of Lamto (Ivory Coast). In: Unesco. *Tropical Grazing Land Ecosystems.* Paris: Unesco/UNEP/FAO, pp. 511–561.

Lamotte, M. (1982). Consumption and decomposition in tropical grassland ecosystems at Lamto, Ivory Coast. In: Huntley, B. J.; Walker, B. H. editors. *Ecology of Tropical Savannas.* Berlin: Springer-Verlag, pp. 415–429.

Lamprey, H. F.; Halevy, G.; Makacha, S. (1974). Interactions between *Acacia,* bruchid seed beetles and large herbivores. *E. Afr. Wildl. J.* 12:81–85.

Landahl, J. T.; Root, R. B. (1969). Difference in the life table of tropical and temperate milkweed bugs *Oncopeltus* (Hemiptera: Lygaeidae). *Ecology* 50:734–737.

Laws, R. M.; Parker, I. S. C.; Johnstone, R. C. B. (1975). *Elephants and Their Habitats.* Oxford: Clarendon Press.

Lawton, J. H.; Strong, D. R., Jr. (1981). Community patterns and competition in folivorous insects. *Am. Nat.* 118:317–338.

Leck, C. F. (1979). Avian extinctions in an isolated tropical wet forest preserve, Ecuador. *The Auk* 96:343–352.

Le Houérou, H. N. (1976a). The nature and causes of desertization. *Arid Lands Newsletter* 3:1–7.

Le Houérou, H. N. (1976b). Rehabilitation of degraded arid lands. *Ecol. Bull.* 24:189–205.

Leigh, E. G. (1975). Population fluctuations, community stability and environmental stability. In: Cody, M. L.; Diamond, J. M. editors. *Ecology and Evolution of Communities.* Cambridge, MA: Belknap Press, pp. 51–73.

Leigh, E. G.; Rand, A. S.; Windsor, D. M. (1982). *The Ecology of a Tropical Forest.* Washington, DC: Smithsonian.

Lepage, M. (1974). *Les termites d'une savane sahélienne (Ferlo Septentrional, Sénégal):* Peuplement, populations, consommation, rôle dans l'écosystem. University of Dijon, Thesis.

Leslie, P. W.; Bindon, J. R.; Baker, P. T. (1984). Caloric requirements of human populations: a model. *Hum. Ecol.* 12:137–162.

Leuthold, W. (1977). Changes in tree populations of Tsavo East National Park, Kenya. *E. Afr. Wildl. J.* 15:61–69.

Levin, D. A.; Turner, B. L. (1977). Clutch size in the Compositae. In: Stonehouse, B.; Perrins, C. editors. *Evolutionary Ecology.* London: Macmillan. pp. 215–222.

Levin, D. L. (1976). Alkaloid-bearing plants: an ecogeographic perspective. *Am. Nat.* 110:261–284.

Levings, S. C.; Windsor, D. M. (1984). Litter moisture content as a determinant of litter arthropod abundance during the dry season on Barro Colorado Island, Panama. *Biotropica* 16:125–131.

Levings, S. C.; Windsor, D. M. (1985). Litter arthropod populations in a tropical deciduous forest: relationships between years and arthropod groups. *J. Anim. Ecol.* 54: 61–69.

Lewis, J. G. E. (1970). The biology of *Scolopendra amazonica* in Nigerian Guinea savanna. *Bull. Mus. National D'Histoire Nat.* 41:85–90.

Lewis, J. G. E. (1974). The ecology of centipedes and millipedes in northern Nigeria. *Symp. zool. Soc. Lond.* 32:423–431.

Lincer, J. L.; Zalkind, D.; Brown, L. H.; Hopcraft, J. (1981). Organochlorine residues in Kenya's Rift Valley lakes. *J. Appl. Ecol.* 18:157–171.

Little, M. A. (1980). Designs for human-biological research among savanna pastoralists. In: Harris, D. R. editor. *Human Ecology in Savanna Environments.* London: Academic Press, pp. 497–503.

Lock, J. M.; Milburn, T. R. (1971). The seed biology of *Themeda Triandra* Forsk, in relation to fire. In: Duffey, E.; Watt, A. S. editors. *The Scientific Management of Animal and Plant Communities for Conservation.* Oxford: Blackwell, pp. 337–349.

Lomnicki, A. (1978). Individual differences between animals and the natural regulation of their numbers. *J. Anim. Ecol.* 47:461–475.

Long, A., Johnson, N. (1981). Forest plantations in Kalimantan, Indonesia. In: Mergen, F. editor: *Tropical Forests: Utilization and Conservation.* New Haven, CN: Yale School of Forestry and Environmental Studies, pp. 77–92.

Lounibos, L. P. (1979). Temporal and spatial distribution, growth and predatory behavior of *Toxorhynchites brevipalpis* (Diptera: Culicidae) on the Kenya coast. *J. Anim. Ecol.* 48:213–236.

Lowman, H. D.; Box, J. D. (1983). Variation in leaf toughness and phenolic content among five species of rain forest trees. *Austral. J. Ecol* 8:17–25.

Maarel, E. van der (1975). Man-made natural ecosystems in environmental management and planning. In: van Dobben, W. H.; Lowe-McConnell, R. H. editors. *Unifying Concepts in Ecology.* The Hague: Junk, pp. 263–274.

MacArthur, J. W. (1975). Environmental fluctuations and species diversity. In: Cody, M. L.; Diamond, J. M. editors. *Ecology and Evolution of Communities.* Cambridge, MA: Belknap, pp. 74–80.

MacArthur, R. H.; Recher, H.; Cody, M. (1966). On the relation between habitat selection and species diversity. *Am. Nat.* 100:319–332.

MacArthur, R. H.; Wilson, E. O. (1967). *The Theory of Island Biogeography.* Princeton, NJ: Princeton University Press.

Madge, D. S. (1965). Leaf fall and litter decomposition in a tropical forest. *Pedobiologia* 5:273–288.

Malaisse, F.; Grégoire, J.; Morrison, R. S.; Brooks, R. R.; Reeves, R. D. (1979). Copper and cobalt in vegetation of Fungurume, Shaba Province, Zaire. *Oikos* 33:472–478.

Margalef, R. (1968). *Perspectives in Ecological Theory.* Chicago: University of Chicago Press.

Markham, R. H.; Babbedge, A. J. (1979). Soil and vegetation catenas on the forest-savanna boundary in Ghana. *Biotropica* 11:224-234.

May, R. M. (1975a). Patterns of species abundance and diversity. In: Cody, M. L.; Diamond, J. M. editors *Ecology and Evolution of Communities.* Cambridge, MA: Belknap, pp. 81–120.

May, R. M. (1975b). Stability in exosystems: some comments. In: van Dobben, W. H.; Lowe-McConnell, R. H. editors: *Unifying concepts in Ecology.* The Hague. Junk, pp. 161–168.

May, R. M. (1981a). Models for single populations. In: May, R. M. editor. *Theoretical Ecology.* Oxford: Blackwell. pp. 5–29.

May, R. M. (1981b). Models for two interacting populations. In: May, R. M. editor. *Theoretical Ecology.* Oxford: Blackwell, pp. 78–104.

May, R. M. (1981c). Patterns in multi-species communities. In: May, R. M. editor. *Theoretical Ecology.* Oxford: pp. 197–227.

May, R. M.; Anderson, R. M. (1979). Population biology of infectious diseases: part II. *Nature* 280:455–461.

Mayr, E. (1969). Bird speciation in the tropics. *Biol. J. Linn. Soc.* 1:1–17.

McCoy, E. D. (1983). *The application of island biogeographic theory to patches of habitat. Biol. Conserv.* 25:53–61.

McEvedy, C.; Jones, R. (1978). *Atlas of World Population History.* Harmondsworth, Middlesex: Penguin Books.

McKey, D. (1974a). Adaptive patterns in alkaloid physiology. *Am. Nat.* 108:305–320.

McKey, D. (1974b). Ant-plants: selective eating of unoccupied *Barteria* by colobus monkeys. *Biotropica* 6:269–270.

McKey, D. (1975). The ecology of coevolved seed dispersal systems. In: Gilbert, L. E.; Raven, P. H. editors. *Coevolution of Animals and Plants.* Austin, TX: University of Texas Press, pp. 159–191.

McKey, D.; Waterman, P. G.; Mbi, C. N.; Gartlan, J. S. (1978). Phenolic content of vegetation in two African rain forests: ecological implications. *Science* 202:61–64.

McNab, B. K. (1963). Bioenergetics and the determination of home range size. *Am. Nat.* 48:133–140.

McNaughton, S. J. (1979). Grassland-herbivore dynamics. In: Sinclair, A. R. E.; Norton-Griffiths, M. editors. *Serengeti.* Chicago: University of Chicago Press, pp. 46–81.

McNaughton, S. J. (1983). Serengeti grassland ecology: the role of composite environmental factors and contingency in community organization. *Ecol. Mon.* 53:291–320.

Meadows, D. H.; Meadows, D. L.; Randers, J.; Behrens, W. W., III. (1972). *The Limits to Growth,* London: Earth Island.

Medway, Lord (1972). Phenology of a tropical rain forest in Malaya. *Biol. J. Linn. Soc.* 4:117–146.

Melillo, J. M.; Palm, C. A.; Houghton, R. A.; Woodwell, G. M.; Myers, N. (1985). A comparison of two recent estimates of destruction of tropical forests. *Env. Conserv.* 12:37–40.

Metz, D. B.; Cawthon, D. A.; Park, T. (1976). An experimental analysis of competitive indeterminacy. *Proc. Natl. Acad. Sci.* 73:1368–1372.

Miller, D. S.; Rivers, J. (1971). Seasonal variations in food intake in two Ethiopian villages. *Proc. Nutr. Soc.* 31:32A–33A.

Milton, K. (1979). Factors influencing leaf choice by howler monkeys: a test of some hypotheses of food selection by generalist herbivores. *Am. Nat.* 114:362–378.

Milton, K. (1980). *The Foraging Strategy of Howler Monkeys.* New York: Columbia University Press.

Milton, K. (1981). Food choice and digestive strategies of two sympatric primate species. *Am. Nat.* 117:496–505.

Milton, K. (1984). Protein and carbohydrate resources of the Maku Indians of Northwestern Amazonia. *Amer. Anthropol.* 86:7–27.

Milton, K.; May, M. L. (1976). Body weight and home range size in primates. *Nature* 259:459–462.

Milton, K; Winsor, D. M.; Morrison, D. W.; Estribi, M. A. (1982). Fruiting phenologies of two neotropical *Ficus* species. *Ecology* 63:752–762.

Misonne, X. (1963). Les rongeurs du Ruwenzori et des régions voisines. *Expl. Parc Natl. Albert Deux Ser.* 14:1–164.

Misra, M. K.; Misra, B. N. (1981). Seasonal changes in leaf area index and chlorophyll in an Indian grassland. *J. Ecol.* 69:797–805.

Molyneux, D. H.; Ashford, R. W. (1983). *The Biology of* Trypanosoma *and* Leishmania, *Parasites of Man and Domestic Animals.* New York: Taylor & Francis.

Monteith, J. L. (1972). Solar radiation and productivity in tropical ecosystems. *J. Appl. Ecol.* 9:747–766.

Montgomery, G. G.; Sunquist, M. E. (1975). Impact of sloths on neotropical forest energy flow and nutrient cycling. In: Golley, F. B.; Medina, E. editors. *Tropical Ecological Systems*. New York: Springer-Verlag, pp. 69–98.

Mooney, H. A. (1972). The carbon balance of plants. *Ann. Rev. Ecol. Syst.* 3:315–346.

Moran, V. C.; Southwood, T. R. E. (1982). The guild composition of arthropod communities in tree. *J. Anim. Ecol.* 51:289–306.

Morris, J. W.; Bezuidenhout, J. J.; Furniss, P. R. (1982). Litter decomposition. In: Huntley, B. J.; Walker, B. H. editors. *Ecology of Tropical Savannas*. Berlin: Springer-Verlag, pp. 535–553.

Morton, J. K. (1966). The role of polyploidy in the evolution of the tropical flora. In: *Chromosomes Today*. Vol. 1. Edinburgh: Oliver & Boyd, pp. 73–76.

Mueller-Dombois, D.; Ellenberg, H. (1974). *Aims and Methods of Vegetation Ecology*. New York: Wiley.

Myers, N. (1979). *The Sinking Ark*. Oxford: Pergamon.

Myers, N. (1980). Kenya's baby boom. *New Sci.* 87:848–850.

Nambudiri, E. M. V.; Tidwell, W. D.; Smith, B. N.; Hebbert, N. P. (1978). A C_4 plant from the Pliocene. *Nature* 276:816–817.

Nergón-Aponte, H.; Jobin, W. R. (1979). Schistosomiasis control in Puerto Rico. Twenty-five years of operational experience. *Am. J. Trop. Med. Hyg.* 28:515–525.

Newberry, D. (1980). Interactions between the coccid *Icerya sechellarum* (Westw.) and its host tree species on Aldabra Atoll. II. *Scaevola taccada* (Gaestn.) Roxb. *Oecologia* 46:180–185.

Newsome, A. E.; Corbett, L. K. (1975). Outbreaks of rodents in semiarid and arid Australia: causes, prevention and evolutionary considerations. In: Prakash, I.; Ghosh, P. K. editors. *Rodents in Desert Environments*. Hague. Junk, pp. 269–275.

Nieuwolt, S. (1965). Evaporation and water balance in Malaya. *J. Trop. Geogr.* 20:34–53.

Nieuwolt, S. (1977). *Tropical Climatology*. London: Wiley.

Noon, B. R.; Dawson, D. K.; Kelly, J. P. (1985). A search for stability gradients in North American breeding birds. *The Auk* 102:64–81.

Norgan, N. G.; Ferro-Lozzi, A.; Durnin, J. V. G. A. (1974). The energy and nutrient intake and energy expenditure of 204 New Guinea adults. *Phil. Trans. Roy. Soc. Lond. B* 268:309–348.

Noy-Meir, I. (1973). Desert ecosystems: environment and producers. *Ann. Rev. Ecol. Syst.* 4:25–51.

Nye, P. H.; Greenland, D. J. (196)0. *The Soil Under Shifting Cultivation. Tech. Comm.* No. 51. Farnham Royal, Buckinghamshire: Commonwealth Agric. Bureaux.

Oates, J. F.; Waterman, P. G.; Choo, G. M. (1980). Food selection by the south Indian leaf-eating monkey *Presbytis johnii* in relation to leaf chemistry. *Oecologia* 45:45–56.

Odend'hal, S. (1972). Energetics of Indian cattle in their environment. *Hum. Ecol.* 1:3–22.

Odum, E. P. (1971). *Fundamentals of Ecology*. Philadelphia: Saunders.

Odum, E. P.; Marshall, S. G.; Marples, T. G. (1965). The caloric content of migrating birds. *Ecology* 46:901–904.

Odum, H. T. (1970). Summary: an emerging view of the ecological system at El Verde. In: Odum, H. T.; Pigeon, R. F. editors. *A Tropical Rain Forest*. US Atomic Energy Commission. pp. I191–I289.

Ohiagu, C. E. (1979). A quantitative study of seasonal foraging by the grass harvesting termite *Trinervitermes geminatus* (Wasmann) (Isoptera, Nasutitermitinae) in southern Guinea savanna, Mokwa, Nigeria. *Oecologia* 40:179–188.

Ojeda, M. M.; Keith, L. B. (1982). Sex and age compostion and breeding biology of cottontail rabbit populations in Venezuela. *Biotropica* 14:99–107.

Opler, P. A. (1983). Nectar production in a tropical ecosystem. In: Bentley, B.; Elias, T. editors. *The Biology of Nectaries*. New York: Columbia University Press, pp. 30–79.

Opler, P. A.; Frankie, G. W.; Baker, H. G. (1980). Comparative phenological studies of treelet and shrub species in tropical wet and dry forests in the lowlands of Costa Rica. *J. Ecol.* 68:167–188.

Orians, G. H. (1969). The number of bird species in some tropical forests. *Ecology* 50:783–801.

Orians, G. H. (1975). Diversity, stability and maturity in natural ecosystems. In: van Dobben, W. H.; Lowe-McConnell, R. H. editors. *Unifying Concepts in Ecology.* The Hague. Junk, pp. 139–150.

Oweyegha-Afunaduula, F. C. (1982). *Vegetation changes in Tsavo National Park (East), Kenya.* University of Nairobi, Thesis.

Park, T. (1948). Experimental studies of interspecies competition. I. Competition between populations of the flour beetles *Tribolium confusum* Duval and *Trobolium castaneum* Herbst. *Ecol. Mon.* 18:265-307.

Pathak, P. S.; Roy, R. D.; Patil, B. D. (1978). Forest grazing: principles, practices and prospects. In: Singh, J. S.; Gopal, B. editors. *Glimpses of Ecology.* Jaipur: International Science Publications, pp. 181–191.

Pearson, D. L. (1980). Patterns of limiting similarity in tropical tiger beetles (Colpeoptera: Cicindelidae). *Biotropica* 12:195-204.

Peet, R. K. (1974). The measurement of species diversity. *Ann. Rev. Ecol. Syst. 4:53*–74.

Pellew, R. A. (1983). The impacts of elephants, giraffe and fire upon *Acacia tortilis* woodlands of the Serengeti. *Afr. J. Ecol.* 21:41–74.

Pemadasa, M. A. (1981a). The mineral nutrition of the vegetation of a montane grassland in Sri Lanka. *J. Ecol.* 69:123–134.

Pemadasa, M. A. (1981b). Cyclic change and pattern in an *Anthocnemum* community in Sri Lanka. *J. Ecol.* 69:565–574.

Pemadasa, M. A. (1983). Effects of added nutrients on the vegetation of two coastal grasslands in the dry zone of Sri Lanka. *J. Ecol.* 71:725–734.

Pemadasa, M. A.; Amarsinghe, L. (1982). The ecology of a montane grassland in Sri Lanka. V. Interference in populations of four major grasses. *J. Ecol.* 70:731–744.

Pemadasa, M. A.; Balasubrmanian, S.; Wijewansa, H. G.; Amarasinghe, L. (1979). The ecology of a salt marsh in Sri Lanka. *J. Ecol.* 67:41–63.

Penning de Vries, F. W. T.; Krul, J. M.; Van Keulen, H. (1979). Productivity of Sahelian rangelands in relation to the availability of nitrogen and phosphorus from the soil. In: **Nitrogen Cycling in West African Ecosystems.** IITA Workshop, MAB Mali/Netherlands: Ibadan.

Perfect, T. J.; Cook, A. G.; Critchley, B. R.; Critchley, U.; Davies, A. L.; Swift, M. J.; Russell-Smith, A.; Yeadon, R. (1979). The effect of DDT contamination on the productivity of a cultivated forest in the subhumid tropics. *J. Appl. Ecol.* 16:705–719.

Peters, W. (1978). Medical aspects–comments and discussion. II. *Symp. Brit. Soc. Parasitol.* 16:25–40.

Petrusewicz, K.; Macfadyen, A. (1970). *Productivity of Terrestrial Animals.* Oxford: Blackwell.

Phillipson, J. (1975). Rainfall, primary production and "carrying capacity" of Tsavo National Park (East), Kenya. *E. Afr. Wildl. J.* 13:171–201.

Pianka, E. R. (1970). On r- and K- selection. *Am. Nat.* 104:592–597.

Pianka, E. R. (1975). Niche relations of desert lizards. In: Cody, M. L.; Diamond, J. M. editors. *Ecology and Evolution of Communities.* Cambridge, MA: Belknap, pp. 292–314.

Pianka, E. R. (1981). Competition and niche theory. In: May, R. M. editor. *Theoretical Ecology.* Oxford: Blackwell. pp. 167–196.

Pielou, E. C. (1975). *Ecological Diversity.* New York: Wiley.

Pimentel, D.; Levin, S. A.; Soans, A. B. (1975). On the evolution of energy balance in some exploiter-victim systems. *Ecology* 56:381–390.

Pimentel, D.; Pimentel, M. (1979). *Food, Energy and Society.* London: Arnold.

Pimm, S. L. (1982). *Food Webs.* London: Chapman & Hall.

Pinero, D.; Sarukhán, J. (1982). Reproductive behaviour and its individual variability in a tropical palm, *Astrocaryum mexicanum. J. Ecol.* 70:461–472.

Plowman, K. P. (1981). Resource utilisation by two New Guinea rain forest ants. *J. Anim. Ecol.* 50:903–916.

Plucknett, D. L.; Smith, N. J. H.; Williams, J. T.; Anishetty, N. M. (1983). Crop germplasm conservation in developing countries. *Science* 220:163–169.

Podoler, H. (1974). Analysis of life tables for a host and parasite *(Plodia-Nemeritis)* ecosystem. *J. Anim. Ecol.* 43:653–670.

Poole, R. W. (1974). *An Introduction to Quantitative Ecology.* New York: McGraw-Hill.

Potter, G. L.; Ellaesser, H. W.; MacCracken, M. C.; Luther, F. M. (1975). Possible climatic impact of tropical deforestation. *Nature* 258:697–698.

Prance, G. T. editor. (1982). *Biological Diversification in the Tropics.* New York: Columbia University Press.

Pratt, D. J.; Greenway, P. J.; Gwynne, M. D. (1966). A classification of East African rangeland, with an appendix on terminology. *J. Appl. Ecol.* 3:369–382.

Pratt, D. J.; Gwynne, M. D. (1977). *Rangeland Management and Ecology in East Africa.* London: Hodder & Stroughton.

Preston, F. W. (1948). The commonness and rarity of species. *Ecology* 29:254–283.

Preston, F. W. (1980). Noncanonical distribution of commonness and rareness. *Ecology* 61:88–97.

Prestwich, G. D.; Bentley, B. L. (1981). Nitrogen fixation by intact colonies of the termite *Nasutitermes corniger. Oecologia* 49:249–251.

Price, P. W. (1975). *Insect Ecology.* New York: Wiley.

Proctor, J.; Anderson, J. M.; Chai, P.; Vallack, H. W. (1983). Ecological studies in four contrasting lowland rain forests in Gunung Mulu National Park, Sarawak. I. Forest environment, structure and floristics. *J. Ecol.* 71:237–260.

Putz, F. E. (1983). Treefall pits and mounds, buried seeds and the importance of soil disturbance to pioneer trees on Barro Colorado Island, Panama. *Ecology* 64:1069–1074.

Putz, F. E. (1984). The natural history of lianas on Barro Colorado Island, Panama. *Ecology* 65:1713–1724.

Raich, J. W. (1983). Effects of forest conversion on the carbon budget of a tropical soil. *Biotropica* 15:177–184.

Ramakrishnan, P. S.; Jeet, N. (1972). Competitive relationships existing between two closely related species of *Argemone* living in the same area. *Oecologia* 9:279–288.

Ramakrishnan, P. S.; Kumar, S. (1971). Productivity and plasticity of wheat and *Cynodon dactylon* (L) Pers. in pure and mixed stands. *J. Appl. Ecol.* 8:85–98.

Ramo, C.; Busto, B. (1984). Nidificacion de los Passeriformes en los llanos de Apure (Venezuela). *Biotropica* 16:59–68.

Randall, R. E. (1970). Vegetation and environment on the Barbados coast. *J. Ecol.* 58:155–172.

Rapp. A. (1976). Introduction. *Ecol. Bull.* 24:11–18.

Rappaport, R. A. (1971). The flow of energy in an agricultural society. *Sci. Am.* 225:116–132.

Redhead, J. F. (1980). Mycorrhiza in natural tropical forests. In: Mikola, P. editor. *Tropical Mycorrhiza Research.* Oxford: Clarendon Press, pp. 127–142.

Regal, P. J. (1982). Pollination by wind and animals: ecology of geographic patterns. *Ann. Rev. Ecol. Syst.* 13:497–524.

Rehr, S. S.; Feeny, P. P.; Janzen, D. H. (1973). Chemical defense in Central American nonant *Acacias. J. Anim. Ecol.* 42:405–416.

Reich, P. B.; Borchert, R. (1984). Water stress and tree phenology in a tropical dry forest in the lowlands of Costa Rica. *J. Ecol.* 72:61–74.

Reichle, D. E. (1971). Energy and nutrient metabolism of soil and litter invertebrates. In: Dubigneaud, P. editor. *Productivity of Forest Ecosystems.* Paris: Unesco, pp 465–477.

Remsen, J. V., Jr.; Parker, T. A., III. (1983). Contribution of river-created habitats to bird species richness in Amazonia. *Biotropica* 15:223–231.

Revelle, R. (1982). Carbon dioxide and world climate. *Sci. Am.* 247:33–41.

Rice, B; Westoby, M. (1983). Plant species richness at the 0.1 hectare level in Australian vegetation compared to other continents. *Vegetatio* 51:129–140.

Richards, P. W. (1952). *The Tropical Rain Forest.* Cambridge: Cambridge University Press.

Ricklefs, R. E. (1977). Environmental heterogeneity and plant species diversity: a hypothesis. *Am. Nat.* 111:376–381.

Ricklefs, R. E. (1979). *Ecology* New York: Chiron Press.

Ricklefs, R. E.; O'Rourke, K. (1975). Aspect diversity in moths: a temperate-tropical comparison. *Evolution* 29:313–324.

Risch, S. J. (1979). A comparison, by sweep sampling, of the insect fauna from corn and sweet potato monocultures and dicultures in Costa Rica. *Oecologia* 42:195–211.

Risch, S. J. (1980). The population dynamics of several herbivorous beetles in a tropical agroecosystem: the effect of intercropping corn, beans and squash in Costa Rica. *J. Appl. Ecol.* 17:593–612.

Rockwood, L. L. (1973a). Distribution, density and dispersion of two species of *Atta* (Hymenoptera: Formicidae) in Guanacaste Province, Costa Rica. *J. Anim. Ecol.* 42:803–817.

Rockwood, L. L. (1973b). The effect of defoliation on seed production of six Costa Rican tree species. *Ecology* 54:1363–1369.

Rockwood, L. L. (1976). Plant selection and foraging patterns in two species of leaf-cutting ants (*Atta*). Ecology 57:48–61.

Rogers, D. (1979). Tsetse population dynamics and distribution: a new analytical approach. *J. Anim. Ecol.* 48:825–849.

Rose, D. J. W. (1972). Dispersal and quality in populations of *Cicadulina* species (Cicadellidae). *J. Anim. Ecol.* 41:589–609.

Rose, D. J. W. (1975). Field development and quality change in successive generations of *Spodoptera exempta* (Wlk), the African armyworm. *J. Appl. Ecol.* 12:727–739.

Rosenzweig, M. L. (1975). On continental steady states of species diversity. In: Cody, M. L.; Diamond, J. M. editors. *Ecology and Evolution of Communities.* Cambridge, MA: Belknap, pp. 121–140.

Roughgarden, J.: Pacala, S.; Rummel, J. (1984). Strong present-day competition between the *Anolis* lizards of St. Maarten (Neth. Antilles). In: Shorrocks, B. editor. *Evolutionary Ecology.* Oxford: Blackwell, pp. 203–220.

Routledge, R. D. (1979). Diversity indices: which ones are admissible? *J. Theor. Biol.* 76:503–515.

Routledge, R. D. (1980). Bias in estimating the diversity of large uncensused communities. *Ecology* 61:276–281.

Rundel, P. W. (1980). The ecological distribution of C_3 and C_4 grasses in the Hawaiian Islands. *Oecologia* 45:354–359.

Russell, E. W. (1973). *Soil conditions and Plant Growth.* 10th ed. London: Longman.

Rypstra, A. L. (1984). A relative measure of predation on web-spiders in temperate and tropical forests. *Oikos* 43:129–132.

Salati, E.; Vose, P. B. (1984). Amazon Basin: a system in equilibrium. *Science* 225:129–138.

Salick, J.; Herrera, R.; Jordan, C. F. (1983). Termitaria: nutrient patchiness in nutrient deficient rain forests. *Biotropica* 15:1–7.

Salt, G. W. (1983). Roles: their limits and responsibilities in ecological and evolutionary research. *Am. Nat.* 122:697–705.

Samways, M. J. (1979). Immigration, population growth and mortality of insects and mites on cassava in Brazil. *Bull. Ent. Res.* 69:491–505.

Sanchez, P. A.; Bandy, D. E.; Villachica, J. H.; Nicholaides, J. J. (1982). Amazon Basin soils: management for continuous crop production. *Science* 216:821–827.

Sanford, R. L., Jr.; Saldarriaga, J.; Clark, K. E.; Uhl, C.; Herrera, R. (1985). Amazon rain forest fires. *Science* 227:53–55.

Sarmiento, G. (1984). *The Ecology of Neotropical Savannas.* Cambridge, MA: Harvard University Press.

Sarukhán, J. (1978). Studies on the demography of tropical trees. In: Tomlinson, P. B.; Zimmerman, M. H. editors. *Tropical Trees as Living Systems.* Cambridge: Cambridge University Press, pp. 163–184.

Sarukhán, J. (1980). Demographic problems in the tropics. In: Solbrig, O. T. editor. *Demography and Evolution in Plant Populations.* Oxford: Blackwell, pp. 161–188.

Schad, G. A.; Rozeboom L. E. (1976). Integrated control of helminths in human populations. *Ann. Rev. Ecol. Syst.* 7:393–420.

Schemske, D. W. (1980a). Floral ecology and hummingbird pollination of *Combretum farinosum* in Costa Rica. *Biotropica* 12:169–181.

Schemske, D. W. (1980b). The evolutionary significance of extrafloral nectar production by *Costus woodsoni* (Zingiberaceae): an experimental analysis of ant protection. *J. Ecol.* 68:959–967.

Schlesinger, W. H. (1977). Carbon balance in terrestrial detritus. *Ann. Rev. Ecol. Syst.* 8:51–81.

Schoener, T. W.; Janzen, D. H. (1968). Notes on environmental determinants of tropical versus temperate insect size patterns. *Am. Nat.* 102:207–224.

Schoener, T. W.; Schoener, A. (1983). Distribution of vertebrates on some very small islands. I. Occurrence sequences of individual species. *J. Anim. Ecol.* 52:209–235.

Scott, J. A.; French, N. R.; Leetham, J. W. (1979). Patterns of consumption in grasslands. In: French, N. R. editor. *Perspectives in Grassland Ecology.* New York: Springer-Verlag, pp. 89–105.

Seely, M. K. (1978). Grassland productivity: the desert end of the curve. *S. Afr. J. Sci.* 74:295–297.

Seely, M. K. (1979). Irregular fog as a water source for desert dune beetles. *Oecologia* 42:213–227.

Siefert, R. P.; Siefert, F. H. (1976). A community analysis of *Heliconia* insect communities. *Am. Nat.* 110:461–483.

Silberbauer, G. (1981). Hunter/gatherers of the central Kalahari. In Harding, R. S. O.; Teleki, G. editors. *Omnivorous Primates, Gathering and Hunting in Human Evolution.* New York: Columbia University Press, pp. 455–498.

Simberloff, D. S. (1978). Colonization of islands by insects: immigration, extinction and diversity. In: Mound, L. A.; Waloff, N. editors. *Diversity of Insect Faunas.* Oxford: Blackwell, pp. 139–153.

Sinclair, A. R. E. (1973). Regulation and population models for a tropical ruminant. *E. Afr. Wildl. J.* 11:307–316.

Sinclair, A. R. E. (1975). Resource limitation in tropical grasslands. *J. Anim. Ecol.* 44:497–520.

Sinclair, A. R. E.; Norton-Griffiths, M. (1982). Does competition of facilitation regulate migrant ungulate populations in the Serengeti? A test of hypotheses. *Oecologia* 53:364–369.

Singh, J. S.; Joshi, M. C. (1979). Primary production. In: Coupland, R. T. editor: *Grassland Ecosystems of the World.* Cambridge: Cambridge University Press, pp. 197–218.

Singh, J. S.; Singh, K. P.; Yadava, P. S. (1979). Ecosystem synthesis. In: Coupland, R. T. editor. *Grassland Ecosystems of the World.* London: Cambridge University Press, pp. 231–239.

Slobodkin, L. B. (1961). *The Growth and Regulation of Animal Populations.* New York: Holt, Rinehart & Winston.

Smith, A. P. (1974). Bud temperature in relation to nyctinastic leaf movement in an Andean giant rosette plant. *Biotropica* 6:263–266.

Smith, A. P. (1979). Function of dead leaves in *Espletia schultzii* (Compositae), an Andean caulescent rosette species. *Biotropica* 11:43–47.

Smith, B. W.; Otto, C. B.; Martin, G. E., III; Boutton, T. W. (1979). Photosynthetic strategies of plants. In: Goodin, J. R.; Northington, D. K. editors. *Arid Land Plant Resources.* Lubbock, TX: Texas Tech University, pp. 474–481.

Snow, A. A. (1982). Pollination intensity and potential seed set in *Passiflora uitifolia. Oecologia* 55:231–237.

Snow, D. W. (1965). A possible selective factor in the evolution of fruiting seasons in tropical forests. *Oikos* 15:274–281.

Snow, D. W. (1981). Tropical frugivorous birds and their food plants: a world survey. *Biotropica* 13:1–14.

Sopher, D. E. (1980). Indian civilization and the tropical savanna environment. In: Harris, D. R. editor. *Human Ecology in Savanna Environments.* London: Academic Press, pp. 185–207.

Southern, H. N. editor (1964). *The Handbook of British Mammals.* Oxford: Blackwell.

Southwood, T. R. E. (1977). Habitat, the templet for ecological strategies. *J. Anim. Ecol.* 46:337–365.

Southwood, T. R. E. (1978). *Ecological Methods.* London: Chapman & Hall.

Southwood, T. R. E. (1981). Bionomic strategies. In: May, R. M. editor. *Theoretical Ecology.* Oxford: Blackwell, pp. 30–52.

Southwood, T. R. E.; Moran, V. C.; Kennedy, C. E. J. (1982). The richness, abundance and biomass of the arthropod communities on trees. *J. Anim. Ecol.* 51:635–649.

Spinage, C. A. (1972). Age estimation in zebra. *E. Afr. Wildl. J.* 10:273–279.

Stanhill, G. (1979). A comparative study of the Egyptian agroecosystem. *Agro-Ecosystems* 5:213–230.

Stanton, N. L. (1979). Patterns of species diversity in temperate and tropical litter mites. *Ecology* 60:295–304.

Stark, N.; Jordan, C. F. (1978). Nutrient retention in the root mat of an Amazonian forest. *Ecology* 59:435–437.

Stiles, F. G. (1975). Ecology, flowering phenology and hummingbird pollination of some Costa Rican *Heliconia* Species. *Ecology* 56:285–301.

Stocker, G. C. (1981). Regeneration of a north Queensland rain forest following felling and burning. *Biotropica* 13:86–92.

Strong, D. R. (1977). Epiphyte load, tree falls, and perennial forest disruption: a mechanism for maintaining higher tree species richness in the tropics without animals. *J. Biogeog.* 4:215–218.

Strong, D. R., Jr.; McCoy, E. D.; Rey, J. R. (1977). Time and the number of herbivore species: the pests of sugarcane. *Ecology* 58:167-175.

Sugden, A. (1982). Long-distance dispersal, isolation and the cloud forest fauna of the Serranía de Macuira, Guajira, Colombia. *Biotropica* 14:208-219.

Sugihara, G. (1980). Minimal community structure: an explanation of species abundance patterns. *Am. Nat.* 116:770–787.

Sutton, S. L.; Hudson, P. J. (1981) Arthropod succession and diversity in umbels of *Cyperus papyrus* L. *Biotropica* 13:117–120.

Swaine, M. D.; Hall, J. B. (1983). Early succession on cleared forest land in Ghana. *J. Ecol.* 71:601–627.

Swift, M. J.; Heal, O. W.; Anderson, J. M. (1979). *Decomposition in Terrestrial Environments.* Oxford: Blackwell.

Swift, M. J.; Russell-Smith, A.; Perfect, T. J. (1981). Decomposition and mineral nutrient dynamics of plant litter in a regenerating bush-fallow in subhumid tropical Nigeria. *J. Ecol.* 69:981–995.

Tanner, E. V. J. (1982). Species diversity and reproductive mechanisms in Jamaican trees. *Biol. J. Linn. Soc.* 18:263–278.

Taylor, L. R. (1978). Bates, Williams, Hutchinson—a variety of diversities. In: Mound, L. A.; Waloff, N. editors. *Diversity of Insect Faunas.* Oxford: Blackwell, pp. 1–18.

Taylor, R. A. J.; Taylor, L. R. (1979). A behavioural model for the evolution of spatial dynamics. In: Anderson, R. M.; Turner, B. D.; Taylor, L. R. editors. *Population Dynamics.* Oxford: Blackwell, pp. 1–27.

Temple, S. A. (1977). Plant-animal mutualism: coevolution with Dodo leads to near extinction of plants. *Science* 197:885–886.

Terborgh, J. (1971). Distribution on environmental gradients: theory and preliminary interpretation of distributional patterns in the avifauna of the Cordillera Vilcabamba, Peru. *Ecology* 52:23–40.

Terhune, E. C. (1977). Prospects for increasing food production in less developed countries through efficient energy utilization. In: Lockeretz, W. editor. *Agriculture and Energy*. New York: Academic Press, pp. 625–637.

Thomas, B. R. (1976). Energy flow at high altitude. In: Baker, P. T.; Little, M. A. editors: *Man in the Andes*. Stroudsburg, PA: Dowden, Hutchinson & Ross, pp. 379–404.

Thompson, J. N. (1982). *Interaction and Coevolution*. New York: Wiley.

Thornton, I. W. B. (1984). Krakatau—the development and repair of a tropical ecosystem. *Ambio* 13:216–225.

Tiezen, L. L.; Senyimba, M. M.; Imbamba, S. K.; Troughton, J. H. (1979). The distribution of C_3 and C_4 grasses and carbon isotope discrimination along an altitudinal and moisture gradient in Kenya. *Oecologia* 37:337–350.

Tilman, D. (1982). *Resource Competition and Community Structure*. Princeton, NJ: Princeton University Press.

Tomlinson, P. B. (1978). Branching and axis differentiation in tropical trees. In Tomlinson, P. B.; Zimmerman, M. H. editors: *Tropical Trees as Living Systems*. London: Cambridge University Press, pp. 187–207.

Turner, F. B.: Jennrich, R. I.; Weintraube, J. D. (1969). Home range and body size of lizards. *Ecology* 50:1076–1081.

Turner, J. R. G. (1971). Studies of Müllerian mimicry and its evolution in burnet moths and heliconid butterflies. In Creed, R. editor. *Ecological Genetics and Evolution*. Oxford: Blackwell, pp. 224–260.

Turner, J. R. G. (1984). Darwin's coffin and Doctor Pangloss—do adaptionist models explain mimicry? In: Shorrocks, B. editor. *Evolutionary Ecology*. Oxford: Blackwell, pp. 313–361.

Udvardy, M. D. F. (1975). *A Classification of the Biogeographic Provinces of the World*. Morges, Switzerland: IUCN Occasional Paper, No. 18.

Uhl, C.; Jordan, C.; Clark, K.; Clark, H.; Herrera, R. (1982). Ecosystem recovery in Amazon caatinga forest after cutting, cutting and burning and bulldozer clearing treatments. *Oikos* 38:313–320.

Unesco. (1978). *Tropical Forest Ecosystems*. Paris: Unesco/UNEP/FAO.

Unesco. (1979). *Tropical Grazing Land Ecosystems*. Paris: Unesco/UNEP/FAO.

Unesco. (1980). *Case Studies on Desertification*. Paris: Unesco/UNEP/FAO.

United Nations. (1981). *Demographic Yearbook*. New York: UN.

Utida, S. (1941). Studies on experimental population of the azuki bean weevil Callosobruchus chinensis (L.) III. The effect of population density upon the mortalities of different stages of the life cycle. *Mem. Coll. Agr. Kyoto Imp. Univ.* 49:21–42.

Utida, S. (1943). Studies on experimental population of the azuki bean weevil *Callosobruchus chinensis* (L.) IX. General considerations and summary of the serial reports I to VIII. *Mem. Coll. Agr. Kyoto Imp. Univ.* 54:23–40.

Uvarov, B. P. (1964). Problems of insect ecology in developing countries. *J. Appl. Ecol.* 1:159–168.

Vandermeer, J. (1984). The evolution of mutualism. In: Shorrocks, B. editor. *Evolutionary Ecology*. Oxford: Blackwell, pp. 221–232.

Varma, B. R.; Chauhan, T. P. S. (1979). Preference for pH of some tropical earthworms. *Geobios* 6:150–153.

Vitousek, P. M. (1984). Litterfall, nutrient cycling and nutrient limitation in tropical forests. *Ecology* 65:285–298.

Wade, N. (1974). Sahelian drought: no victory for western aid. *Science* 185:234–237.

Wagner, F. H. (1981). Population dynamics. In: Goodall, D. W.; Perry, R. A. editors. *Arid-land Ecosystems: Structure, Functioning and Management*. Vol. 2. Cambridge: Cambridge University Press, pp. 125–168.

Walker, B. H.; Ludwig, D.; Holling, C. S.; Peterman, R. M. (1981). Stability of semiarid savanna grazing systems. *J. Ecol.* 69:473–498.

Wallace, L. L. (1981). Growth, morphology and gas exchange of mycorrhizal and nonmycorrhizal *Panicum coloratum* L., a C_4 grass species, under different clipping and fertilization regimes. *Oecologia* 49:272–278.

Walter, H. (1971). *Ecology of Tropical and Subtropical Vegetation.* Edinburgh: Oliver & Boyd.

Walter, H. (1979). *Vegetation of the Earth.* New York: Springer-Verlag.

Waterman, P. G. (1983). Distribution of secondary metabolites in rain forest plants: toward an understanding of cause and effect. In: Sutton, S. L.; Whitmore, T. C.; Chadwick, A. C. editors. *Tropical Rain Forest: Ecology and Management.* Oxford: Blackwell, pp. 167–179.

Weathers, W. W. (1979). Climatic adaptation in avian standard metabolic rate. *Oecologia* 42:81–89.

Webb, D. S. (1978). The history of savanna vertebrates in the New World. II. South America and the great interchange. *Ann. Rev. Ecol. Syst.* 9:393–426.

Webb, L. J.; Tracy, J. C.; Haydock, K. P. (1967). A factor toxic to seedlings of the same species, associated with the living roots of the nongregarious subtropical rain forest tree *Grevillea robusta. J. Appl. Ecol.* 4:13–25.

Weiner, J. S. (1980). Work and well-being in savanna environments: physiological considerations. In: Harris, D. R. editor. *Human Ecology in Savanna Environments.* London: Academic Press, pp. 421–437.

Weir, J. S. (1971). The effects of creating additional water supplies in a central African National Park. In: Duffey, E.: Watt, A. S. editors. *The Scientific Management of Animal and Plant Communities for Conservation.* Oxford: Blackwell, pp. 367–385.

Weir, J. S. (1973). Air flow, evaporation and mineral accumulation in mounds of *Macrotermes subhyalinus* (Rambur). *J. Anim. Ecol.* 42:509–520.

Western, D. (1982). Patterns of depletion in a Kenya rhino population: are there lessons for nature reserve design? *Biol. Conserv.* 24:147–156.

Western, D.; Henry, W. (1979). Economics and conservation in third world national Parks. *Bioscience* 29:414–418.

Western, D; Ssemakulä, J. (1981). The future of savannah ecosystems: ecological islands or faunal enclaves? *Afr. J. Ecol.* 19:7–19.

Western, D.; Van Praet, C. (1973). Cyclical changes in the habitat and climate of an East African ecosystem. *Nature* 241:104–106.

Whalley, W. B.; Smith, B. J. (1981). Mineral content of harmattan dust from northern Nigeria examined by scanning electron microscope. *J. Arid Env.* 4:21–29.

Wheeler, E. F. (1980). Nutritional status of savanna peoples. In: Harris, D. R. editor. *Human Ecology in Savanna Environments.* London: Academic Press, pp. 439–455.

Wheelwright, N. T.; Orians, G. H. (1982). Seed dispersal by animals: contrasts with pollen dispersal, problems of terminology and contraints on coevolution. *Am. Nat.* 119:402–413.

White, T. C. R. (1976). Weather, food, and plagues of locusts. *Oecologia* 22:119–134.

White, T. C. R. (1978). The importance of a relative shortage of food in animal ecology. *Oecologia* 33:71–86.

Whitmore, T. C. *Tropical Rain Foreets of the Far East.* Oxford: Oxford University Press.

Whitmore, T. C. (1982). On pattern and process in forests. In: Newman, E. I. editor. *The Plant Community as a Working Mechanism.* Oxford: Blackwell. pp. 45–59.

Whittaker, R. H. (1975a). *Communities and Ecosystems.* New York: MacMillian.

Whittaker, R. H. (1975b). The design and stability of plant communities. In: van Dobben, W. H.; Lowe-McConnell, R. H. editors. *Concepts in Ecology.* The Hague: Junk, pp. 169–181.

Whitwell, A. C.; Phelps, R. J.; Thomson, W. R. (1974). Further records of chlorinated hydrocarbon pesticide residues in Rhodesia. *Arnoldia* 6:1–8.

Wickler, W. (1968). *Mimicry in Plants and Animals.* London: Weidenfeld and Nicholson, World University Library.

Wiebes, J. T. (1979). Coevolution of figs and their insect pollinators. *Ann. Rev. Ecol. Syst.* 10:1–12.

Wiegert, R. G. (1968). Thermodynamic considerations in animal nutrition. *Amer. Zool.* 8:71–81.

Wilcox, B. A.; Murphy, D. D. (1985). Conservation strategy: the effects of fragmentation on extinction. *Am. Nat.* 125:879–887.

Wilson, W. L.; Johns, A. D. (1982). Diversity of selected animal species in undisturbed forest, selectively logged forest and plantations in East Kalimantan, Indonesia. *Biol. Conserv.* 24:205–218.

Wint, G. R. W. (1983). Leaf damage in tropical rain forest canopies. In: Sutton, S. L.; Whitmore, T. C.; Chadwick, A. C. editors. *Tropical Rain Forest: Ecology and Management.* Oxford: Blackwell, pp. 229–239.

Winterhalder, B.; Larsen, R.; Thomas, R. B. (1974). Dung as an essential resource in a highland Peruvian community. *Hum. Ecol.* 1:89–104.

Wolda, H. (1982). Seasonality of Homoptera on Barro Colorado Island. In: Leigh, E. G.; Rand, A. S.; Windsor, D. M. editors. *The Ecology of a Tropical Forest.* Washington, DC: Smithsonian, pp. 319–330.

Wolda, H. (1983). Spatial and temporal variation in abundance in tropical animals. In: Sutton, S. L.; Whitmore, T. C.; Chadwick, A. C. editors. *Tropical Rain Forest: Ecology and Management.* Oxford: Blackwell. pp. 93–105.

Wood, B. J. (1971). Development of integrated control programs for pests of tropical perennial crops in Malaysia. In: Huffaker, C. B. editor. *Biological Control.* New York: Plenum pp. 422–457.

Wood, T. G. (1978). Food and feeding habits of termites. In: Brian, M. V. editor. *Production Ecology of Ants and Termites.* Cambridge: Cambridge University Press, pp. 55–80.

Wood, T. G.; Sands, W. A. (1978). The role of termites in ecosystems. In: Brian, M. V. *Production Ecology of Ants and Termites.* Cambridge: Cambridge University Press, pp. 245–292.

Wrangham, R. W. and Waterman, P. G. (1981). Feeding behaviour of vervet monkeys on *Acacia tortilis* and *Acacia xanthophloea:* with special reference to reproductive strategies and tannin production. *J. Anim. Ecol.* 50:715–731.

Wright, S. J. (1983). The dispersion of eggs by a bruchid beetle among *Scheela* palm seeds and the effect of distance to the parent plant. *Ecology* 64:1016–1021.

Young, A. M. (1982). Population Biology of Tropical Insects. New York: Plenum.

Index